AF839195

VOLUME EIGHTY FIVE

ADVANCES IN QUANTUM CHEMISTRY

Jack Sabin, Scientist and Friend

EDITORIAL BOARD

Remigio Cabrera-Trujillo (UNAM, México)
Hazel Cox (UK)
Frank Jensen (Aarhus, Denmark)
Mel Levy (Durham, NC, USA)
Jan Linderberg (Aarhus, Denmark)
Svetlana A. Malinovskaya (Hoboken, NJ, USA)
William H. Miller (Berkeley, CA, USA)
John W. Mintmire (Stillwater, OK, USA)
Manoj K. Mishra (Mumbai, India)
Jens Oddershede (Odense, Denmark)
Josef Paldus (Waterloo, Canada)
Pekka Pyykko (Helsinki, Finland)
Mark Ratner (Evanston, IL, USA)
Dennis R. Salahub (Calgary, Canada)
Henry F. Schaefer III (Athens, GA, USA)
John Stanton (Austin, TX, USA)
Alia Tadjer (Sofia, Bulgaria)
Harel Weinstein (New York, NY, USA)

VOLUME EIGHTY FIVE

ADVANCES IN QUANTUM CHEMISTRY

Jack Sabin, Scientist and Friend

Edited by

JENS ODDERSHEDE

Department of Physics, Chemistry and Pharmacy,
University of Southern Denmark,
Campusvej 55, DK-5230 Odense M,
Denmark

ERKKI J. BRÄNDAS

Department of Chemistry,
Uppsala University,
Uppsala, Sweden

ELSEVIER

ACADEMIC PRESS
An imprint of Elsevier

Academic Press is an imprint of Elsevier
50 Hampshire Street, 5th Floor, Cambridge, MA 02139, United States
525 B Street, Suite 1650, San Diego, CA 92101, United States
The Boulevard, Langford Lane, Kidlington, Oxford OX5 1GB, United Kingdom
125 London Wall, London, EC2Y 5AS, United Kingdom

First edition 2022

Copyright © 2022 Elsevier Inc. All rights reserved.

No part of this publication may be reproduced or transmitted in any form or by any means, electronic or mechanical, including photocopying, recording, or any information storage and retrieval system, without permission in writing from the publisher. Details on how to seek permission, further information about the Publisher's permissions policies and our arrangements with organizations such as the Copyright Clearance Center and the Copyright Licensing Agency, can be found at our website: www.elsevier.com/permissions.

This book and the individual contributions contained in it are protected under copyright by the Publisher (other than as may be noted herein).

Notices
Knowledge and best practice in this field are constantly changing. As new research and experience broaden our understanding, changes in research methods, professional practices, or medical treatment may become necessary.

Practitioners and researchers must always rely on their own experience and knowledge in evaluating and using any information, methods, compounds, or experiments described herein. In using such information or methods they should be mindful of their own safety and the safety of others, including parties for whom they have a professional responsibility.

To the fullest extent of the law, neither the Publisher nor the authors, contributors, or editors, assume any liability for any injury and/or damage to persons or property as a matter of products liability, negligence or otherwise, or from any use or operation of any methods, products, instructions, or ideas contained in the material herein.

ISBN: 978-0-323-99188-9
ISSN: 0065-3276

For information on all Academic Press publications
visit our website at https://www.elsevier.com/books-and-journals

Publisher: Zoe Kruze
Acquisitions Editor: Sam Mahfoudh
Developmental Editor: Tara Nadera
Production Project Manager: Vijayaraj Purushothaman
Cover Designer: Vicky Pearson

Typeset by STRAIVE, India

ELSEVIER Book Aid International Working together to grow libraries in developing countries
www.elsevier.com • www.bookaid.org

Contents

Contributors

Javier Aizpurua
Centro de Física de Materiales, Centro Mixto CSIC-UPV/EHU and Donostia International Physics Center (DIPC), Donostia-San Sebastián, Spain

Tomasz J. Antosiewicz
Department of Physics and Gothenburg Physics Center, Chalmers University of Technology, Göteborg, Sweden; Faculty of Physics, University of Warsaw, Warsaw, Poland

S. Peter Apell
Department of Physics and Gothenburg Physics Center, Chalmers University of Technology, Göteborg, Sweden; Centro de Física de Materiales, Centro Mixto CSIC-UPV/EHU and Donostia International Physics Center (DIPC), Donostia-San Sebastián, Spain

Remigio Cabrera-Trujillo
Instituto de Ciencias Físicas, Universidad Nacional Autónoma de México, Cuernavaca, Morelos, Mexico; Department of Physics, University of Gothenburg, Gothenburg, Sweden

Sylvio Canuto
Instituto de Física, Universidade de São Paulo, São Paulo, SP, Brazil

Kaline Coutinho
Instituto de Física, Universidade de São Paulo, São Paulo, SP, Brazil

Salvador A. Cruz
Departamento de Física, Universidad Autónoma Metropolitana-Iztapalapa, Mexico City, Distrito Federal, Mexico

James Dufty
Department of Physics, University of Florida, Gainesville, FL, United States

F. Flores
Departamento Física Teórica de la Materia Condensada and Condensed Matter Physics Center (IFIMAC) Universidad Autónoma de Madrid, Madrid, Spain

E.B. Genio
KETD East Qianjin Road Enterprise Science & Technology Park, Kunshan, China

E.C. Goldberg
Instituto de Física del Litoral, CONICET-UNL, Santa Fe, Argentine

Pedro L. Grande
Ion Implantation Laboratory, Institute of Physics, Federal University of Rio Grande do Sul, Porto Alegre, RS, Brazil

Maria Silvia Gravielle
Instituto de Astronomía y Física del Espacio (UBA-CONICET), Buenos Aires, Argentina

J.A. Hamida
Department of Physics, University of Florida, Gainesville, FL, United States

B. Andes Hess Jr.
Department of Chemistry, Vanderbilt University, Nashville, TN, United States

Evanildo G. Lacerda Jr.
Instituto de Física, Universidade de São Paulo, São Paulo, SP, Brazil

Morten Matras
CEO, Leaddoubler.com, Odense, Denmark

Alejandra M.P. Mendez
Institute of Astronomy and Space Physics, CONICET and University of Buenos Aires, Buenos Aires, Argentina

David A. Micha
Quantum Theory Project, Departments of Chemistry and of Physics, University of Florida, Gainesville, FL, United States

J.W. Mintmire
Department of Physics, Oklahoma State University, Stillwater, OK, United States

Jorge E. Miraglia
Instituto de Astronomía y Física del Espacio (UBA-CONICET), Buenos Aires, Argentina

Darío M. Mitnik
Institute of Astronomy and Space Physics, CONICET and University of Buenos Aires, Buenos Aires, Argentina

Claudia C. Montanari
Institute of Astronomy and Space Physics, CONICET and University of Buenos Aires, Buenos Aires, Argentina

Eduardo C. Montenegro
Instituto de Física, Universidade Federal do Rio de Janeiro, Rio de Janeiro, Brazil

Gautam Mukhopadhyay
Department of Physics and Gothenburg Physics Center, Chalmers University of Technology, Göteborg, Sweden; Department of Physics, Indian Institute of Technology Bombay, Powai, Mumbai, India

K.A. Muttalib
Department of Physics, University of Florida, Gainesville, FL, United States

Jens Oddershede
Department of Physics, Chemistry and Pharmacy, University of Southern Denmark, Campusvej, Denmark

J.V. Ortiz
Department of Chemistry and Biochemistry, Auburn University, Auburn, Alabama, United States

John W. Perram
Mærsk McKinney Moller Institute, University of Southern Denmark, Odense, Denmark

M. Pilla
Qorvo Inc., Richardson, TX, United States

Stephan P.A. Sauer
Department of Chemistry, University of Copenhagen, Copenhagen, Denmark

Lucas Sigaud
Instituto de Física, Universidade Federal Fluminense, Niterói, Brazil

Lea Northcote Sørensen
Department of Chemistry, University of Copenhagen, Copenhagen, Denmark

N.S. Sullivan
Department of Physics, University of Florida, Gainesville, FL, United States

M.S. Tacca
Instituto de Física del Litoral, CONICET-UNL, Santa Fe, Argentine

Maarten Vos
Department of Materials Physics, Research School of Physics, Australian National University, Canberra, ACT, Australia

Jeffrey Wrighton
Department of Physics, University of Florida, Gainesville, FL, United States

Preface

Prof. John R. Sabin

Professor John R. Sabin from the Quantum Theory Project at the University of Florida has decided to step down as the series editor of *Advances in Quantum Chemistry* after many years of service to this book series and the quantum chemistry community. Jack Sabin to his friends—and he has very many—has been part of the series for longer than most of us can remember. He became part of the crew involved with the series soon after he joined the faculty at the University of Florida in 1971. In the early days of the Sanibel Symposia, Jack was handling the manuscripts from symposium participants that went into the *Advances in Quantum Chemistry* Symposium issues. Later he became more and more involved in business of the regular series as well. During these many years, a countless number of manuscripts have passed through his hands, and he has been instrumental in disseminating the results of quantum chemistry research to several generations of scientists. So, it is not just an editor stepping down from the front page of the series, it is a part of the series's DNA that we now must get on without. This volume will only be a small token of our appreciation for his service to all of us, but it will give us an opportunity to pay tribute to him for being both a scientist and a friend at the same time.

Erkki J. Brändas

Jack Sabin has been and still is an enthusiastic and devoted supporter of Scandinavia, its language(s), traditions, and way of life. He arrived as a newly graduated PhD and holder of an NIH postdoctoral fellowship for 1966–67, becoming a young member of the growing Uppsala Quantum Chemistry Group. As a maturing graduate student in 1966, I met Jack, arriving in his new Swedish Volvo, with an already unquenchable thirst for science and learning Scandinavian living and culture. He acclimated quickly and was a perfect fit for the erudite versatility and international experience weathered at the department. This was the birth and initiation of a distinguished academic career as physicist, teacher, supervisor, science editor, and university administrator, but he also stands for significantly much more! Sailing, philately, genealogy, to mention a few, display his adroit interests in depth, for instance, reflected in his compilation/edition *Adventures in Civilization*, a fascinating and touching family history tracing his ancestry back to more than 400 years ago. Unfortunately for Sweden, he was hijacked to Denmark to enjoy the enchanted realm of Nordic summers and in particular the Danish gemüt. During the more than 50 years of friendship and editorial camaraderie, at both Wiley and Elsevier, complex interactive social and scientific editing has been a joy, multiply rewarding and diversifying—a seminal occupation in consensus and compromise. For that, I am genuinely grateful.

Jens Oddershede

Jack Sabin is a prominent scientist. I knew so already when I first got in touch with him in the early 1970s where he was visiting Aarhus University in Denmark. After 1978, when I moved to the University of Southern Denmark in Odense, Jack and I have been engaged in close scientific collaboration, with him visiting Odense at least once every year, a tradition we are still upholding. Over the years, we have published 90 joint papers, the last being published in 2021, and most of them with several other coauthors. A scientific collaboration can only last so many years if we both enjoy it and both contribute. Jack was the driving force in the search for new and interesting applications of electronic structure methods. He saw things I had not noticed, he knew the literature and the new trends very well, and he had a clear view of the potential and the limitations in any application.

Even though he has a PhD in experimental physical chemistry, he is an eminent physicist, much better than me, who has had a formal education in physics! Over the years, we moved more and more into physics applications, primarily into heavy ion stopping, a trend that was surely initiated and driven by him and a field in which he became a true expert.

Jack is a team player. Wherever he is, he creates a positive spirit. He is the "favorite visitor" at my department in Odense among the faculty, staff, students, and coworkers, and he is a master in collecting the right team behind a new project. He is the kind of person one always wishes to have visiting your research group. When I was spending more and more time first as dean and later as university president, he was the one who kept me alive scientifically. Without him, my scientific career would have ended there.

Even though Jack and I have had many papers together, his total scientific output exceeds 250, so he is also engaged in many other sides of quantum theory other than the ones we have in common. John R. Sabin has an extensive network all over the world, and the many contributions and laudations in this volume bear witness to a highly esteemed scientific fellow and friend.

Thank you, Jack, for your many contributions to quantum chemistry and for your services to this series for so many years, and thank you for being who you are. I am proud you are my friend.

JENS ODDERSHEDE and ERKKI J. BRÄNDAS

CHAPTER ONE

Energy deposition in a many-atom system with dissipative dynamics

David A. Micha*
Quantum Theory Project, Departments of Chemistry and of Physics, University of Florida, Gainesville, FL, United States
*Corresponding author: e-mail address: micha@qtp.ufl.edu

Contents

Abstract

The focus in this presentation is the formulation of a general treatment of energy deposition applicable to target excitation by both fast and slow projectiles, including the dissipative dynamics of a non-equilibrium medium by means of a reduced statistical density operator treatment, which can be implemented computationally to provide measurable properties. The treatment combines two formalisms to provide a general theory of energy deposition in a target which interacts with a surrounding medium. The first formalism involves expressing inelastic collision cross sections in terms of time-correlation functions of scattering operators, usually given in the literature for a medium at thermal equilibrium. It is used to derive expressions for fast electronic excitation of the target, and for slow multicenter scattering by target ions. The second formalism introduces a reduced statistical density operator dependent only on variables of a primary region of interest, and accounting for fluctuation and dissipation phenomena arising from interactions of the target with its medium, or secondary region. The combined treatment leads to energy deposition rates corrected to include a term for disturbance by the projectile of the target original equilibrium, and a term accounting for dissipation phenomena in the target.

Advances in Quantum Chemistry, Volume 85
ISSN 0065-3276
https://doi.org/10.1016/bs.aiq.2022.03.001
Copyright © 2022 Elsevier Inc.
All rights reserved.

1. Introduction

When fast charged particles, with energies from a few eV up to hundreds of thousands of eV, traverse matter composed of atoms, molecules or biological compounds, they deposit energy along their trajectories and lose some of their kinetic energy. Electron beams and beams of atomic ions such as protons, He^{2+} and Li^{+}, have been used in instruments to measure energy deposition and extract from it a variety of properties of interest in applications. Among them, electron and ion projectiles are used in energy-loss spectroscopies to provide information on quantum state-to-state transition probabilities in beam targets. An important application is energy deposition in medical treatments with by ion beams used in applications to destroy harmful tissues.

The mechanism of energy transfer between beam particles and medium particles changes as the charged particles in a beam loose energy and slow down, and that requires different treatments of their collision dynamics. Quantal treatments can be derived and applied to situations where beam particle velocities vary from values much higher than velocities of electrons and ions in the medium, all the way down to small projectile velocity values comparable to velocities of electrons in atomic shells and to velocities of recoiling ions in molecules or solids.

The treatment of energy deposition by fast charged projectiles was developed by H. Bethe in seminal work,[1] generalized and expanded by various authors[2–5] and applied to many target systems. These treatments link the energy deposition rates to quantal state-to-state oscillator strengths, which can be obtained from detailed calculations of electronic structure of target species, including electron correlation.[6,7] In particular, this has led to many results obtained for atoms, molecules and biological compounds.[8–10]

The energy deposition rate of a charged beam into a target, or stopping power $S(\vec{v})$ for a projectile of mass m_0 and velocity $\vec{v}$, is related for large velocities to the generalized oscillator strength (or GOS) $F_{an}(\vec{q})$ for a transition between target quantum states a and n of a target with N electrons at positions $\vec{r}_j, j = 1 - N$, in a framework of atomic ions at locations $\vec{R}_\mu$. This depends on the momentum transfer during collision $\hbar\vec{q} = \vec{p}_n - \vec{p}_a$, with $\vec{p}_n(\vec{v})$ the projectile momentum for state n, a function of the initial projectile velocity. For a target originally in state $a = 0$, the relation is (in Hartree units with $\hbar = e = m_e = 1$ $a.\ u.$)

$$S_0(\vec{v}) = \frac{4\pi Z_1{}^2}{v^2}\sum\nolimits_{n>0}\int_{q_{min}}^{q_{max}}\frac{dq}{q}F_{0n}(\vec{q})$$

$$F_{0n}(\vec{q}) = \frac{2(E_n - E_0)}{q^2}\left|\left\langle 0\right|\sum\nolimits_{j=1}^{N}\exp\left(-i\vec{q}.\vec{r}_j\right)\left|n\right\rangle\right|^2$$

In the limit of semiclassical scattering of the projectile at sufficiently higher velocities (or short wavelengths) it is found that $\sum_{n\neq0}F_{0n}(\vec{q}) = N$ for variable $\vec{q}$, a useful Bethe sum rule.[5]

At high projectile velocities, the main contributions to energy loss are due to excitation or ionization of electronic states. In a quantal treatment, the GOS can be re-expressed to involve the time dependence of electronic density operators and their time-correlation. At low velocities, energy transfer occurs through close-range collisions involving interactions of the beam particles with closed electron-shell structures, in a medium with delocalized electrons. The following treatment is based on our formal theory of collisional energy transfer based on time-correlation functions (TCFs) of collision operators, which is quite general and can account for varying energy transfer regimes as collision velocities change.[11,12] It will be used to briefly describe as special cases electronic excitation at high velocities, and multiple scattering by target ions at low velocities.

Furthermore, the TCF treatment is expanded to include additional dissipative medium effects when a target is at thermal equilibrium to start with, and is then excited by energy deposition. Collisions between a projectile and a target species occur in a medium which may be away from thermal equilibrium due to the presence of electronic and lattice excitations created by concurrent collisional events, leading to fluctuation and dissipation phenomena. However, most present treatments of excitation and recoil during energy deposition rely on thermal averages of quantal operators and their time correlation, on the assumption that the medium is at thermal equilibrium. Here it is shown that an extension of those treatments, to include dissipative medium phenomena due to the presence of electronic and ion recoil excitations, can be done introducing instead averages over reduced density matrices (RDMs). A variety of RDM treatments have been developed in the literature to deal with fluid dynamics, molecular spectroscopy and kinetics, and condensed matter phenomena[13–18] and are employed here for energy deposition.

We have developed a treatment of the dissipative dynamics of phenomena involving molecules and materials in terms of reduced density operators (RDOps), and have implemented calculations of optical properties of semiconductors using density functional theory and ab initio molecular dynamics.[19–23] The time-evolution of RDOps can be derived from equations of motion which include both delayed and instantaneous dissipation rates. The focus in this presentation is instead on the formulation of a general treatment of energy deposition including the dissipative dynamics of a non-equilibrium medium by means of a similar RDOp treatment, which can be implemented computationally to provide measurable properties, but does not include applications here.

Fig. 1 below is a pictorial description of an energy deposition event, with a beam of charged particles of initial velocity v interacting with a target containing an assembly of molecules (in shaded shapes) in a bath of solvent species. The beam emerges with velocity $v' < v$. The whole system is partitioned into a primary region P containing the assembly of molecules and adjacent interacting solvent, while the remaining solvent is labeled as the secondary S region. The beam excites the molecular assembly and also some electrons and ions in the medium, which lead to fluctuation and dissipation phenomena in the primary region.

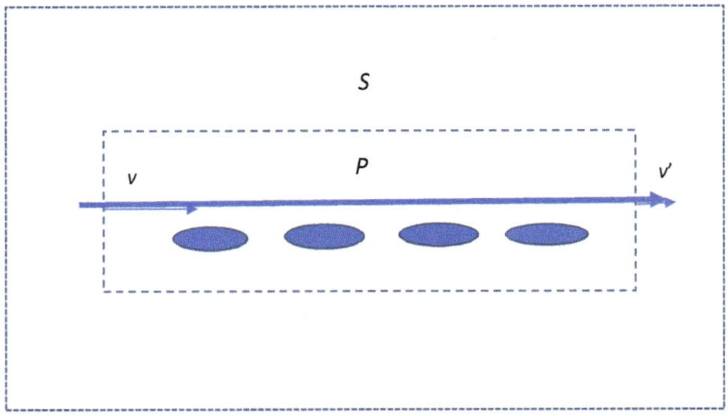

Fig. 1 Pictorial description of an energy deposition event, with a beam of charged particles of initial velocity *v* interacting with a target containing an assembly of molecules (in shaded shapes) in a bath of solvent species. The whole system is partitioned into a primary region *P* containing the assembly of molecules and adjacent interacting solvent, while the remaining solvent is labeled as the secondary *S* region.

2. Energy deposition by atomic ions from time-correlation functions of collisional operators

2.1 Formalism of collisional TCFs

A general formulation relating cross sections to collisional time-correlation functions (or TCFs) can employs the transition operator $\widehat{T}^{(+)}(E)$ of scattering theory, where (+) indicates outgoing waves and an argument with $E+i\epsilon=E^{(+)}$, $\epsilon\rightarrow 0$,[24]

$$\widehat{T}^{(+)}(E) = V + V\widehat{G}_0^{(+)}(E)\widehat{T}^{(+)}(E), \quad \widehat{G}_0^{(+)}(E) = (E^{(+)} - \widehat{H}_0)$$

for a system Hamiltonian $\widehat{H} = \widehat{H}_0 + V$, including the Hamiltonian $\widehat{H}_0 = \widehat{K}_{rel} + \widehat{H}_{int}$ of non-interacting beam and internal motions of system particles, plus a potential energy V which accounts for the interactions of the beam particles with the system particles. Here $\widehat{G}_0^{(+)}(E)$ is the Green function (or propagator) for free motion. States $|\vec{k}\rangle$ for beam particle satisfy $\widehat{K}_{rel}|\vec{k}\rangle = E_k|\vec{k}\rangle$ with beam momentum $\vec{p}_0 = m_0\vec{v} = \hbar\vec{k}$ and kinetic energy $E_k=\hbar^2k^2/(2m_0)$, and internal motion states from $\widehat{H}_{int}|\alpha\rangle = E_\alpha|\alpha\rangle$ give total energies $E=E_k+E_\alpha$. From this, the rate of transition between states $|\vec{k},\alpha\rangle$ and $|\vec{k}',\alpha'\rangle$, with $\widehat{H}_0|\vec{k},\alpha\rangle = (E_k + E_\alpha)|\vec{k},\alpha\rangle$, is given by

$$R\left(\vec{k},\alpha\rightarrow\vec{k}',\alpha'\right) = (2\pi/\hbar)\left|\left\langle\vec{k}',\alpha'\right|\widehat{T}^{(+)}\left|\vec{k},\alpha\right\rangle\right|^2\delta(E-E')$$

for state-to-state changes of the relative momentum $\hbar\vec{k}$ and internal system state α.

This rate can be transformed into a TCF using that $\delta(E-E')=(2\pi\hbar)^{-1}\int_{-\infty}^{\infty}dt\exp\{i(E-E')t/\hbar]$ and that $\exp\left(i\,\widehat{H}_{int}t/\hbar\right)|\alpha\rangle = \exp\left(i\,E_\alpha t/\hbar\right)|\alpha\rangle$, which lead to

$$R\left(\vec{k},\alpha\rightarrow\vec{k}',\alpha'\right) = \hbar^{-2}\int_{-\infty}^{\infty} dt\,\exp\left(i\frac{\varepsilon t}{\hbar}\right)\left\langle\alpha'\left|\widehat{T}_{\vec{k}'\vec{k}}(t)\right|\alpha\right\rangle^*\left\langle\alpha'\left|\widehat{T}_{\vec{k}'\vec{k}}(0)\right|\alpha\right\rangle$$

$$\widehat{T}_{\vec{k}'\vec{k}}(t) = \exp\left(i\frac{\widehat{H}_{int}t}{\hbar}\right)\widehat{T}_{\vec{k}'\vec{k}}\exp\left(-i\frac{\widehat{H}_{int}t}{\hbar}\right)$$

where $\widehat{T}_{\vec{k}'\vec{k}}=\langle\vec{k}'|\widehat{T}^{(+)}|\vec{k}\rangle$ is yet an operator in the space of internal states of the system, evolving in time with $\widehat{H}_{int}$. Also, $\varepsilon=E_k-E_{k'}$ is the amount of beam kinetic energy (or energy deposition) transferred into the system. When a target system has a statistical distribution w_α of internal states, the quantal and statistical rate, indicated with a double parenthesis, is

$$R\left(\vec{k}\rightarrow\vec{k}'\right)=\sum\nolimits_{\alpha}w_\alpha\sum\nolimits_{\alpha'}R\left(\vec{k},\alpha\rightarrow\vec{k}',\alpha'\right)$$
$$=\hbar^{-2}\int_{-\infty}^{\infty}dt\,\exp\left(i\frac{\varepsilon t}{\hbar}\right)F\left(t;\vec{k}',\vec{k}\right)$$
$$F\left(t;\vec{k}',\vec{k}\right)=\sum\nolimits_{\alpha}w_\alpha\left\langle\alpha\left|\widehat{T}_{\vec{k}'\vec{k}}(t)^{\dagger}\widehat{T}_{\vec{k}'\vec{k}}(0)\right|\alpha\right\rangle=\left\langle\left\langle\widehat{T}_{\vec{k}'\vec{k}}(t)^{\dagger}\widehat{T}_{\vec{k}'\vec{k}}(0)\right\rangle\right\rangle,$$

with this result obtained by using the completeness property of the set of internal states. Therefore the statistical rate of energy deposition is given quite generally as a Fourier transform $\int_{-\infty}^{\infty}dt\,\exp(i\varepsilon t/\hbar)\,F(t;\vec{k}',\vec{k})=\widetilde{F}(\varepsilon/\hbar;\vec{k}',\vec{k})$ of a statistical TCF of scattering operators. Alternatively, an inelastic differential cross section is given from the rate per unit beam flux, introducing the orientation solid angle Ω of $\vec{k}'$ relative to the beam direction $\vec{k}$, as.[11]

$$\frac{d^2\sigma}{d\varepsilon\,d\Omega}=\left(\frac{2\pi}{\hbar}\right)^4 m_0{}^2\frac{k'}{k}R\left(\vec{k}\rightarrow\vec{k}'\right)$$

This compact expression contains as special cases energy deposition treatments for fast and slow ion beams traversing a target with delocalized electrons in a lattice of atomic ions. The special cases can be derived from expressions for the scattering operator in different limits.

2.2 Special cases of collisional TCFs

For a fast charged projectile of charge C_0 at position $\vec{R}_0(t)=\vec{b}+\vec{v}t$ interacting with target electrons of charge c_e, it is usually accurate to use the Born approximation, with $\widehat{T}_{\vec{k}'\vec{k}}\cong V_{\vec{k}'\vec{k}}$, a Coulomb matrix element. This is $V_{\vec{k}'\vec{k}}=C_0c_e\langle\vec{k}'|\int d^3r\,\rho(\vec{r})|\vec{R}_0-\vec{r}|^{-1}|\vec{k}\rangle$ for a medium with a density of electrons $\widehat{\rho}_{el}(\vec{r})=\sum_j\delta(\vec{r}-\vec{r}_j)=(2\pi\hbar)^{-3}\int d^3q\,\exp(i\vec{q}.\vec{r})\widehat{\rho}_{\vec{q}}$, where $\widehat{\rho}_{\vec{q}}=\sum_j\exp(-i\vec{q}.\vec{r}_j)$ is a transformed density amplitude. This gives

$$\widehat{T}_{\vec{k}'\vec{k}}(t) = \frac{C_0 c_e}{(2\pi\hbar)^3} \int d^3q \left\langle \vec{k}' \left| \int d^3r \exp\left(i\,\vec{q}\,.\,\vec{r}\right) \left| \vec{R}_0(t) - \vec{r} \right|^{-1} \right| \vec{k} \right\rangle \widehat{\rho}_{\vec{q}}(t)$$
$$= \int d^3q \, A_{\vec{k}'\vec{k}}(\vec{q}\,, t) \widehat{\rho}_{\vec{q}}(t)$$

showing that the transition operator changes linearly with the density amplitude, so that the scattering cross section depends on the Fourier transform of $\langle\langle \widehat{\rho}_{\vec{q}}(t)\widehat{\rho}_{\vec{q}'}(0)\rangle\rangle$, the statistical density-density TCF closely related to the oscillator strength of the electron system.[16,25,26] The calculation of generalized oscillator strengths for energy deposition in molecular assemblies and biomolecules using the many-electron polarization propagator in the random-phase approximation, has been the subject of several papers.[6,7,9,10]

Replacing this $\widehat{T}_{\vec{k}'\vec{k}}(t)$ in the rate obtained from collisional TCFs, one finds for fast projectiles

$$\frac{d^2\sigma}{d\varepsilon\, d\Omega} = \left(\frac{2\pi}{\hbar}\right)^4 \frac{m_0{}^2}{\hbar^2} \frac{k'}{k} \int d^3q \int d^3q' \int_{-\infty}^{\infty} dt$$
$$\exp\left(i\frac{\varepsilon t}{\hbar}\right) A_{\vec{k}'\vec{k}}(\vec{q}, t)^* \left\langle \left\langle \widehat{\rho}_{\vec{q}}(t)^\dagger \widehat{\rho}_{\vec{q}'}(0) \right\rangle \right\rangle A_{\vec{k}'\vec{k}}\left(\vec{q}\,', 0\right)$$

which involves a weighted Fourier transform of the density-density TCF, or generalized oscillator strength for excitations of the electrons in the target.

For a projectile with low velocities compared to medium electron velocities, a collision process involves close-range interactions of the projectile ion with ions in the target, mediated by short-range ion-ion potential energies. For large molecular targets, this may be conveniently treated as interactions of the projectile p with molecular fragments (or n-atom clusters) C, with interaction potentials V_{0C}, and a total interaction $V = \sum_C V_{0C}$. A treatment of the multiple collisions involved in this case can be described starting with the partial transition operators $\widehat{T}_{0C} = V_{0C} + V_{0C}\widehat{G}_0\widehat{T}_{0C}$ for a single encounter. A final-channel decomposition of the total multiple transition operator is then $\widehat{T} = \sum_C \widehat{T}^{(C)}$, with $\widehat{T}^{(C)}$ describing a transition with the last interaction occurring through V_{0C}, and given by[11,12,24]

$$\widehat{T}^{(C)} = \widehat{T}_{0C} + \widehat{T}_{0C}\widehat{G}_0 \sum\nolimits_{C' \neq C} \widehat{T}^{(C')}$$

The energy deposition rate is given using these operators as

$$R\left(\vec{k} \rightarrow \vec{k}'\right) = \hbar^{-2} \sum\nolimits_{C,C'} Re\left[\widetilde{F}^{(C,C')}\left(\varepsilon/\hbar; \vec{k}', \vec{k}\right)\right]$$

from $F^{(C,\,C')}(t; \vec{k}', \vec{k}) = \langle\langle \widehat{T}^{(C)}_{\vec{k}'\vec{k}}(t)^{\dagger}\, \widehat{T}^{(C')}_{\vec{k}'\vec{k}}(0)\rangle\rangle$, correlating two final cluster interactions. Calculations of this rate expression can be simplified to account for shadow and eclipse effects during collisions along the projectile trajectory. Within the target structure, interior clusters can be shadowed by exterior ones as seen from the projectile source, and exterior clusters can be eclipsed as seen from a detector location. An approximation can be developed from the formalism of multiple scattering[11,12] to write that

$$F^{(C,\,C')}\left(t; \vec{k}', \vec{k}\right) \cong w^{(C-)}\left(\vec{k}'\right)\left\langle\left\langle (T_{0C})_{\vec{k}'\vec{k}}(t)^{\dagger} (T_{0C'})_{\vec{k}'\vec{k}}(0)\right\rangle\right\rangle w^{(C'+)}\left(\vec{k}\right)$$

with functions $w^{(C\pm)}(\vec{k})$ describing shadow and eclipse effects. These can be important in energy deposition into polymers and biomolecules composed of large molecular fragments.[10,26,27] These fragments (or clusters) can be peptides in a protein, or nucleotides in DNA or RNA. Further useful approximations can be developed for multiple scattering phenomena, including treatments of combined slow and fast degrees of freedom, and simplifications of T_{0C} for impulsive collisions.[11,12]

3. Energy deposition in a dissipative medium using a reduced density operator

3.1 Dissipative dynamics after initial thermal equilibrium

To begin with, a target system may be assumed to be at thermal equilibrium for given pressure and temperature $T = (k_B\,\beta)^{-1}$ with k_B the Boltzmann constant and β a thermal inverse energy. In this case the weight factors w_α for its states α can be obtained from the Boltzmann distribution, as $w_\alpha = \exp(-\beta\, E_\alpha)/\sum_\mu \exp(-\beta\, E_\mu)$, with $H_{int}|\alpha\rangle = E_\alpha|\alpha\rangle$. Introducing this distribution in the statistical-quantum average $\langle\langle \ldots \rangle\rangle$ for the deposition rate and combining exponentials containing H_{int}, the rate for a thermalized target is found to be the Fourier transform of

$$F_{thrm}\left(t; \vec{k}', \vec{k}\right) = F\left(t - i\hbar\beta; \vec{k}', \vec{k}\right)$$

which describes energy deposition for a target that remains at thermal equilibrium on the average.

However, it can be expected that in some experimental conditions the passage of projectile charges through the target will excited its electronic and ionic structure due to electronic density fluctuations and ion recoils. This departure from thermal equilibrium involves fluctuation and dissipation phenomena which can be treated in terms of a von Neumann statistical density operator $\widehat{\Gamma}(t)$.[13,17–20,28]

In more detail, the treatment involves classifying the whole target as containing interacting primary (or *P*) and secondary (or *S*) regions, with *P* treated in detail based on its atomic structure and *S* treated as a medium with given statistical properties. In the present treatment of energy deposition, the *P* region contains the target species plus neighboring species from its environment, while *S* contains all other species in the medium. Degrees of freedom of particles in the two regions can be identified, and their distributions of initial values found from an experimental setup. A reduced density operator (or RDOp) $\widehat{\rho}(t) = \langle tr_S[\widehat{\Gamma}]\rangle_{in}$ can be defined with a trace over *S* variables indicating these are summed over in $\widehat{\Gamma}$ and the pointed bracket meaning an statistical average over initial conditions. The equation of motion (or EOM) for $\widehat{\rho}$ follows from that of $\widehat{\Gamma}$ by taking traces of the latter EOM and defining effective operators in the *P* subspace. Hamiltonians for *P* and *S* regions are $\widehat{H}_P$ and $\widehat{H}_S$ and their interaction Hamiltonian is $\widehat{H}_{PS}$. State transitions $|\vec{k},\alpha\rangle$ to $|\vec{k}',\alpha'\rangle$ of interest occur in the *P* region, with $\widehat{H}_{int} = \widehat{H}_P$ and $V = \widehat{H}_{0P}$ for the interaction potential of the projectile with the *P* region. The total Hamiltonian contains the kinetic energies and interaction terms in $\widehat{H} = \widehat{K}_{rel} + \widehat{H}_P + \widehat{H}_S + \widehat{H}_{PS} + \widehat{H}_{0P} + \widehat{H}_{0S}$, where now $\widehat{K}_{rel}$ is the projectile kinetic energy along a prescribed trajectory interacting with the *P* region, and only indirectly with the *S* region. Here details of the treatment are omitted, but follow the developments and notation in previous publications.[21,29] The EOM for $\widehat{\rho}(t)$ is found to be

$$\frac{\partial\widehat{\rho}}{\partial t} = -\frac{i}{\hbar}\left[\widehat{F}_P + \widehat{H}_0 + \widehat{H}_{0P}, \widehat{\rho}(t)\right] + \widehat{\mathcal{A}}_P\widehat{\rho}(0) - \frac{1}{\hbar^2}\int_0^t dt'\widehat{\mathcal{M}}_P(t,t')\widehat{\rho}(t')$$

to be solved for an initial RDOp $\widehat{\rho}(t_{in}) = \widehat{\rho}_{in}$, its initial statistical distribution of *P*-states. Here $\widehat{F}_P = \widehat{H}_P + \langle \widehat{V}_P^{(S)}(t)\rangle_{in}$ where $\widehat{V}_P^{(S)}(t) = tr_S[\widehat{H}_{PS}\widehat{\Gamma}(t)]$, the $\widehat{\mathcal{A}}_P\widehat{\rho}(0)$ term describes a rate of fluctuations, and $\widehat{\mathcal{M}}_P(t,t')$ is the kernel of a delayed dissipative rate. Fluctuation and dissipation rates can be obtained from a partitioning formalism and are found to contain TCFs of the S region electronic and ionic densities.[17] In most cases of practical interest for the interpretation of experiments, time scales are such that fluctuations average to zero, and dissipation is instantaneous so one can write that

$$\left\langle \frac{1}{\hbar^2}\int_0^t dt'\widehat{\mathcal{M}}_P(t,t')\widehat{\rho}(t') \right\rangle_{in} \cong \widehat{\mathcal{R}}_P(t)\widehat{\rho}(t)$$

with $\widehat{\mathcal{R}}_P$ an instantaneous dissipative superoperator at time t. This has components $\widehat{\mathcal{R}}_P = \widehat{\mathcal{R}}_P^{(el)} + \widehat{\mathcal{R}}_P^{(ion)}$ coming from electronic and ionic excitations in the whole system. For the usual time scale of experimental measurements, $\widehat{\mathcal{R}}_P$ can be taken as being constant over time. Electronic dissipation can be constructed from Lindblad vibronic transition amplitudes in the P region induced by interactions with the S region, and ionic dissipation follows from a Redfield treatment of bilinear P-S couplings of ionic displacements.[18] Terms with $\widehat{H}_0$ disappear when the projectile trajectory is treated classically, as assumed in what follows.

The essential change in the formalism that includes dissipative phenomena consists in replacing the thermal average in $F_{thrm}(t;\vec{k}',\vec{k}) = \sum_\alpha w_\alpha(\beta)\langle\alpha|\widehat{T}_{\vec{k}'\vec{k}}(t)^\dagger\widehat{T}_{\vec{k}'\vec{k}}(0)|\alpha\rangle$, where $w_\alpha(\beta)$ is a Boltzmann probability from $\widehat{\rho}^{(thrm)} = exp(-\beta\widehat{H}_P)/tr_P[exp(-\beta\widehat{H}_P)]$, with a function describing the total system of primary plus secondary regions by means of

$$F_{tot}\left(t;\vec{k}',\vec{k}\right) = tr_P\left[\widehat{\rho}(t)\widehat{T}_{\vec{k}'\vec{k}}(t)^\dagger\widehat{T}_{\vec{k}'\vec{k}}(0)\right]$$

involving the trace over all P region variables of the RDOp coming from the EOM with fluctuation and dissipation rates. The transition operators $\widehat{T}_{\vec{k}'\vec{k}}$ are constructed in the P-subspace, insofar this has been chosen to include all the system interactions with the projectile.

3.2 Effects of medium dissipative dynamics

The above EOM can be solved for steady-state or pulse beam conditions involving the projectile interaction with its target. What follows addresses the steady-state situation in some detail. The formalism is made more compact by introducing Liouvillian superoperators as in $\widehat{\mathcal{L}}_x\widehat{\rho} = -(i/\hbar)\left[\widehat{F}_x, \widehat{\rho}\right]$. The EOM is then

$$\frac{d\widehat{\rho}}{dt} = \widehat{\mathcal{K}}(t)\widehat{\rho}(t), \ \widehat{\mathcal{K}} = \widehat{\mathcal{L}}_P + \widehat{\mathcal{L}}_{0P}(t) - \widehat{\mathcal{R}}_P$$

containing Liouvillians for the P region and for the projectile interacting with P. A steady-state solution $\widehat{\rho}^{(ss)}$ for a beam of given velocity $\vec{v}$ satisfies $0 = \widehat{\mathcal{K}}(t)\widehat{\rho}^{(ss)}(t)$, while the thermal equilibrium Boltzmann RDOp of P satisfies $0 = \widehat{\mathcal{L}}_P\widehat{\rho}^{(thrm)}(\beta)$.

Further development of the theory is best based on the introduction of a unitary transformation $\widehat{U}_{\vec{v}}(t) = \exp\left[-i \ \ \vec{v}t.(\widehat{\vec{p}})/\hbar\right]$ of the projectile variables, describing a translation by $\vec{v}t$ of the reference frame so that it moves with the initial constant velocity of the beam particle, generated by the momentum operator of the projectile. For an operator $\widehat{A}\left[\vec{R}_0(t)\right]$ dependent on the position $\vec{R}_0$ of the beam particle, one finds that[30] $\widehat{U}_{\vec{v}}(t)^{\dagger}\widehat{A}(\vec{R}_0)\widehat{U}_{\vec{v}}(t) = \widehat{A}(\vec{R}_0 - \vec{v}t)$, with the projectile position shifted in the new frame. This transformation can be applied to both sides of the EOM, further introducing a new P-region RDOp in $\widehat{\rho}(t) = \widehat{U}_{\vec{v}}(t)^{\dagger}\widehat{\sigma}(t)\widehat{U}_{\vec{v}}(t)$. This gives

$$\frac{d\widehat{\sigma}}{dt} = \left[\widehat{\mathcal{L}}_P + \widehat{\mathcal{L}}_{0P}(0) - \widehat{U}_{\vec{v}}(t)\,\widehat{\mathcal{R}}_P\widehat{U}_{\vec{v}}(t)^{\dagger}\right]\widehat{\sigma}(t)$$

This shows that the new dissipation superoperator may change with time in the new frame, insofar the particle velocity is changing along its trajectory. However, this dependence is indirect, since it occurs through the P-S interactions, and the new dissipation can be assumed to stay independent of time in a (TA) *translation approximation* so that $\widehat{\sigma}^{(TA)}(t)$ evolves under a time-independent Liouvillian, and its steady-state form is given by $0 = \left[\widehat{\mathcal{L}}_P + \widehat{\mathcal{L}}_{0P}(0) - \widehat{\mathcal{R}}_P\right]\widehat{\sigma}^{(ss)}$, where $\widehat{\sigma}^{(ss)}$ is independent of time and simpler to calculate. In the original frame, $\widehat{\rho}^{(ss)}(t) = \widehat{U}_{\vec{v}}(t)^{\dagger}\widehat{\sigma}^{(ss)}\widehat{U}_{\vec{v}}(t)$ within the TA approximation.

It remains to show how the steady-state and thermal equilibrium RDOps are related. For equilibrium we have that $0 = \widehat{\mathcal{L}}_P\widehat{\sigma}^{(thrm)}$ and that $\widehat{\rho}^{(thrm)}(t) = \widehat{U}_{\vec{v}}(t)^{\dagger}\widehat{\sigma}^{(thrm)}\widehat{U}_{\vec{v}}(t) = \widehat{\sigma}^{(thrm)}$. The connection can be done by means of perturbation theory writing $\mathcal{K}_P = \widehat{\mathcal{L}}_P + \mathcal{K}'_P$, with $\mathcal{K}'_P = \widehat{\mathcal{L}}_{0P}(0) - \widehat{\mathcal{R}}_P$ a perturbation Liouvillian and, to first order, $\widehat{\sigma}^{(ss)} = \widehat{\sigma}^{(thrm)} + \widehat{\sigma}'$. This gives

$$\widehat{\mathcal{L}}_P \widehat{\sigma}' = -\mathcal{K}'_P \widehat{\sigma}^{(thrm)}$$

to first order, with the righthand side known and containing both the projectile-P interaction rate and the dissipative rate. This equation can be solved using the eigenoperators and eigenvalues of $\widehat{\mathcal{L}}_P$ as a basis set for expansion of $\widehat{\sigma}'$. Here this is done only in outline, with details that can be found in our previous publications on the eigenvalue procedure used for relaxation phenomena.[21,31] A basis set $\{\widehat{\sigma}_A\}$ of orthogonal operators and their non-zero eigenvalues λ_A are generated from

$$\widehat{\mathcal{L}}_P \widehat{\sigma}_A = -\lambda_A \widehat{\sigma}_A \,, \quad tr_P\left[\overline{\widehat{\sigma}_B}\,\widehat{\sigma}_A\right] = \delta_{AB}$$

with the bar signifying the adjoint operator, and used to expand

$$\widehat{\sigma}' = \sum\nolimits_A \Lambda_A \widehat{\sigma}_A, \quad \Lambda_A = \lambda_A{}^{-1}\, tr_P\left(\overline{\widehat{\sigma}_A}\widehat{\mathcal{K}'}\widehat{\sigma}^{(thrm)}\right)$$

which shows two contributions to the departure from thermal equilibrium in $\widehat{\sigma}' = \widehat{\sigma}^{(0P)} + \widehat{\sigma}^{(diss)}$, with terms in $\Lambda_A = \Lambda_A^{(0P)} + \Lambda_A^{(diss)}$ coming from the projectile-P interaction rate, and from the dissipation rate. These two expansion coefficients are

$$\Lambda_A^{(0P)} = \lambda_A{}^{-1} tr_P\left[\overline{\widehat{\sigma}_A}\,\widehat{\mathcal{L}}_{0P}(0)\widehat{\sigma}^{(thrm)}\right]$$
$$\Lambda_A^{(diss)} = -\lambda_A{}^{-1} tr_P\left(\overline{\widehat{\sigma}_A}\,\widehat{\mathcal{R}}_P\widehat{\sigma}^{(thrm)}\right)$$

which can be calculated from the given trace expressions.

The perturbation results are incorporated into the deposition rate using the corrected RDOp within the trace in $R_{tot}(\vec{k} \to \vec{k}') = F_{tot}(t; \vec{k}', \vec{k}) = tr_P[\widehat{\rho}(t)\widehat{T}_{\vec{k}'\vec{k}}(t)^{\dagger}\widehat{T}_{\vec{k}'\vec{k}}(0)]$, with $\widehat{\rho}(t) = \widehat{\rho}^{(thrm)}(t) + \widehat{\rho}'(t)$ and $\widehat{\rho}'(t) = \widehat{U}_{\vec{v}}(t)^{\dagger}\widehat{\sigma}'\widehat{U}_{\vec{v}}(t) = \widehat{\rho}_{\vec{v}}^{(0P)}(t) + \widehat{\rho}_{\vec{v}}^{(diss)}(t)$. The trace simplifies insofar the transition amplitudes $\widehat{T}_{\vec{k}'\vec{k}}$ commute with $\widehat{U}_{\vec{v}}(t)$, which then can be removed with $\widehat{\rho}(t)$ replaced in the trace by $\widehat{\sigma} = \widehat{\sigma}^{(thrm)} + \widehat{\sigma}'$ and this given by a sum over A-eigenoperators $\widehat{\sigma}_A$. The rate function for the P-region in interaction with the S-region finally becomes

$$F_{tot}\left(t;\vec{k}',\vec{k}\right)=F_{thrm}\left(t;\vec{k}',\vec{k}\right)+F_{0P}\left(t;\vec{k}',\vec{k}\right)+F_{diss}\left(t;\vec{k}',\vec{k}\right)$$

$$F_{thrm}\left(t;\vec{k}',\vec{k}\right)=tr_P\left[\widehat{\sigma}^{(thrm)}\widehat{T}_{\vec{k}'\vec{k}}(t)^{\dagger}\widehat{T}_{\vec{k}'\vec{k}}(0)\right]$$

$$F_{0P}\left(t;\vec{k}',\vec{k}\right)=\sum\nolimits_A\Lambda_A^{(0P)}tr_P\left[\widehat{\sigma}_A\widehat{T}_{\vec{k}'\vec{k}}(t)^{\dagger}\widehat{T}_{\vec{k}'\vec{k}}(0)\right]$$

$$F_{diss}\left(t;\vec{k}',\vec{k}\right)=\sum\nolimits_A\Lambda_A^{(diss)}tr_P\left[\widehat{\sigma}_A\widehat{T}_{\vec{k}'\vec{k}}(t)^{\dagger}\widehat{T}_{\vec{k}'\vec{k}}(0)\right]$$

These expressions can be further developed as done in Section 2.2, to describe fast beam excitation of electronic motions, now in interaction with the *S*-region, with a non-equilibrium generalized oscillator strength. Also, slow beam particles undergoing multiple scattering by clusters in the *P*-region while they interact with the *S*-region follow from binary rates $F^{(C,C')}(t;\vec{k}',\vec{k})$ obtained from non-equilibrium TCFs of *C*-cluster transition operators.

3.3 Outline of treatments for calculations of energy deposition including medium dissipative dynamics

In applications of the above expressions to specific energy deposition events, it is possible to do calculations of the associated rates introducing expansion basis sets $\{|a\rangle\}$ to construct matrix representations of operators and super-operators (or Liouvillians) and to obtain the needed eigenmatrices $\boldsymbol{\sigma}_A = [\langle a|\widehat{\sigma}_A|a'\rangle]$ and eigenvalues λ_A.

The choice of basis set is dictated by the type of structures found in the *P*-region. If these are relatively small molecules like N_2 or H_2O, it is possible to construct a basis set $\{|a\rangle=|I\rangle\otimes|\boldsymbol{\nu}_I\rangle\}$ of vibronic states, with $|I\rangle$ a molecular electronic state and $|\boldsymbol{\nu}_I\rangle$ a vibrational state in the *P*-region. Then the first rate term, $F_{thrm}(t;\vec{k}',\vec{k})$, follows for fast projectiles from the generalized oscillator (GOS) function of the molecular electrons, constrained by a Frank-Condon factor from vibrational state overlaps[4,9,10] However, calculation of the two following rate terms F_{0P} and F_{diss} involve also a more general, non-equilibrium GOS containing matrix elements of type $\langle a|\widehat{\rho}_{\vec{q}}(t)\widehat{\rho}_{\vec{q}'}(0)|a'\rangle$. For slower projectile velocities, involving non-equilibrium TCFs of transition operators for cluster scattering, relevant matrix elements are similarly of the form $\langle a|\widehat{T}^{(C)}_{\vec{k}'\vec{k}}(t)^{\dagger}\widehat{T}^{(C')}_{\vec{k}'\vec{k}}(0)|a'\rangle$

When the *P*-region contains a large structure with many atoms, such as a polymer, a biomolecule, or a solid surface, the large number of active electronic states requires other practical choices for the basis set and for vibrational motions. This can be computationally implemented in a treatment which introduces the one-electron density operator $\widehat{\sigma}^{(1)}(t)$ and its EOM including the dissipation rate, involving also one-electron effective Hamiltonians and dissipative rate superoperator. Here relevant operators and states are expanded in a basis set of one-electron orbitals $\{|a\rangle = |j\rangle\}$ generated with a density functional for the *P*-region. The electronically non-adiabatic transitions induced by vibrational motions can be described in terms of ab initio molecular dynamics. This approach has been applied to the treatment of optical properties of semiconductor surfaces[21,29] and could be useful in applications to energy deposition in the presence of dissipative media.

In this approach it is necessary to generate one-electron density matrices $\boldsymbol{\sigma}^{(1)} = [\langle j|\widehat{\sigma}^{(1)}|j'\rangle]$ and eigenmatrices $\boldsymbol{\sigma}_A{}^{(1)} = [\langle j|\widehat{\sigma}_A{}^{(1)}|j'\rangle]$. For high velocity projectiles, the density amplitudes $\widehat{\rho}_{\vec{q}}$ are one-electron operators which can be expanded in the basis set of electron orbitals. At low projectile velocities, the projectile-cluster transition operator can be approximated as $\widehat{T}_{0C} \cong \widehat{t}_{0C}$, for a projectile-(cluster electron) interaction and calculated also in the basis set of orbitals. The generation of an orbital basis set within density functional theory, and of vibronic transitions from ab-initio dynamics, require careful choices of density functionals needed to generate correct electronic density of states, and have been the subject of RDOp treatments of optical properties in many-atom systems, in our previous work.[21,32]

4. Conclusion

The above treatment has combined two formalisms to provide a general theory of energy deposition in a target which interacts with a surrounding medium. The first formalism involves expressing inelastic collision cross sections in terms of time-correlation functions of scattering operators, usually given in the literature for a medium at thermal equilibrium. The second formalism introduces a reduced statistical density operator dependent only on variables of a primary region of interest, and accounting for fluctuation and dissipation phenomena arising from interactions of the target with its medium.

The combined treatment leads to energy deposition rates corrected to include a term for disturbance by the projectile of the target original equilibrium, and a term accounting for dissipation phenomena in the target.

Whether these two corrections are important in an application would depend on experimental conditions, involving the projectile mass and charge, its velocity, and whether the target temperature is, to begin with, close to room temperature or larger.

Acknowledgments

This work has been partly supported by the NSF of the USA and by the Dreyfus Foundation. It is a pleasure to dedicate this publication to John R. Sabin, long-time colleague and friendly companion, and great conversationalist at many lunch occasions.

Work partly supported by the Chemistry Division of the National Science Foundation of the USA, with previous grants NSF 1011967 and NSF 1445825, by the Dreyfus Foundation, and by the University of Florida High Performance Computing facility.

References

1. Bethe, H. A. Theory of the Passage of Fast Corpuscular Rays Through Matter (Translation From the German). *Ann. Phys. (Leipzig)* **1930**, *5*, 325–400.
2. Bloch, F. The Slowdown of Rapidly Moving Particles in Their Passing Throughsolid Matter. *Ann. Phys. (Leipzig)* **1933**, *16*, 285–320.
3. Inokuti, M. Inelastic Collisions of Fast Charged Particles With Atoms and Molecules: The Bethe Theory Revisited. *Rev. Mod. Phys.* **1971**, *43*, 297–347.
4. Inokuti, M.; Itikawa, Y.; Turner, J. E. Addenda: Inelastic Collisions of Fast Charged Particles With Atoms and Molecules the Bethe Theory Revisited. *Rev. Mod. Phys.* **1978**, *50*, 23.
5. Bethe, H. A.; Jackiw, R. *Intermediate Quantum Mechanics*, 2nd ed.; W. A. Benjamin, Inc.: New York, 1968.
6. Oddershede, J.; Sabin, J. R. Orbital and Whole Atom Stopping Power and Shell Corrections for Atoms With Z< 36. *At. Dat. Nucl. Dat. Tables* **1984**, *31*, 275–297.
7. Cabrera-Trujillo, R.; Sabin, J. R.; Oddershede, J.; Sauer, P. A. The Bethe Sum Rule and Basis Set Selection in the Calculation of Generalized Oscillator Strengths. *Adv. Quantum Chem.* **1999**, *35*, 175–192.
8. Sabin, J. R.; Oddershede, J. Shell Corrections to Electronic Stopping Powers from Orbital Mean Excitation Energies. *Phys. Rev. A* **1982**, *26*, 3209.
9. Sauer, S. P. A.; Sabin, J. R.; Oddershede, J. Calculated Mean Excitation Energies for Some Small Molecules. *Nucl. Instr. Meth. B* **1995**, *100*, 458–463.
10. Sauer, S. P. A.; Oddershede, J.; Sabin, J. R. Mean Excitation Energies for Biomolecules-Glycine to DNA. *Adv. Quantum Chem.* **2011**, *62*, 215–242.
11. Micha, D. A.; Vilallonga, E. F. The Collisional Time-Correlation Function Approach to Molecular Energy Transfer. In *Adv. Chem. Phys.*; Prigogine, I., Rice, S. A., Eds.; 84; J. Wiley and Sons Inc.: New York, 1993; pp. 1–72.
12. Vilallonga, E.; Micha, D. A. The Calculation of Time-Correlation Functions for Molecular Collisions. *Phys. Rep.* **1992**, *212*, 329–390.
13. Kubo, R.; Toda, M.; Hashitsume, N. *Statistica Physics II. Non-equilibrium Statistical Mechanics*; Springer-Verlag: Heidelberg, Germany, 1978.
14. Schatz, G. C.; Ratner, M. A. *Quantum Mechanics in Chemistry*; Prentice-Hall, Inc.: Englewood Cliffs, NJ, 1993.
15. Blum, K. *Density Matrix Theory and Applications*, 2nd ed.; Plenum Press: New York, 1996.
16. Mukamel, S. *Principles of Nonlinear Optical Spectroscopy*; Oxford University Press: Oxford, England, 1995.

17. May, V.; Kuhn, O. *Charge and Energy Transfer Dynamics in Molecular Systems*, 2nd ed.; Wiley-VCH Verlag: Weinheim, Germany, 2004.
18. Nitzan, A. *Chemical Dynamics in Condensed Phases*; Oxford University Press: Oxford, England, 2006.
19. Micha, D. A. Density Matrix Theory and Computational Aspects of Quantum Dynamics in an Active Medium. *Int. J. Quantum Chem.* **2000**, *80*, 394–405.
20. Micha, D. A. Density Matrix Calculations of Gaseous and Adsorbate Dynamics in Electronically Excited Molecular Systems. *Int. J. Quantum Chem.* **2006**, *106*, 3371–3382.
21. Micha, D. A. Generalized Response Theory for a Photoexcited Many-Atom System. In *Advances in Quantum Chemistry*; 71; Elsevier-Academic Press: Amsterdam, Holland, 2015; pp. 195–220.
22. Hembree, R.; Vazhappilly, T.; Micha, D. A. Quantum Confinement Effects on Electronic Photomobilities at Nanostructured Semiconductor Surfaces: Si(111) Without and With Adsorbed Ag Clusters. *J. Chem. Phys.* **2017**, *147*, 224703.
23. Vazhappilly, T.; Micha, D. A. Optical properties of the TiO2(110) surface with adsorbed Ag atoms relevant to photocatalysis and photovoltaics. In *Computational Photocatalysis: Modeling of Photophysics and Photochemistry at Interfaces*; Kilin, D., Kilina, S., Han, Y., Eds.; American Chemical Society: Washington DC, 2019; pp. 47–66.
24. Rodberg, L. S.; Thaler, R. M. *Introduction to the Quantum Theory of Scattering*; Academic Press: New York, 1967.
25. McQuarrie, D. A. *Statistical Mechanics*; Harper & Row Publ: New York, 1973.
26. Micha, D. A. *Molecular Interactions. Concepts and Methods*; John Wiley & Sons: Hoboken, NJ, USA, 2020.
27. Atkins, P.; de Paula, J. *Physical Chemistry*, 9th ed.; WHFreeman and Co: New York, 2010.
28. Von Neumann, J. *Mathematische Grundlagen Der Quantenmechanik* (Transl. by R. T. Beyer, Princeton Univ. Press, New Jersey), 1932; Springer-Verlag: Berlin, 1955.
29. Kilin, D. S.; Micha, D. A. Photovoltage of Nanostructures on Si Surfaces: Ab Initio Results. *J. Phys. Chem. C* **2009**, *113*, 3530.
30. Messiah, A. *Quantum Mechanics*, 1; North-Holland: Amsterdam, 1962.
31. Kilin, D. S.; Micha, D. A. Relaxation of Photoexcited Electrons at a Nanostructured Si(111) Surface. *J. Phys. Chem. Lett.* **2010**, *1*, 1073.
32. Vazhappilly, T.; Han, Y.; Kilin, D. S.; et al. Electronic Relaxation of Photoexcited Open and Closed Shell Adsorbates on Semiconductors: Ag and Ag2 on TiO2. *J. Chem. Phys.* **2022**, *156*, 104705. https://doi.org/10.1063/5.0082748.

CHAPTER TWO

Shape-sensitive inelastic scattering from metallic nanoparticles

S. Peter Apell[a,b,*], Gautam Mukhopadhyay[a,c], Tomasz J. Antosiewicz[a,d], and Javier Aizpurua[b]

[a]Department of Physics and Gothenburg Physics Center, Chalmers University of Technology, Göteborg, Sweden
[b]Centro de Física de Materiales, Centro Mixto CSIC-UPV/EHU and Donostia International Physics Center (DIPC), Donostia-San Sebastián, Spain
[c]Department of Physics, Indian Institute of Technology Bombay, Powai, Mumbai, India
[d]Faculty of Physics, University of Warsaw, Warsaw, Poland
*Corresponding author: e-mail address: apell@chalmers.se

Contents

Abstract

We investigate the interplay of shape changes and localized surface plasmons in small metal particles with the potential of a large enhancement of the inelastic signal from the particles' own vibrations. The framework is a geometrical one where we study the change in geometric factors during the vibrational movement. The resulting cross-section is found to be of a detectable order of magnitude, however, much smaller than the elastic cross-section.

1. Raman scattering from single metal particles

Raman scattering is a powerful tool to understand the interplay between vibrational and electronic degrees of freedom in a system. For many years, this has been a very useful tool in molecular physics to identify specific vibrational fingerprints of target molecules now reaching the single molecule

Advances in Quantum Chemistry, Volume 85
ISSN 0065-3276
https://doi.org/10.1016/bs.aiq.2022.04.001
Copyright © 2022 Elsevier Inc.
All rights reserved.

detection limit.[1–5] From the inelastically scattered light, which is vibrationally shifted with respect to the incoming light, we can infer important properties of the system at hand. One of the major goals in this area has for many years been to be able to monitor the Raman scattering using individual molecules to probe a particular local environment. A ubiquitous problem in this case has been to get enough field strength to pick up an informative signal, often handicapped by signals averaged over many molecules and hence many local environments. Considerable progress has therefore been made in this area using the excitation of local electromagnetic resonances in order to enhance both the incoming and the scattered radiation fields. In particular, small metal particles are very interesting from this point of view with their characteristic surface plasmon modes which are tunable to a particular energy by simply changing their shape, size, or local dielectric environment, and produce a very large and localized field enhancement which dramatically boosts the inelastic scattered signal. The advent of new fabrication and detection techniques has allowed for, in principle, accurate measurement of the Mie scattering spectrum of a single particle. Hence, one should also be able to pick up the inelastic Mie scattering part, especially if it can be enhanced by the aforementioned local surface plasmon/polariton resonances.

Inelastic scattering from surfaces with or without molecules has a long history.[6–12] It was followed up focusing on details in single particle or molecular vibrations[13–20] and was preceded by colloidal studies.[21–23] A special direction of research in the inelastic scattering of metallic nanoparticles has been the study of vibrations of elongated metallic particles.[24–29] In recent years,[30–40] the focus has shifted toward increasingly more elaborated systems such as arrays of particles, higher-order multipoles, and the influence of the surrounding medium. In this context, we should remind the reader about a development in cluster physics[41–43] approaching from below in size the development indicated above. The study of vibrational modes of metallic nanoparticles in Raman spectroscopy has been complemented during the last years with time-domain studies of the vibrations[44,45] as well as with transient absorption spectroscopy.[46]

Typical vibrational energies are in the range of a few meV for nanosized particles. Classic elasticity theory is often used where the frequency of vibration is related to the corresponding sound velocity and in the simplest models inversely proportional to size of the particle.[47] For a spherical particle, one distinguishes between spheroidal modes (purely radial displacements) and torsional modes (purely tangential displacements).

Duval early on pointed out[48,49] that $L = 0$, 2 spheroidal modes are Raman active, but all torsional ones are Raman inactive. Bachelier and Mlayah[50] investigated in what way a vibration of a particle could couple to surface plasmon polaritons and found two contributions. One is a modulation of the interband susceptibility via deformation potential changes due to purely radial vibrations. The other is a more surface localized effect where quadrupolar vibrations modulate the surface polarization. A change of particle shape gives a change in the surface polariton polarization vectors. Pure transverse and pure radial vibrations are however not active since such modes do not change the shape of the particle and hence to lowest order only quadrupolar spheroidal vibrations contribute.

Several studies have been performed on rods where a breathing mode dominates, a pure radial mechanical deformation of the rod with no change in length. Also extensional modes induced by laser heating have been demonstrated where the rod waist oscillates out and in changing both length and width. If the rod has indentations, they can act as acoustic hotspots with both surface plasmons and strong shape deformations in the same place.[51] Recently, more complex configurations of metallic coupled nanoparticles have been shown to produce selection-rule breaking in inelastic light scattering.[40] For both rods and particles, a thermal expansion can trigger anisotropic shape oscillations implying the ionic background is moving, hence interacting with the electronic system and probing its responsiveness of setting up local plasma oscillations.

In what follows, we investigate the Raman scattering properties of individual metal particles with an emphasis on shape effects. We make realistic estimates of anticipated cross-sections to validate the feasibility of experimentally detecting them. It turns out that a type of quadrupolar geometric factors together with the more well-known dipolar geometric factors completely characterize the shape-related part of the vibrational response. This in turn can get greatly enhanced provided one can meet the condition to set up a local surface plasmon excitation in the particle. We use a simple classical electromagnetic field description throughout the chapter.

2. Raman cross-section

For a polarizable object irradiated with light of frequency ω_{inc}, there will be an induced dipole moment $\vec{\mu}$ proportional to the amplitude $\vec{E}$ of the total electric field acting:

$$\mu_i(t) = \sum_j \alpha_{ij} \cdot E_j \cos(\omega_{\text{inc}} t), \tag{1}$$

where i, j denote appropriate coordinates. The "constant" of proportionality, α_{ij}, is the polarizability tensor of the particle and contains all its symmetry properties. If the particle performs an oscillatory motion, in our case an acoustic vibration with normal mode frequency ω_{nom}, we may write to lowest order

$$\alpha_{ij}(t) = \alpha_{ij}^0 + \sum_k \left.\frac{\partial \alpha_{ij}}{\partial q_k}\right|_o q_k \cos(\omega_{\text{nom}} t) + \cdots \tag{2}$$

Here, α_{ij}^0 is the polarizability tensor for the equilibrium situation and q_k is a normal mode coordinate introducing an index k pertaining to the polarization of the normal mode. There is of course a whole set of selection rules depending on the symmetry of the mode k, the field acting, as well as different types of normal modes. However, we address these issues when pertinent further down. Combining Eqs. (1) and (2) and, for illustrational purposes, restricting to a situation where the polarizability tensor is diagonal (but not necessarily isotropic), we have

$$\begin{aligned}\mu_i(t) &= \alpha_{ii}^0 E_i \cos(\omega_{\text{inc}} t) \\ &+ \frac{1}{2}\sum_k E_i \left.\frac{\partial \alpha_{ii}}{\partial q_k}\right|_o q_k \{\cos[(\omega_{\text{inc}} - \omega_{\text{nom}})t], + \cos[(\omega_{\text{inc}} + \omega_{\text{nom}})t]\} + \cdots\end{aligned} \tag{3}$$

The first term corresponds to the standard elastic Rayleigh scattering from an oscillating dipole (same frequency as ω_{inc}) and the second term gives rise to Raman scattering at the so-called Stokes and anti-Stokes frequencies with a cross-section σ_R being proportional to the square of the change in the polarizability as the particle is vibrating. For our purposes, we use the simplest possible form for the cross-section according to Ref. 52 (c being the velocity of light)

$$\sigma_R \cong \left(\frac{\omega_{\text{inc}}}{c}\right)^4 \left|\sum_k q_k \left.\frac{\partial \alpha_{ii}}{\partial q_k}\right|_o\right|^2, \tag{4}$$

neglecting any shift of the incoming and outgoing frequencies because of the vibrational mode.

3. Particle polarizability

In order to proceed, we now introduce a generic polarizability for our particle which illustrates in a realistic way how shape and also the size of the particle influence the Raman cross-section. To this order, we use the dipolar polarizability for an ellipsoidal particle (p) of volume V in a surrounding medium (d)[53]

$$\alpha_{ii}(\omega, L) = V\epsilon_d \frac{\epsilon_p(\omega) - \epsilon_d(\omega)}{\epsilon_d(\omega) + L_i[\epsilon_p(\omega) - \epsilon_d(\omega)]}, \tag{5}$$

where ϵ_p and ϵ_d are the dielectric permeability of the particle and surrounding medium, respectively. For the simplest order of magnitude estimates we will use $\epsilon_d = \epsilon_0$, the free space value, and for ϵ_p the simplest possible free electron like Drude form for a metallic particle.

L_i in the denominator of Eq. (5) is a geometric factor often mistakenly called depolarization factor (see p. 147 of Bohren and Huffman[53]) and corresponding to the direction i of the applied field. It depends only on the shape of the particle.[53,54] In general, $L_1 + L_2 + L_3 = 1$, and they can be evaluated analytically for the case of spheroids (prolate: $a_1 = a_2 < a_3$; oblate $a_1 = a_2 > a_3$) where a_1, a_2, and a_3 are the lengths of the semi-axes of the ellipsoid along x, y, and z, respectively. For what comes below, we only need L_3 ($L_1 = L_2 = (1 - L_3)/2$)

$$\begin{aligned} L_3 &= \frac{a_1 a_2 a_3}{2} \int_0^\infty ds \frac{1}{f(s)(a_3^2 + s)} \\ &= \frac{e^2 - 1}{e^2}\left[1 + \frac{1}{2e} \ln \frac{1-e}{1+e}\right] \quad \text{when } a_1 = a_2, \end{aligned} \tag{6}$$

defining $e^2 = 1 - a_1^2/a_3^2$ and $f(s) \equiv \sqrt{(a_1^2 + s)(a_2^2 + s)(a_3^2 + s)}$. In the spherical limit ($e = 0$), $L_3 = 1/3$, and it goes to zero when $e \rightarrow 1$ (needle). Oblate expressions can be obtained in a straightforward manner from the literature.[53]

When the particle is executing a vibrational motion (whose basic mechanisms we will not consider in detail in this chapter), it has the potential, depending on symmetry, to influence both the size and shape of the particle. This in turn will change the volume as well as the geometric factors involved in the expression for the polarizability in Eq. (5). Since we are mainly considering a resonant Raman scattering situation to have as large effect as possible, shape effects (L) dominate size effects (V) as shown and discussed further in Appendix. There we also address three effects which go beyond Eq. (5): radiation damping, dynamic depolarization, and change in volume influencing the plasma frequency of the particle, while we at this stage continue to analyze the simpler situation to see if the resulting cross-section is large enough to be detected in a reasonable experimental setup.

We want to find how the polarizability of the particle changes with respect to a vibrationally induced deformation of the particle, in a particular direction. We therefore rewrite according to

$$\left.\frac{\partial \alpha_{ii}}{\partial q_j}\right|_o = \left.\frac{\partial \alpha_{ii}}{\partial L_i}\frac{\partial L_i}{\partial a_j}\frac{\partial a_j}{\partial q_j}\right|_o . \tag{7}$$

It is straightforward to find $\partial \alpha_{ii}/\partial L_i = -\alpha_{ii}^2/V$. For the last derivative in Eq. (7), we use the following lowest-order Taylor expansion for the semi-axis j

$$a_j \equiv a_j^o + q_j \left.\frac{\partial a_j}{\partial q_j}\right|_o , \tag{8}$$

only allowing for a change in direction i from a normal coordinate change in direction j. While the focus is on a single normal mode only, in practice these oscillations can affect all relevant dipolar and quadrupolar quantities by virtue of changing the L-shape factor and/or particle volume.

Introducing a relative displacement $\delta_j \equiv (a_j - a_j^o)/a_j^o$ and a dimensionless shape change tensor:

$$\Gamma_{ij} \equiv a_j^o \frac{\partial L_i}{\partial a_j}, \tag{9}$$

we can write for our Raman cross-section:

$$\sigma_R \cong \left(\frac{\omega_{\text{inc}}}{c}\right)^4 \left|\sum_k \left.\frac{\partial \alpha_{ii}}{\partial q_k}\right|_o\right|^2 = \left(\frac{\omega_{\text{inc}}}{c}\right)^4 \left|\frac{1}{V}\alpha_{ii}^2 \sum_j \Gamma_{ij}\delta_j\right|^2 . \tag{10}$$

The implication of the findings above is that if we have incoming light close to a surface collective electronic resonance of the particle, we anticipate an enormous enhancement provided the resonance is reasonably sharp. A more detailed treatment gives that "one" polarizability should be evaluated at the incoming photon energy and the other "one" at the vibrationally shifted energy, potentially yielding a smaller overall enhancement in general.

Now most surface plasmons are much broader than the few meV of the vibrational excitation for the situation at hand, so we do not see this as any major weakening of the possible enhancement.

4. Shape factor for Raman scattering

We now focus on the shape-change tensor Γ_{ij} as defined in Eq. (9), i.e., by taking the relevant derivatives of the geometric factor L_i pertaining to the direction i with respect to a change of particle dimension in the direction j. In order to facilitate this, we first need to introduce a dimensionless factor L_{ij} carrying the quadrupolar character of the distortion giving rise to the Raman scattering. Following the structure of L_i, we define L_{ij} as

$$L_{ij} = \frac{a_1 a_2 a_3 a_i^2}{2} \int_0^\infty \frac{ds}{f(s)(a_i^2 + s)(a_j^2 + s)}, \quad f(s) = \sqrt{(a_1^2 + s)(a_2^2 + s)(a_3^2 + s)}. \tag{11}$$

We will call L_{ij} the geometric distortion tensor. The symmetry of the problem ($a_1 = a_2$) makes it enough to know L_{11}, L_{13}, and L_{33}. Now we have that $L_{12} = L_{21} = L_{22} = L_{11}$ and $L_{31} = L_{32} = L_{13}/(1 - e^2) = L_{23}/(1 - e^2)$. e was defined in connection with Eq. (6). However like all L_i, L_{ij} fulfill sum-rules[55] so $L_{11} = (1 - L_{13})/4$ and $L_{33} = (1 - 2L_{13}/(1 - e^2))/3$ which means it is enough to know L_{13} to generate all the other L_{ij}'s. In fact, using yet another sum-rule $4L_{11} + L_{13}/(1 - e^2) = 3L_1 = 3(1 - L_3)/2$ we can even generate L_{13} from L_3 alone

$$L_{13} = \frac{1 - e^2}{2e^2}(1 - 3L_3) = \frac{1 - e^2}{2e^4}\left[3 - 2e^2 + \frac{3}{2e}(1 - e^2)\ln\frac{1 - e}{1 + e}\right]. \tag{12}$$

L_{13} starts out from 1/5 in the spherical limit ($e = 0$) and goes to 0 in the needle limit ($e = 1$). For most of the shape range, it has very little variation. The same is true for the other L_{ij}'s which are also of about the same order of magnitude.

From the definition of Γ_{ij} in Eq. (9), combined with $L_1 + L_2 + L_3 = 1$, it follows immediately that $\sum_i \Gamma_{ij} = 0$. This together with the general symmetry properties at hand and the sum-rules implies that $\Gamma_{22} = \Gamma_{11}$, $\Gamma_{12} = \Gamma_{21}$, $\Gamma_{32} = \Gamma_{31} = -(\Gamma_{11} + \Gamma_{12})$, and $-\Gamma_{33}/2 = \Gamma_{23} = \Gamma_{13}$. Hence, it is enough to calculate Γ_{13}. Performing the derivative in Eq. (9), we get

$$\Gamma_{13} = L_1 - L_{31} = \frac{3(e^2-1)}{2e^4}\left[1 + \frac{1-e^3/3}{2e}\ln\frac{1-e}{1+e}\right]. \tag{13}$$

It starts out from $-2/15$ for a sphere ($e = 0$) and goes toward 0 in the needle limit ($e = 1$). Overall this means that all the Γ_{ij} factors are of the order 0.1 for our estimates below.

5. Estimates and discussion

From Eq. (10), we can derive that for a good quality resonance in the polarizability the Raman scattering cross-section is proportional to the fourth power of the corresponding quality factor of the resonance (resonance frequency divided by damping). This is important since we found a very small value of $\Gamma_{ij} \simeq 0.1$ above and we will find that δ_j is even smaller. The general elastic properties of metals and the surface localization of the distortion due to the vibration in itself imply that δ_j should be rather small. A simple estimate is $\delta_j \approx l_j/a^o_j$ where l_j is the lattice constant in direction j, as a measure of the maximum deflection possible. Using the Lindeman melting criterion, a 10% change would be too much for the system. We therefore take the conservative value $\delta_j \approx 0.01$ for our estimates in what follows. Notice also that if there is a relative stretch δ_j which is the same in all directions as for a breathing mode, however large amplitude it has, our theory conforms with the earlier findings that this mode is not Raman active expressed as $\sum_i \Gamma_{ij} = 0$.

We now have the ingredients necessary for estimating the cross-section. With a Drude dielectric function characterized by plasma frequency ω_p and scattering rate δ, we can write

$$\sigma_R = \left(\frac{\omega_{\text{inc}}}{c}\right)^4 \left|\alpha_{ii}^2 \sum_j \frac{\Gamma_{ij}\delta_j}{V}\right|^2 \approx \frac{V_{\text{eff}}^2}{l_o^4}, \tag{14}$$

where we have introduced an effective volume $V_{\text{eff}} = V\Gamma\delta$, which for a particle of typical dimension of 20 nm and values for Γ and δ above will be around $(4\ \text{nm})^3$. This should be compared with the square of the electromagnetic length-scale $l_o \equiv c\delta/\omega_p^2 \approx 100\text{nm}$, yielding a typical cross-section of 10^{-22}m^2.

This is a value which carries a positive possibility to explore given typical Raman cross-sections for molecules. The problem is of course that the vibrations are in the meV range which makes it difficult for a Raman spectrometer to see the difference between the inelastic Raman and the elastic Rayleigh scattering. On top of this, we have the traditional problems with small metal particles with respect to size distribution and interactions with substrate or between the particles. Using our model estimate, we have the following ratio for Raman to Rayleigh

$$\frac{\sigma_R}{\sigma_{\mathrm{Ray}}} = \left| \alpha_{ii} \sum_j \frac{\Gamma_{ij}\delta_j}{V} \right|^2 \approx 10^{-6} \left(\frac{\omega_o}{L\delta} \right)^2 \approx 10^{-3}, \tag{15}$$

with the quality factor of the order of 10. Thus, the Raman signal is within the elastic signal and it is also three orders of magnitude smaller. However, the experimental techniques in the area, see for instance Ref. 26, give hope that such a signal can be detected and shed light on the possible enhancement of a Raman signal due to the localized surface plasmons excited on the particle. A final point which impacts the visibility of the signal is related to temperature. The mechanical vibrations considered here and, hence, their inelastic cross-sections are strongly influenced by temperature. The phonon population at room temperature for typical vibrations in the order of meV is well below 1. Therefore, the ratio between anti-Stokes and Stokes is always around 10^{-2} to 10^{-1} and both contributions will not be equal.

Appendix

In order to evaluate the derivative of the polarizability with respect to the direction of a deformation with a volume change of the particle, we write

$$\sum_j q_j \frac{\partial \alpha_{ii}}{\partial q_j}\bigg|_o = \frac{\partial \alpha_{ii}}{\partial V} \left(\sum_j \frac{\partial V}{\partial a_j} q_j \frac{\partial a_j}{\partial q_j}\bigg|_o \right). \tag{A.1}$$

The factor within parenthesis can be dealt with first. Using Eq. (8), the last derivate in Eq. (A.1) can be combined with q_j to form $a_j\delta_j$ for the result

$$\sum_j q_j \frac{\partial \alpha_{ii}}{\partial q_j}\bigg|_o = V \frac{\partial \alpha_{ii}}{\partial V} \sum_j \delta_j. \tag{A.2}$$

To get to this, we also used that $a_j \partial V / \partial a_j = V$ for the spheroids we are considering. Thus in what follows, it is enough to pay attention to the volume derivative of the polarizability. In principle, we have three contributions:

The *first* is the explicit volume dependence in the polarizability itself whereby $\partial a_{ii}/\partial V = \alpha_{ii}/V$. It is seen to be of first order in the polarizability as compared to the treatment in the main body of the chapter. Close to resonance we thus expect this contribution to be small in comparison, all else being equal.

The *second* contribution comes from a volume dependence in L itself when we include radiation damping since this is known to be important when particles get larger. For a simple estimate, we do the customary replacement $\alpha \to \alpha/(1 - 2ik^3\alpha/3)$ (see Ref. 12), where k is the wave number of the incoming radiation. We then find that the result above is reduced even further by a factor $(1-2ik^3\alpha/3)^2$.

The *third* contribution is the possible dependence a volume change has on the plasma frequency of the free electron gas. To this end, we get (L being the geometric factor in direction i)

$$\frac{\partial \alpha_{ii}}{\partial V} \equiv L\left(\frac{\alpha_{ii}}{V}\right)^2. \tag{A.3}$$

It is of second order in the polarizability as the main contribution in this chapter and in size it is L as compared to Γ, i.e., of the same order of magnitude.

Acknowledgments

S.P.A. acknowledges hospitality and support from CFM/DIPC, San Sebastian, Spain.

I (S.P.A.) met Jack Sabin for the first time in the highly inspiring Physics milieu in the Basque town of Donostia nearly 30 years ago. With the enduring work of Professor Pedro Echenique and his colleagues one had for a number of years built up a world class international environment. Little did I realize at the time that all of us who met there contributed to the whole atmosphere. Different areas of physics research, different ways of doing physics and cultural differences in general contributed to this.

Jack is the kind of person you very easily get to talk to and quickly get involved in common subjects. I especially remember how we combined his and the Gainesville Quantum Theory Project people's knowledge in the chemistry aspect of nature with my own expertise in surface electrodynamics to write our Viagrapaper as Jack called it. His humor and wit is always an inspiration. The actual journal title was *Molecular shape, capacitance, and chemical hardness* and appeared in the Zerner festschrift in this very journal. Altogether we have written 16 papers. I now realize that Jack at an early stage could see in what direction to proceed and what expertise to involve.

This volume is to Jack as the scientist. He has also been a highly influential mentor in my private life where I have been torn between an academic and an administrative career. He influenced my decisions at many critical moments—especially pushing the idea of grasping the chance when it comes along. With children about the same age, we shared a lot of inspiration and problem solving when it comes to deal with real-life issues.

And as the saying goes, "behind every great man there is even a greater woman." It is not possible writing a tribute to Jack without mentioning his wonderful and talented wife Birgit.

Since the Sabins spend the summers in Nyborg, Denmark, and we live just across the strait in Sweden my family share the good summer meetings we've had. At these meetings, I also visited Jack at the University in Odense where he had a desk and I learned another truth from hi "Don't be a door mat to someone else if you want to do good and independent research." At the University, and also at various conferences I went to, I was often approached by well-established figures in the field giving the advice—"Young man look out who you associate yourself with since it can and will affect your reputation." I must therefore say that I'm both grateful and proud to have been able to collaborate with Jack during the years.

I will conclude with one final memory from when there was a conference in honor of his 70th birthday at a wonderful place just outside of Sønderborg in the very south of Denmark. To this very day, I cannot but remember one remark he made in his personal speech at that conference—"When you wake up at my age and it does not hurt in your body anywhere—you're most likely dead." Reaching 70 myself this year this example of Jack's witty humor always with a twist—almost like a punch in your face—symbolizes also the way I remember his way of doing physics. There was always, even in a very old and well-established topic as Stopping Power, a mind-blowing blow somewhere in the text.

Peter Apell

References

1. Moskovits, M. Surface-Enhanced Spectroscopy. *Rev. Mod. Phys.* **1985**, *57* (3), 783–826. https://doi.org/10.1103/RevModPhys.57.783.
2. Nie, S.; Emory, S. R. Probing Single Molecules and Single Nanoparticles by Surface-Enhanced Raman Scattering. *Science* **1997**, *275* (5303), 1102–1106. https://doi.org/10.1126/science.275.5303.1102.
3. Kneipp, K.; Wang, Y.; Kneipp, H.; Perelman, L. T.; Itzkan, I.; Dasari, R. R.; Feld, M. S. Single Molecule Detection Using Surface-Enhanced Raman Scattering (SERS). *Phys. Rev. Lett.* **1997**, *78* (9), 1667–1670. https://doi.org/10.1103/PhysRevLett.78.1667.
4. Xu, H.; Bjerneld, E. J.; Käll, M.; Börjesson, L. Spectroscopy of Single Hemoglobin Molecules by Surface Enhanced Raman Scattering. *Phys. Rev. Lett.* **1999**, *83* (21), 4357–4360. https://doi.org/10.1103/PhysRevLett.83.4357.
5. Zhang, R.; Zhang, Y.; Dong, Z. C.; Jiang, S.; Zhang, C.; Chen, L. G.; Zhang, L.; Liao, Y.; Aizpurua, J.; Luo, Y.; Yang, J. L.; Hou, J. G. Chemical Mapping of a Single Molecule by Plasmon-Enhanced Raman Scattering. *Nature* **2013**, *498*, 82–86. https://doi.org/10.1038/nature12151. https://www.nature.com/articles/nature12151.
6. Mills, D. L.; Chen, Y. J.; Burstein, E. Raman Scattering of Light by Polaritons in Thin Films; Surface Polaritons and Size Effects. *Phys. Rev. B* **1976**, *13* (10), 4419–4438. https://doi.org/10.1103/PhysRevB.13.4419.
7. Gersten, J. I.; Weitz, D. A.; Gramila, T. J.; Genack, A. Z. Inelastic Mie Scattering From Rough Metal Surfaces: Theory and Experiment. *Phys. Rev. B* **1980**, *22* (10), 4562–4571. https://doi.org/10.1103/PhysRevB.22.4562.
8. Weitz, D. A.; Gramila, T. J.; Genack, A. Z.; Gersten, J. I. Anomalous Low-Frequency Raman Scattering From Rough Metal Surfaces and the Origin of Surface-Enhanced Raman Scattering. *Phys. Rev. Lett.* **1980**, *45* (5), 355–358. https://doi.org/10.1103/PhysRevLett.45.355.

9. Wang, D. S.; Kerker, M. Enhanced Raman Scattering by Molecules Adsorbed at the Surface of Colloidal Spheroids. *Phys. Rev. B* **1981**, *24* (4), 1777–1790. https://doi.org/10.1103/PhysRevB.24.1777.
10. Laor, U.; Schatz, G. C. The Role of Surface Roughness in Surface Enhanced Raman Spectroscopy (SERS): The Importance of Multiple Plasmon Resonances. *Chem. Phys. Lett.* **1981**, *82* (3), 566–570. https://doi.org/10.1016/0009–2614(81)85442-5.
11. Arya, K.; Zeyher, R. Theory of Surface-Enhanced Raman Scattering From Molecules Adsorbed at Metal Surfaces. *Phys. Rev. B* **1981**, *24* (4), 1852–1865. https://doi.org/10.1103/PhysRevB.24.1852.
12. Wokaun, A.; Gordon, J. P.; Liao, P. F. Radiation Damping in Surface-Enhanced Raman Scattering. *Phys. Rev. Lett.* **1982**, *48* (14), 957–960. https://doi.org/10.1103/PhysRevLett.48.957.
13. Montagna, M.; Dusi, R. Raman Scattering From Small Spherical Particles. *Phys. Rev. B* **1995**, *52* (14), 10080–10089. https://doi.org/10.1103/PhysRevB.52.10080.
14. Rao, A. M.; Richter, E.; Bandow, S.; Chase, B.; Eklud, P. C.; Williams, K. A.; Feng, S.; Subbaswamy, K. R.; Menon, M.; Thess, A.; Smalley, R. E.; Dresselhaus, G.; Dresselhaus, M. S. Diameter-Selective Raman Scattering From Vibrational Modes in Carbon Nanotubes. *Science* **1997**, *275* (5297), 187–191. https://doi.org/10.1126/science.275.5297.187.
15. Oldenburg, S. J.; Hale, G. D.; Jackson, J. B.; Halas, N. J. Light scattering from dipole and quadrupole nanoshell antennas. *Appl. Phys. Lett.* **1999**, *75* (8), 1063–1065. https://doi.org/10.1063/1.124597.
16. Krenn, J. R.; Schider, G.; Rechberger, W.; Lamprecht, B.; Leitner, A.; Aussenegg, F. R.; Weeber, J. C. Design of Multipolar Plasmon Excitations in Silver Nanoparticles. *Appl. Phys. Lett.* **2000**, 77 (21), 3379–3381. https://doi.org/10.1063/1.1327615.
17. Xu, H.; Aizpurua, J.; Käll, M.; Apell, P. Electromagnetic Contributions to Single-Molecule Sensitivity in Surface-Enhanced Raman Scattering. *Phys. Rev. E* **2000**, *62* (3), 4318–4324. https://doi.org/10.1103/PhysRevE.62.4318.
18. Perner, M.; Gresillon, S.; März, J.; von Plessen, G.; Feldmann, J.; Porstendorfer, J.; Berg, K. J.; Berg, G. Observation of Hot-Electron Pressure in the Vibration Dynamics of Metal Nanoparticles. *Phys. Rev. Lett.* **2000**, *85* (4), 792–795. https://doi.org/10.1103/PhysRevLett.85.792.
19. Hernández-Rosas, J.; Picquart, M.; Haro-Poniatowski, E.; Kanehisa, M.; Jouanne, M.; Morhange, J. F. Elastic Vibrations of Spheroidal Nanometric Particles. *J. Phys. Condens. Matter* **2003**, *15* (44), 7481–7487. https://doi.org/10.1088/0953-8984/15/44/004.
20. Weigel, T.; Schulte, J.; Schweiger, G. Inelastic Scattering by Particles of Arbitrary Shape. *J. Opt. Soc. Am. A* **2006**, *23* (11), 2797–2802. https://doi.org/10.1364/JOSAA.23.002797. http://opg.optica.org/josaa/abstract.cfm?URI=josaa-23-11-2797.
21. Rzepka, E.; Taurel, L.; Lefrant, S. First-Order Raman Scattering Induced by Na and Ag Colloids in NaCl, NaBr and NaI. *Surf. Sci.* **1981**, *106* (1), 345–349. https://doi.org/10.1016/0039-6028(81)90221-1.
22. Shaw, K. D.; Garrell, R. L.; Krimm, S. Surface Enhanced Raman Spectroscopy of Halide Ions on Colloidal Silver: Simulation of Silver-Halide Ion Spectra. *Surf. Sci.* **1983**, *124* (2), 625–637. https://doi.org/10.1016/0039-6028(83)90816-6.
23. Mariotto, G.; Montagna, M.; Viliani, G.; Duval, E.; Lefrant, S.; Rzepka, E.; Maï, C. Low-Energy Raman Scattering from Silver Particles in Alkali Halides. *Europhys. Lett. (EPL)* **1988**, *6* (3), 239–243. https://doi.org/10.1209/0295-5075/6/3/009.
24. Kornyushin, Y.; Amusia, M. Y. On the Frequencies of Collective Electron Oscillations in Nanotubes. *Contemp. Phys.* **2001**, *42* (3), 159–165. https://doi.org/10.1080/00107510110048057.

25. Hu, M.; Wang, X.; Hartland, G. V.; Mulvaney, P.; Juste, J. P.; Sader, J. E. Vibrational Response of Nanorods to Ultrafast Laser Induced Heating: Theoretical and Experimental Analysis. *J. Am. Chem. Soc.* **2003**, *125* (48), 14925–14933. https://doi.org/10.1021/ja037443y.
26. Margueritat, J.; Gonzalo, J.; Afonso, C. N.; Mlayah, A.; Murray, D. B.; Saviot, L. Surface Plasmons and Vibrations of Self-Assembled Silver Nanocolumns. *Nano Lett.* **2006**, *6* (9), 2037–2042. https://doi.org/10.1021/nl061237h.
27. Maier, S. A. Plasmonic Field Enhancement and SERS in the Effective Mode Volume Picture. *Opt. Express* **2006**, *14* (5), 1957–1964. https://doi.org/10.1364/OE.14.001957. http://opg.optica.org/oe/abstract.cfm?URI=oe-14-5-1957.
28. Petrova, H.; Lin, C. H.; de Liejer, S.; Hu, M.; McLellan, J. M.; Siekkinen, A. R.; Wiley, B. J.; Marquez, M.; Xia, Y.; Sader, J. E.; Hartland, G. V. Time-Resolved Spectroscopy of Silver Nanocubes: Observation and Assignment of Coherently Excited Vibrational Modes. *J. Chem. Phys.* **2007**, *126* (9), 094709. https://doi.org/10.1063/1.2672907.
29. Burgin, J.; Langot, P.; Arbouet, A.; Margueritat, J.; Gonzalo, J.; Afonso, C. N.; Vallée, F.; Mlayah, A.; Rossell, M. D.; Van Tendeloo, G. Acoustic Vibration Modes and Electron-Lattice Coupling in Self-Assembled Silver Nanocolumns. *Nano Lett.* **2008**, *8* (5), 1296–1302. https://doi.org/10.1021/nl073123r.
30. Haynes, C. L.; McFarland, A. D.; Zhao, L.; Van Duyne, R. P.; Schatz, G. C.; Gunnarsson, L.; Prikulis, J.; Kasemo, B.; Käll, M. Nanoparticle Optics: The Importance of Radiative Dipole Coupling in Two-Dimensional Nanoparticle Arrays. *J. Phys. Chem. B* **2003**, *107* (30), 7337–7342. https://doi.org/10.1021/jp034234r.
31. Mock, J. J.; Smith, D. R.; Schultz, S. Local Refractive Index Dependence of Plasmon Resonance Spectra From Individual Nanoparticles. *Nano Lett.* **2003**, *3* (4), 485–491. https://doi.org/10.1021/nl0340475.
32. Payne, E. K.; Shuford, K. L.; Park, S.; Schatz, G. C.; Mirkin, C. A. Multipole Plasmon Resonances in Gold Nanorods. *J. Phys. Chem. B* **2006**, *110* (5), 2150–2154. https://doi.org/10.1021/jp056606x.
33. Félidj, N.; Grand, J.; Laurent, G.; Aubard, J.; Lévi, G.; Hohenau, A.; Galler, N.; Aussenegg, F. R.; Krenn, J. R. Multipolar Surface Plasmon Peaks on Gold Nanotriangles. *J. Chem. Phys.* **2008**, *128* (9), 094702. https://doi.org/10.1063/1.2839273.
34. Esteban, R.; Vogelgesang, R.; Dorfmüller, J.; Dmitriev, A.; Rockstuhl, C.; Etrich, C.; Kern, K. Direct Near-Field Optical Imaging of Higher Order Plasmonic Resonances. *Nano Lett.* **2008**, *8* (10), 3155–3159. https://doi.org/10.1021/nl801396r.
35. Mahajan, S.; Cole, R. M.; Soares, B. F.; Pelfrey, S. H.; Russell, A. E.; Baumberg, J. J.; Bartlett, P. N. Relating SERS Intensity to Specific Plasmon Modes on Sphere Segment Void Surfaces. *J. Phys. Chem. C* **2009**, *113* (21), 9284–9289. https://doi.org/10.1021/jp900661u.
36. Dorfmüller, J.; Vogelgesang, R.; Weitz, R. T.; Rockstuhl, C.; Etrich, C.; Pertsch, T.; Lederer, F.; Kern, K. Fabry-Pérot Resonances in One-Dimensional Plasmonic Nanostructures. *Nano Lett.* **2009**, *9* (6), 2372–2377. https://doi.org/10.1021/nl900900r.
37. Guzatov, D. V.; Klimov, V. V.; Pikhota, M. Y. Plasmon Oscillations in Ellipsoid Nanoparticles: Beyond Dipole Approximation. *Laser Phys.* **2010**, *20*, 85–99. https://doi.org/10.1134/S1054660X09170083.
38. Ng, M. Y.; Chang, Y. C. Laser-Induced Breathing Modes in Metallic Nanoparticles: A Symmetric Molecular Dynamics Study. *J. Chem. Phys.* **2011**, *134* (9), 094116. https://doi.org/10.1063/1.3563803.
39. López-Ríos, T. Enhanced Raman Scattering Mediated by Long Wave Vector Surface Plasmon Polaritons. *Phys. Rev. B* **2012**, *85* (12), 125438. https://doi.org/10.1103/PhysRevB.85.125438.

40. Portalès, H.; Goubet, N.; Casale, S.; Xu, X. Z.; Ariane, M.; Mermet, A.; Margueritat, J.; Saviot, L. Inelastic Light Scattering by Long Narrow Gold Nanocrystals: When Size, Shape, Crystallinity, and Assembly Matter. *ACS Nano* **2021**, *14* (4), 4395–4404. https://doi.org/10.1021/acsnano.9b09993.
41. Lipparini, E.; Stringari, S. Collective Excitations in Deformed Alkali Metal Clusters. *Z. Phys. D At. Mol. Clusters* **1991**, *18*, 193–201. https://doi.org/10.1007/BF01437447.
42. Lyalina, A.; Matveentsev, A.; Solov'yov, I. A.; Greiner, W. On the Applicability of Deformed Jellium Model to the Description of Metal Clusters. *Eur. Phys. J. D* **2003**, *24*, 15–18. https://doi.org/10.1140/epjd/e2003-00119-4.
43. Moroz, A. Depolarization Field of Spheroidal Particles. *J. Opt. Soc. Am. B* **2009**, *26* (3), 517–527. https://doi.org/10.1364/JOSAB.26.000517. http://opg.optica.org/josab/abstract.cfm?URI=josab-26-3-517.
44. van Dijk, M. A.; Lippitz, M.; Orrit, M. Detection of Acoustic Oscillations of Single Gold Nanospheres by Time-Resolved Interferometry. *Phys. Rev. Lett.* **2005**, *95* (26), 267406. https://doi.org/10.1103/PhysRevLett.95.267406.
45. Crut, A.; Maioli, P.; Del Fatti, N.; Vallée, F. Acoustic Vibrations of Metal Nano-Objects: Time-Domain Investigations. *Phys. Rep.* **2015**, *549*, 1–43. https://doi.org/10.1016/j.physrep.2014.09.004.
46. Staleva, H.; Hartland, G. V. Vibrational Dynamics of Silver Nanocubes and Nanowires Studied by Single-Particle Transient Absorption Spectroscopy. *Adv. Funct. Mater.* **2008**, *18* (23), 3809–3817. https://doi.org/10.1002/adfm.200800605.
47. Gouadec, G.; Colomban, P. Raman Spectroscopy of Nanomaterials: How Spectra Relate to Disorder, Particle Size and Mechanical Properties. *Prog. Cryst. Growth Charact. Mater.* **2007**, *53* (1), 1–56. https://doi.org/10.1016/j.pcrysgrow.2007.01.001.
48. Duval, E.; Boukenter, A.; Champagnon, B. Vibration Eigenmodes and Size of Microcrystallites in Glass: Observation by Very-Low-Frequency Raman Scattering. *Phys. Rev. Lett.* **1986**, *56* (19), 2052–2055. https://doi.org/10.1103/PhysRevLett.56.2052.
49. Duval, E. Far-Infrared and Raman Vibrational Transitions of a Solid Sphere: Selection Rules. *Phys. Rev. B* **1992**, *46* (9), 5795–5797. https://doi.org/10.1103/PhysRevB.46.5795.
50. Bachelier, G.; Mlayah, A. Surface Plasmon Mediated Raman Scattering in Metal Nanoparticles. *Phys. Rev. B* **2004**, *69* (20), 205408. https://doi.org/10.1103/PhysRevB.69.205408.
51. Large, N.; Saviot, L.; Margueritat, J.; Gonzalo, J.; Afonso, C. N.; Arbouet, A.; Langot, P.; Mlayah, A.; Aizpurua, J. Acousto-Plasmonic Hot Spots in Metallic Nano-Objects. *Nano Lett.* **2009**, *9* (11), 3732–3738. https://doi.org/10.1021/nl901918a.
52. Tang, J.; Albrecht, A. C. Developments in the Theories of Vibrational Raman Intensities. In *Raman Spectroscopy*; Szymanski, H. A., Ed.; Plenum: New York, 1970; pp. 33–68.
53. Bohren, C.; Huffman, D. *Absorption and Scattering of Light by Small Particles*; John Wiley and Sons, Inc.: New York, 1983.
54. Levin, M. L.; Muratov, R. Z. On the Internal Potential of Heterogeneous Ellipsoids. *Astrophys. J.* **1971**, *166*, 441–445.
55. Stevenson, A. F. Electromagnetic Scattering by an Ellipsoid in the Third Approximation. *J. Appl. Phys.* **1953**, *24* (9), 1143–1151. https://doi.org/10.1063/1.1721462.

CHAPTER THREE

Artificial intelligence and E-learning

Morten Matras[a] and John W. Perram[b,*]
[a]CEO, Leaddoubler.com, Odense, Denmark
[b]Mærsk McKinney Moller Institute, University of Southern Denmark, Odense, Denmark
*Corresponding author: e-mail address: johnperram06@yahoo.com.au

Dedication: This paper is dedicated to John's long friendship with Jack Sabin going all the way back to the late seventies when they met at the new, small Odense University. He last met Jack in Nyborg in 2019, just before the Covid closed us all down.

Contents

Abstract

Since the invention of computers, one of the aims has been to get a machine to resemble human intelligence. Following the first application in decryption, this aim became the automation of mathematical knowledge as this was considered to be the best structured branch of science. Early results came in the form of symbolic algebra packages such as REDUCE and *Mathematica*. The arrival of the internet and the increasing speed of computers greatly increased the possibility of AI into areas such as e-learning and robotics. In this article, we describe our experiences with building a fully automated course in classical mechanics.

Advances in Quantum Chemistry, Volume 85
ISSN 0065-3276
https://doi.org/10.1016/bs.aiq.2022.03.002
Copyright © 2022 Elsevier Inc.
All rights reserved.

1. Introduction

Two of the overused current expressions are Artificial Intelligence and E-learning. So let us try to define what these terminologies mean. If we start with Artificial Intelligence or AI, let us agree that it must involve a digital computer. So the beginning of AI must have occurred after the opening of the first digital computer, the Colossus, around 1942. The aim was to aid the decryption of German intelligence messages during the second world war and the aim seems to have been achieved. In any case, decryption has for centuries been one of the activities of mathematics, so that using a computer to aid it must certainly have qualified the activities as AI.

Soon after the end of the war when programmable computers first appeared in Manchester and Boston, the users, who were mainly mathematicians, formed the academic subject called Artificial Intelligence.

Like AI, e-learning can be no older than the digital computer. The arrival of cheap software to produce full screen videos enabled the production of downloadable videos which can be seen on YouTube. Soon after came learning management systems whose main role was to restrict access to on-line learning material. However, an investigation of the real economics of education revealed that the real riches are provided by examinations, the indicator that the technology has actually been transferred. The work in this paper is, we believe, the first to document a fully automatic method of examining in the mathematical sciences.

2. Early history of computing

The first and best documented application of digital computers involved the decoding of German ENIGMA codes during the second world war. The Bletchley Park computer was probably the first digital computer, the application successful and the persecution of the principal investigator Alan Turing the worst since Galileo. A good picture of his history and attainments can be found in his Wikipedia biography.[1]

So, when the discipline of computer science was founded after the second world war and one of its central research areas was artificial intelligence. A good article about its history since then can be found in Wikipedia.[2] In the next section, we discuss how our careers have meshed with its changing history.

3. Our experience with ai and e-learning

3.1 John Perram

I have been a user of computers since 1965 when I attempted to carry out calculations on the shape of mercury drops. The attempts were not very successful and, soon after, I moved to Manchester to start a PhD study in the Department of Mathematics. There, I had access to the Atlas computer, then the most powerful machine in the world. Aided by Joan Walsh, I succeeded in simulating the behavior of a drop emerging from a capillary tube. Following advice, I then abandoned that project and spent the next 3 years to studying the statistical mechanics of liquid water. The main results of the unsuccessful project were later published.[3]

This got me interested in the theory of polar fluids and their dielectric constants. After 2 years postdoctoral work in Manchester, I returned to the newly established Department of Applied Mathematics in the Research School of Physical Sciences at the Australian National University where I began to work with my PhD student Lee White on doing Monte Carlo simulations of polar fluids. I then began work on the perturbation theory of molecular fluids. The conclusion of this work saw me as Professor of Applied Mathematics at the University of Odense where I continued to work on these topics.

I first met AI when I used the program REDUCE to carry out the complicated symbolic manipulations required in the perturbation theory of liquid mixtures. However a calculation I carried out after meeting Joel Lebowitz on a rapid overlap algorithm for ellipsoids[4] in 1982 was later to prompt my entry into AI without my realizing it at the time.

One of my duties in Odense was to chair the committee for the appointment of a professor of computer science. The position was offered to Dr. John Campbell from the University of Exeter in 1985. When I interviewed him in Odense. I told him about the ellipsoid algorithm. He told me that I should apply it to robotics, one of the new areas of artificial intelligence. At that time, AI was concerned with speeding up algorithms to solve such problems as the traveling salesman problem or trajectory planning for multi axis robots where the essential problem was seen as collision avoidance. I figured that my work on the dynamics of complex molecules and their implementation on parallel computers[5] qualified me to work in the area.

Unfortunately for us, he accepted the chair of artificial intelligence at University College, London. However, I kept contact with him and was involved in the EU programs in parallel computing and multi-agent systems, as well as spending the 1987–88 academic year on sabbatical year with him.

My attempts to establish a research presence in robotics were initially unsuccessful until my famous meeting with Torben Andensen of Odense Steel Shipyard in 1988 where, on a visit to Odense University, he asked me if anyone there knew anything about classical mechanics. My answer that I did started the process which led to the formation of the Mærsk McKinney Moller Institute and the establishment of Odense as a major European centre of robotics research and business. It is a great tribute to Torben, who passed away recently, that he realized the essential barrier to real robotics.

3.2 Morten Matras

Morten Matras was a computer science/applied mathematics student at SDU who completed his master project on Exercise Publisher[6] in 2001, supervised by Professor Bo Nørregåd Jørgensen of the Maersk McKinney Institute. During his studies, he was fascinated by the economics of education and how it might be improved.

His thesis differentiates between fact-based studies such as the humanities and biological sciences and algorithmic studies such as mathematics, physics, some areas of chemistry and computer science. Roughly stated, the language of the first group was natural language and of the second group, mathematics.

He also identified evaluation as one of the principal costs of education. The value of an education is first realized when the student passes an examination on its material. For fact-based educations, the main form of evaluation has been the multiple choice test where candidates have to choose the answer from 5 alternatives. John Perram remembers his first year biology examination for which he achieved a pass by answering the few questions he knew and reducing the rest to two or three possible answers by removing the impossible choices.

For algorithmic studies, the current method was the written examination where students were invited to write short reports on problems motivated by the syllabus. Unless absolutely correct, assigning marks to such reports has always been time consuming.

He established contact with John Perram in 2001 when Perram returned from sabbatical in Melbourne. At that time, many universities were attempting to establish e-learning using packages such as BlackBoard.[7] Perram was appointed as the Science Faculty's part time director of e-learning whose initial job was to choose a package to implement at SDU.

When they first met, Morten realized that *Mathematica*, the language used by Perram to implement he used to carry out his experiments on e-learning was an ideal solution to his idea that marks could be given for on-line projects by allowing students to purchase hints for the solution of their projects. Full marks would be awarded to projects which did not need the hint server. As *Mathematica* could carry out mathematics, it could be used to fully automate the hint server. They agreed to collaborate on integrating Exercise Publisher fully in an online version of the course in classical mechanics.

The savings from Perram's appointment were used to hire Morten as an amanuensis to implement the ideas of Exercise Publisher in the evaluation of the course in Classical Mechanics he had taken over and intended to turn into a piece of e-learning.

The question of security and the suppression of cheating has always been a problem for the evaluation of projects. In particular the possibility of the writer of a project report collaborating with other students in the solution of a project has attracted a lot of attention.

4. Classical mechanics as E-learning

The course at SDU in classical mechanics had previously been taught in a traditional way by Henrik Gordon Petersen. That is, with lectures, tutorials and a written examination. A traditional lecture of such material consists of the teacher speaking to the class while writing the equations describing the content on a board or overhead projector. The writing of equations forces the eyes of the students to follow the development of the argument. This advance over static text cannot be overestimated.

One way to implement this activity electronically is to prepare lectures as executable code embedded in text using the text interface of programs such as *Mathematica*[8] or Maple.[9] That is, the main steps of the argument are stored as executable code which are then executed during the production of the video of the lecture in the classroom. The students can then experience the lecture and download it and the original code from the learning management system. Such a file can be converted to a the video of a classical lecture[10] using a program such as Camtasia Studio[11] which can be loaded up onto the LMS.[7] or a platform such as YouTube.[12] Once there, anyone can download it and enjoy most of the material of a university course.

So one of the main functions of an e-learning program is to restrict access to such material to fee paying students. When we start to think of knowledge as e-learning we realize that the product being paid for is the evaluation that such knowledge has actually been transferred.

Written examinations are a crude way of doing this, mainly due to the short time available for students to carry them out. Much better is one or more off-line projects for which Exercise Publisher gives a basis for quantitative evaluation. The results of applying all this to the course in classical mechanics is described in the next section.

5. Results

The first phase of the course is to decide how it will be evaluated. We decided that the evaluation would consist of two parts. In the first part, the students will be asked to complete 2 project reports in the form of *Mathematica* notebooks. The projects were described in some detail and the text and formulas for each part were entered into Exercise Publisher as a hint. Students carrying out the project could purchase a hint for a part they couldn't complete. One of the things which went wrong here was that groups of students collaborating on a project could purchase a hint which could be shared between them. There are techniques in software engineering which would enable this behavior to be detected, but in the first instance we agreed to ignore it.

The next phase of implementing the course consisted of generating the first lecture by converting the first segment of the course notes into *Mathematica* notebooks. When we did this, we noticed that some operations were repeated, so students were encouraged to store such operations in their tool box. The tool box would then be the subject of a final oral examination, We noted that this oral examination could be held remotely using a tool such as Skype. Many and more powerful tools such as Zoom, Webex and Teams for holding remote on line examinations are now available.

When it was time for the students to carry out their first project, a standard solution to the problems was generated and the individual parts were stored as hints in Exercise Publisher, each with a cost in percent of full marks. We decided not to impose any sanctions on students who had shared hints, but instead judged their projects traditionally.

At the end of the course, students had to take part in an oral examination of their tool box. This was done with an external examiner, either Associate Professor Poul Hjort or Professor Steen Markvorsen, both of the Department of Mathematics at the Technical University of Denmark. Students were given 10 min to begin to solve one of a series of projects using their tool box.

The first time the course ran, the number of students was small, mainly Chinese students in the International Masters Program. These performed uniformly well. For students from Applied Mathematics and Computer Systems Engineering, students either performed very well or dropped out and failed to deliver projects. Finally, in 2004, for some reason the course was dropped.

In 2006, after Perram had retired and Morten terminated, the course had to be run once more to allow students from the previous study plan to complete their course. There quite a lot of these (about 60) and Perram ran the course from Sydney in Australia with Lars Peter Ellekilde as tutor. Perram moved to Odense in time to carry out the oral examination with Professor Markvorsen as external censor, The percentage of students passing this course (70%) was almost the same as when it was held in a traditional manner.

6. Perspectives

Classical Mechanics is, of course, only one topic in the world of mathematics, physics and chemistry. Neither of us is a quantum chemist, so we leave it up to the other contributors to this volume to think about what other aspects of computer science could be used to heighten our understanding and visualization of it and similar branches of science.

E-learning and artificial intelligence are constantly evolving and adapting to improve everyday performance. Although our experience has essentially been devoted to classical mechanics and computer science, interesting and efficient applications of machine learning are currently providing assistance to quantum chemistry research, the primary focus of the present book series. For a recent review, see Dral.[13]

References

1. https://en.wikipedia.org/wiki/*Alan_Turing*.
2. https://en.wikipedia.org/wiki/*Artificial_intelligence*.
3. Perram, J. W.; Hayter, J. B.; Hunter, R. J. *J. Electroanal. Chem.* **1973**, *42*, 291–298.
4. Perram, J. W.; Wertheim, M. S. *J. Comput. Phys.* **1985**, *58*, 409–416.
5. Petersen, H. G.; Perram, J. W. *Mol. Phys.* **1989**, *67*, 849–860.
6. Andersen, M. *Exercise Publisher: Evolving E-learning (MSc thesis)*, Odense University, 2001; pp. 1–135.
7. Blackboard. www.blackboard.com.
8. Mathematica, Wolfram Corporation, www.wolfram.com.
9. Maple. www.maplesoft.com.

10. Being renewed. An equivalent lecture which shows how mathematics teaching can be computerised can be found at www.https://www.youtube.com/watch?v=uEC5dmx6wmw.
11. *Camtasia*, is a visualization program produced by www.techsmith.com. It produces high quality movies from a screen video on a PC.
12. Storage site for videos www.youtube.com.
13. Dral, P. O. Quantum chemistry assisted by machine learning. *Adv. Quant. Chem.* **2020**, *891*, 291–324.

CHAPTER FOUR

Structure and correlations for harmonically confined charges

Jeffrey Wrighton and James Dufty*
Department of Physics, University of Florida, Gainesville, FL, United States
*Corresponding author: e-mail address: dufty@phys.ufl.edu

Contents

Abstract

Coulomb charges confined by a harmonic potential display a rich structure at strong coupling, both classical and quantum. A simple density functional theory is reviewed showing the essential role of correlations in forming shell structure and order within the shells. An overview of previous comparisons with molecular dynamics and Monte Carlo simulations is summarized and extended. It is shown that correlations for the fluid phase (shell structure only) are well approximated by those for the uniform one-component plasma even at very strong coupling. A corresponding representation of the correlations for the ordered phase is still an open question. The confirmed success for the classical density functional theory is important for the subsequent representation of the quantum case. Here, a mapping of the quantum description onto an equivalent classical description with effective potentials allows direct application of the classical methods, both theory and simulation. This is particularly relevant at low but finite temperatures where quantum simulation methods are compromised. The special case of Coulomb charges in a harmonic trap is the simplest example of more complex systems of experimental interest where confinement and strong coupling play an essential role (e.g., quantum dots, ions in complex traps, electrons on a helium surface, dusty Yukawa plasmas, ultracold neutral plasmas).

Advances in Quantum Chemistry, Volume 85
ISSN 0065-3276
https://doi.org/10.1016/bs.aiq.2022.04.003
Copyright © 2022 Elsevier Inc.
All rights reserved.

1. Introduction

The problem considered here is the structure and correlation among N equal Coulomb charges confined by an external harmonic potential. It is a generalization of the Thomson problem[1] for charges confined to the surface of a sphere, posed 117 years ago, to three dimensions and finite temperatures. The ground state for the harmonic confinement is well studied, exposing a rich shell structure (distribution of particles localized about well-defined radii with localization on each radius similar to those of the single sphere Thomson problem). Within classical mechanics, these ground-state results have been quantified in detail via shell models, molecular dynamics (MD) simulation, and Monte Carlo (MC) simulation.[2–7] They are also realized experimentally for dusty plasmas. The corresponding results for confined charges at finite temperatures are the extension described here. We provide here a summary of our theoretical work in collaboration with the Bonitz group at the Institute fur Theoretische Physik und Astrophysik, Christian-Albrechts Universitat, Kiel.[8–11] In addition, we describe the method for extension to quantum theory and list some remaining outstanding problems.

The primary effect of temperature on the classical ground-state shell models is to broaden the sharp shell structure and smooth their angular distribution due to thermal motion. The governing parameters are the Coulomb coupling constant Γ (ratio of Coulomb to thermal energies of a pair) and the average number of charges $\bar{N}$. The number of shells is fixed by $\bar{N}$, while their relative resolution (sharpness) is determined by Γ. The zero temperature ground-state results from shell models are recovered in the limit of large Γ. A simple approximate density functional theory described below is able to capture these results quantitatively, in comparison with those from MC simulations. It is based on approximating correlations among the charges in the trap by those in the uniform one-component plasma (OCP). The close relationship of correlations in these two quite different systems has been confirmed by MD simulation.[10,11]

At still larger Γ, corresponding to lower temperatures, the rotational invariance of the fluid phase is broken and the particles within each shell become localized about sites close to those of the Thomson problem for a single shell. Those localized domains are approximated here by Gaussian distributions centered at these sites, and the correlations among them are calculated showing good agreement with results from MC simulation.

The effects studied here result from strong coupling conditions for which there are relatively few theoretical methods available. In the classical case, the density functional model described below is confirmed by MD and MC simulations. The latter simulations are not available for the quantum case, and the quantum density functional model has problems at finite temperatures. However, it has been shown that the quantum system can be mapped onto a corresponding classical system with quantum effects embedded in effective Coulomb and trap potentials.[12] Applications to the OCP (jellium) show good agreement with quantum Monte Carlo results.[13,14] This approach has been applied subsequently to the case of quantum charges in a harmonic trap[15,16] as described below. In this way, a broad scope of confined Coulomb systems of interest can be addressed. Fig. 1 gives a simple overview of the parameter space.

It is a pleasure to dedicate this contribution to our friend and colleague of many years, Professor John (Jack) Sabin. Jack has been an inspiration for all

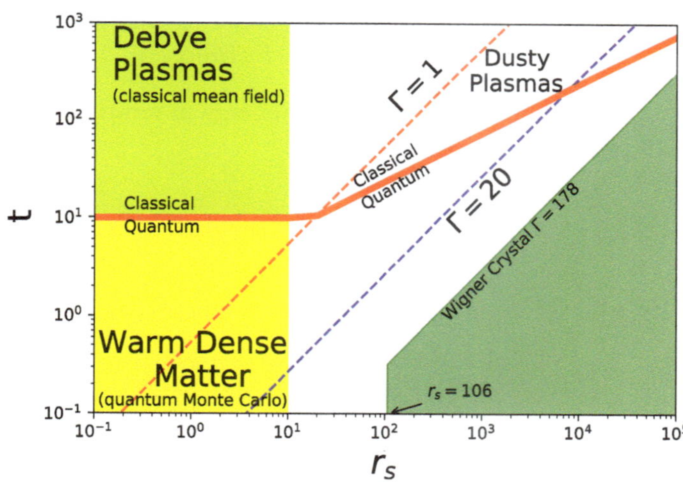

Fig. 1 Overview of the relevant parameter space. Here, r_s is the Wigner–Seitz radius in terms of the Bohr radius of the confined particles ($r_s = r_0/a_b$), and t is the temperature relative to the ideal gas Fermi temperature per particles ($t = k_BT/\varepsilon_F$).[16] Here, $a_b = \hbar^2/me^2$ and $\varepsilon_F = \hbar^2(3\pi^2\bar{n})^{2/3}/2m$, where $\bar{n}$ is the average density from (6). The Fermi energy is used for scaling as it is relevant to the low-temperature quantum domain and thus t is a measure of the onset of quantum effects. Note that the classical phase for dusty plasmas shown in this figure represents the fluid phase.

that is expected of those with an academic career, exemplifying the best in teaching, research, and administration. His cheerful nature and good will have brightened the lives of all who knew him well.

2. Density functional theory

Consider N particles of charge q at equilibrium in a harmonic trap at inverse temperature β. The free energy is denoted $F(\beta|n)$, indicating that it is a function of β and a functional of the nonuniform density $n(\mathbf{r})$. The equilibrium density profile is determined from

$$\frac{\delta F(\beta|n)}{\delta n(\mathbf{r})} = \mu - V(r), \qquad V(r) = \frac{1}{2} m\omega^2 r^2. \tag{1}$$

The potential $V(r)$ is the confining harmonic trap. The free energy functional can be separated into that for a system without interactions, $F_0(\beta|n)$, and a remainder $F_{ex}(\beta|n)$ containing all effects of Coulomb interactions among the particles

$$F(\beta|n) = F_0(\beta|n) + F_{ex}(\beta|n). \tag{2}$$

A formal representation for the excess free energy in terms of pair correlations can also be written exactly

$$F_{ex}(\beta|n) = -\int_0^1 dy(1-y) \int d\mathbf{r} d\mathbf{r}' n(\mathbf{r}) n(\mathbf{r}') \beta^{-1} c^{(2)}(\mathbf{r}, \mathbf{r}'|yn), \tag{3}$$

where $c^{(2)}(\mathbf{r}, \mathbf{r}'|n)$ is the direct pair correlation function

$$\beta^{-1} c^{(2)}(\mathbf{r}, \mathbf{r}'|n) \equiv -\frac{\delta^2 F_{ex}(\beta|n)}{\delta n(\mathbf{r}) \delta n(\mathbf{r}')}. \tag{4}$$

In this way, Eq. (1) is an equation for the local density in terms of the pair correlations of the direct correlation function[9]

$$\frac{\delta F_0(\beta|n)}{\delta n(\mathbf{r})} = \mu - V(r) + \int_0^1 dy \int d\mathbf{r}' n(\mathbf{r}') \beta^{-1} c^{(2)}(\mathbf{r}, \mathbf{r}'; yn). \tag{5}$$

The average density $\bar{n}$ is defined by

$$\bar{n} = \frac{\bar{N}}{V}, \qquad \bar{N} = \int d\mathbf{r} n(\mathbf{r}). \tag{6}$$

The system is self-confined with spherical symmetry. The maximum radius R is the point at which Coulomb repulsion force on a particle is balanced by the harmonic trap confinement

$$\bar{N}\frac{q^2}{R^2} = m\omega^2 R, \qquad V = \frac{4}{3}\pi R^3. \tag{7}$$

The mean distance between particles r_0 is introduced by

$$\frac{4}{3}\bar{n}\pi r_0^3 = 1. \tag{8}$$

Scaling the coordinates with respect to r_0 in the above equations gives the dimensionless form

$$\frac{\delta F_0^*(n^*)}{\delta n^*(\mathbf{r}^*)} = \beta\mu - \frac{1}{2}\Gamma r^{*2} + \int_0^1 d\gamma \int d\mathbf{r}'^* n^*(\mathbf{r}'^*) c^{(2)}(\mathbf{r}^*, \mathbf{r}'^*; \gamma n^*), \tag{9}$$

where

$$F_0^*(n^*) = \beta F_0(\beta|n); \qquad n^*(\mathbf{r}^*) = n(\mathbf{r}) r_0^3, \qquad \Gamma = \beta m\omega^2 r_0^2 = \beta\frac{q^2}{r_0}. \tag{10}$$

The parameter Γ is the Coulomb coupling constant (Coulomb energy of a pair at the average distance relative to the thermal energy β^{-1}). The constant $\beta\mu$ can be eliminated in terms of $\bar{N}$.

Up to this point, the results apply for both quantum and classical mechanics. The classical case is considered more explicitly next.

3. Classical mechanics

Within classical statistical mechanics, $F_0(\beta|n)$ can be written exactly as a functional of the density

$$F^{(0)}(\beta|n) = -\frac{1}{\beta}\int d\mathbf{r} n(r)\left(1 - \ln\left(n(r)\lambda^3\right)\right). \tag{11}$$

Here, $\lambda = \left(h^2\beta/2\pi m\right)^{1/2}$. Then Eq. (9) becomes[9]

$$\ln\left(n^*(\mathbf{r}^*)\right) = \ln\left(\left(\frac{\lambda}{r_0}\right)^3 e^{\beta\mu}\right) - \frac{1}{2}\Gamma r^{*2} + \int_0^1 d\gamma \int d\mathbf{r}'^* n^*(\mathbf{r}'^*) c^{(2)}(\mathbf{r}^*, \mathbf{r}'^*; \gamma n^*). \tag{12}$$

The constant first term on the right side can be eliminated in terms of $\bar{N}$. Consequently, the dimensional density profile and associated free energy density profile depend only on the two parameters $\bar{N}$ and Γ.

The solutions to (12) are expected to confirm the following qualitative behavior observed from ground-state energy functions,[8] and MC and MD simulations.[7] For given $\bar{N}$, the profiles have a strong dependence on the coupling strength Γ. At very small values, the density profile is rotationally symmetric and monotonically decreasing to zero from a maximum at $r^* = 0$. At increasing values, the radial dependence develops local maxima, referred to as shells. This is illustrated in Fig. 2. The number of shells increases with $\bar{N}$ and their width sharpens with increasing Γ. The shell populations are greater for increasing radius and grow linearly with $\bar{N}$.

Eventually, at sufficiently large Γ rotational symmetry is broken. The uniform distribution of particles within each shell distorts to local domains for the associated population. Their locations are close to those of the Thomson problem—the ground-state configuration for a given number of charges confined to a sphere. In the following, the extent to which approximations to the density functional theory (12) captures these features is described.

3.1 Fluid phase

The approximate determination of the pair correlations described by the direct correlation functional in (12) is motivated as follows. First, it is shown

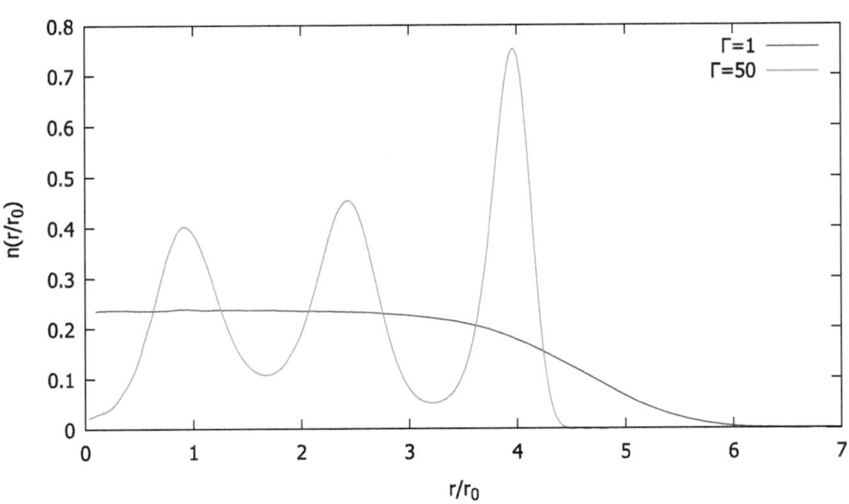

Fig. 2 Formation of shell structure in a harmonic trap as the coupling constant Γ increases. Monte Carlo simulation of 100 particles in a trap.

elsewhere that the functional $F_{ex}(\beta|n)$ is exactly equal to the corresponding system with a uniform neutralizing background (inhomogeneous jellium), i.e., charges in a uniform neutralizing background with the same harmonic potential[17]

$$F_{ex}(\beta|n) = F_{jex}(\beta|n). \tag{13}$$

The advantage of this is that the OCP has a finite uniform limit in the absence of the harmonic confinement, which is the uniform OCP. It has been observed elsewhere[11] that the distribution of pairs within the trap *without reference to their center of mass* position is almost identical to those of the OCP (see Fig. 3). Therefore, as an approximation for the fluid phase, OCP correlations have been used,

$$c^{(2)}(\mathbf{r},\mathbf{r}';\lambda n)\big|_{\lambda n=\overline{n}} = c^{(2)}(\mathbf{r},\mathbf{r}';\overline{n}) = c^{(2)}_{OCP}(|\mathbf{r}-\mathbf{r}'|;\overline{n}). \tag{14}$$

Evaluation of the direct correlation function for the OCP in (14) still poses a formidable many-body problem at strong coupling. However, it is a well-studied problem and an excellent approximation, the *adjusted hypernetted chain approximation* (AHNC), is known.[20,21] The solutions to (12) with these two approximations for $c^{(2)}(\mathbf{r},\mathbf{r}';n)$ give all of the above expected properties for the density profile quantitatively in comparison to MD and MC results, across the entire domain of Γ and $\bar{N}$. An example is illustrated in Fig. 4.

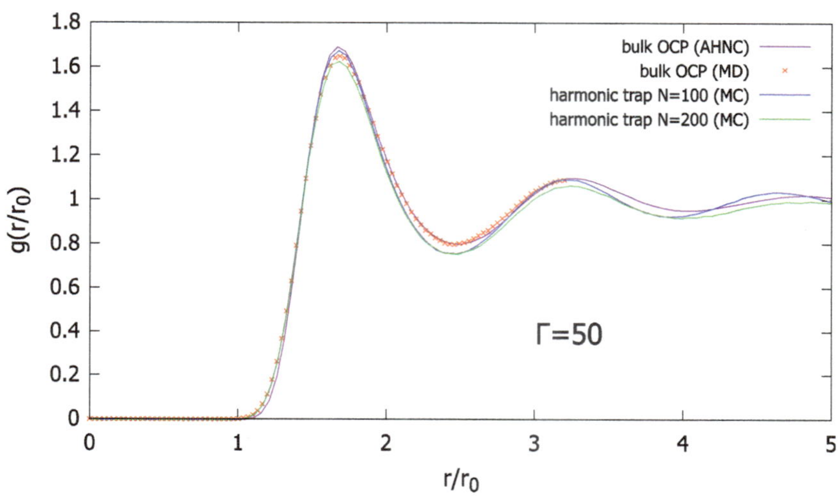

Fig. 3 Comparison of the pair distribution of particles in a one-component plasma (OCP) and harmonic trap. MD calculations for the OCP were performed using Sarkas.[18,19]

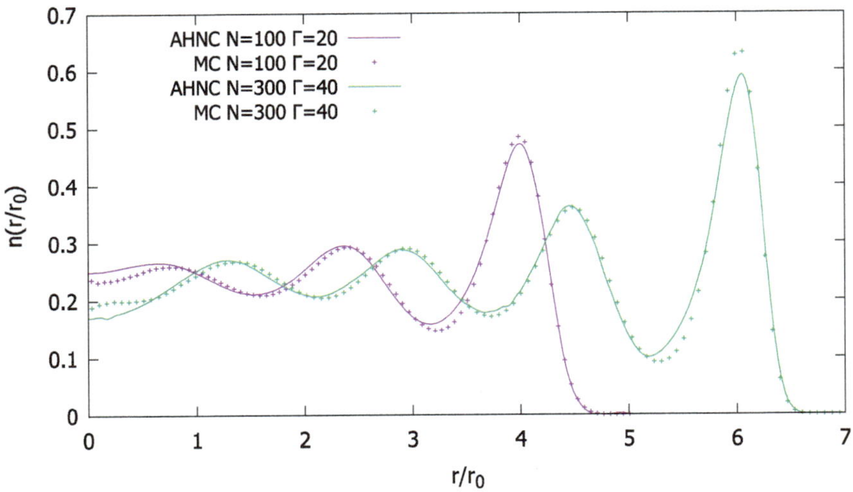

Fig. 4 Comparison of density profiles for AHNC and MC, for $N = 100$ and $\Gamma = 20$, and for $N = 300$ and $\Gamma = 40$.

3.2 Ordered phase

In the fluid phase, the particles are uniformly distributed throughout each shell. As the coupling increases eventually the particles enter an ordered state where rotational symmetry is broken within the shell. Fig. 5 shows the angular correlations within the outer shell from a MD simulation, for three values of the coupling constant corresponding to fluid and ordered phases. The system consisted of $N = 38$ charges, with 32 in the outer shell. The pair correlation function $g(\theta)$ is the probability to find a particle displaced on the shell by an angle θ from an arbitrary reference particle. As in a uniform fluid, the peaks represent nearest neighbor, next nearest neighbor, etc. At the lower values of Γ, there is not much qualitative difference in the angular correlations, but the correlation peaks are narrowing and some structure is starting to develop in the form of shoulders in the middle peaks. For significantly larger values of Γ however, a definite structure appears within each peak of the correlation graph. In particular, the two broad peaks at $\Gamma = 100$ that appeared at 70 and 110 degrees have condensed into triplets at $\Gamma = 1000$. In addition, the two peaks at 40 and 140 degrees are showing the beginning formation of a doublet structure with the presence of shoulders at $\Gamma = 1000$. This set of doublet and triplet features persists at much higher Γ, without the appearance of any more peaks.

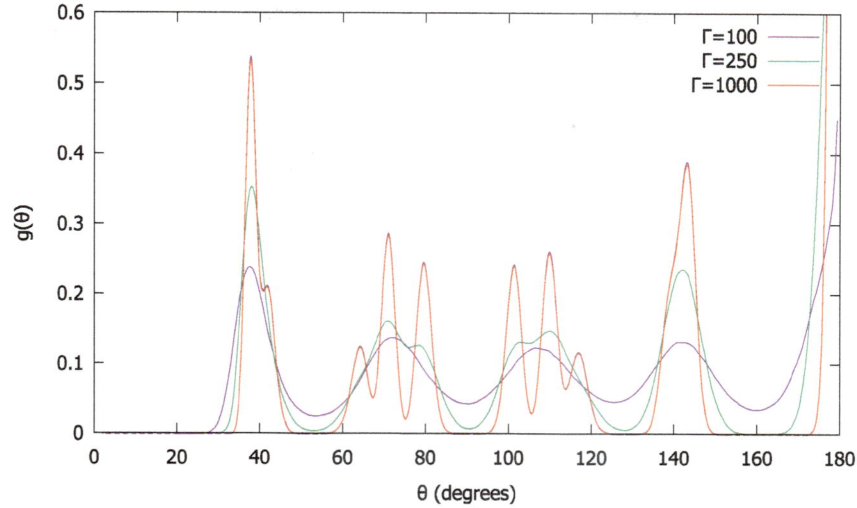

Fig. 5 Angular correlations within a single shell from MD simulation. The harmonic trap contained 38 particles, with 32 in the outer shell. MD simulations were performed with LAMMPS.[22,23]

These features can be modeled using a thin-shell model where the particle configuration results from the Thomson problem.[1] The context of the original Thomson paper lies in a mistaken view of atomic structure at the time it was written, addressing the question of the ground-state energy of charges fixed to a sphere. It is thus related to the questions here of effectively having multiple spheres representing the shell structure and the distribution of charges within those spheres. At very low temperature (large Γ), simulations and other studies show that the density changes from a uniform distribution to have features very close to those of the Thomson problem. As shown below, the number of localized sites and their locations are quite similar. Differences can be associated with the widths of the shells and interactions between different shells. The specific ordering depends on the number of charges because of the spherical geometry. In Fig. 6, the Thomson configuration for a randomly chosen particle from a system containing $N = 32$ charges is shown along with constant-angle planes showing the angular displacement of the other particles. To compare more directly, the angular correlations of all 32 charges in the Thomson problem were calculated. A plot of the angular displacement of all pairs is shown in Fig. 7. The charges occur in two different angular

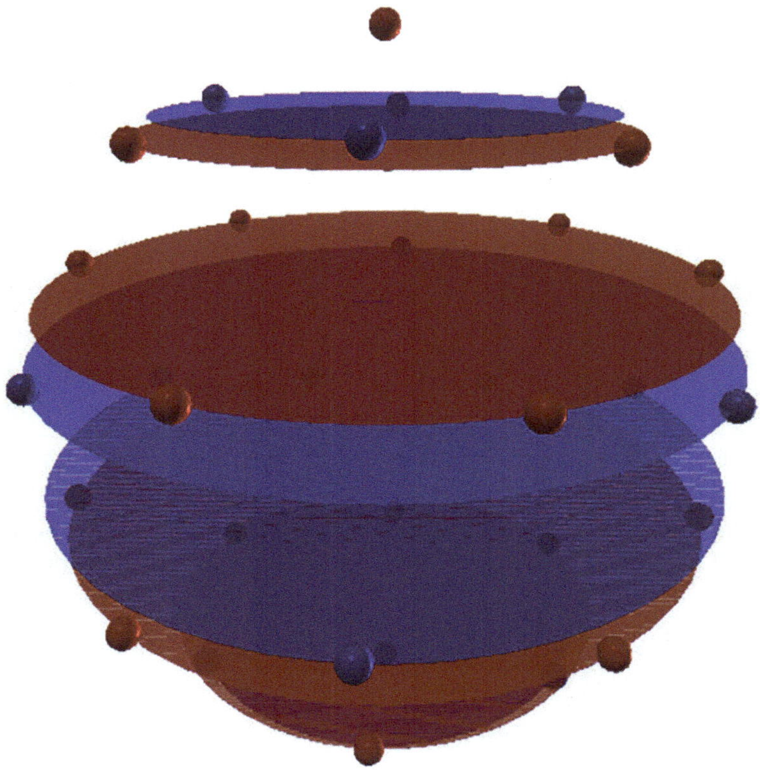

Fig. 6 A system of 32 charges in the Thomson problem. *Horizontal circles* connect those charges that are at constant angle θ from the chosen particle at the *top* of the figure.

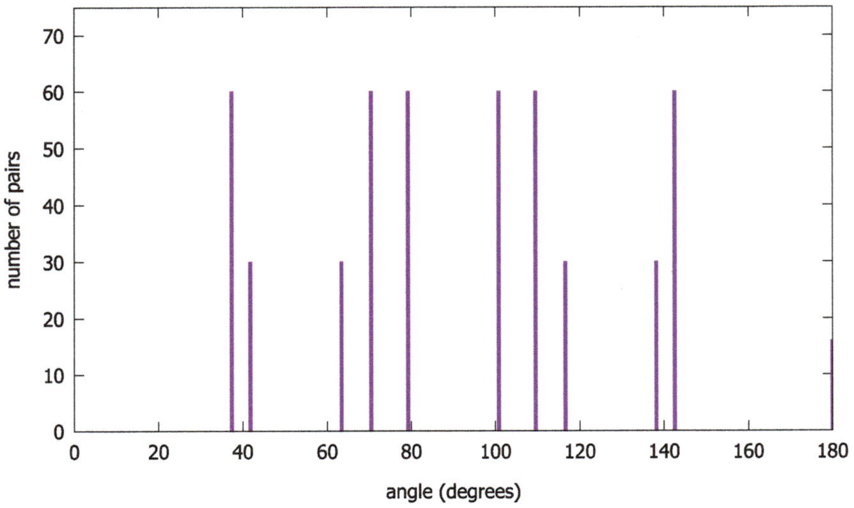

Fig. 7 Number of each angle present between each pair of particles in the Thomson system for $N = 32$ charges.

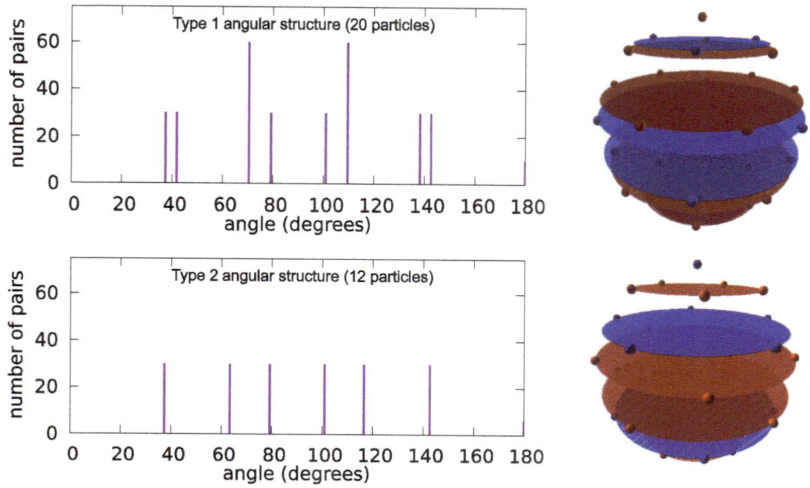

Fig. 8 Angular distribution for the two types of angular configuration for $N = 32$ charges in the Thomson problem. The first type of angular configuration has particles at eight specific angles between $\theta = 0$ degrees and $\theta = 180$ degrees. There are 20 particles in this configuration. The remaining 12 particles are in a configuration with 6 specific angles to the other particles. Plots show the total number of bonds that each population contributes for the entire system; their sum gives the results in Fig. 7.

correlation structures. These are shown in Fig. 8. The combination of these two populations accounts for the specific angular correlation structure for the case of $N = 32$ charges.

To account for thermal effects, the charges were modeled using a Gaussian function along the sphere of the form

$$f(\theta) = \sqrt{\frac{\alpha}{\pi}} \exp\left(-\alpha(\theta - \theta_0)^2\right) \tag{15}$$

where θ_0 is the angle corresponding to the Thomson site and α is a parameter that increases with decreasing temperature. Fig. 9 shows how this model reproduces the correct splitting, from the four broad peaks at small Γ which condense to the characteristic doublets and triplets at higher Γ. Fig. 10 shows the effect of the fitting parameter α showing how it models the effect of thermal motion. This supports the idea of considering the Thomson sites on a sphere to be analogous to a fundamental lattice for the ordered state, to the extent that the shell can be approximated as thin.

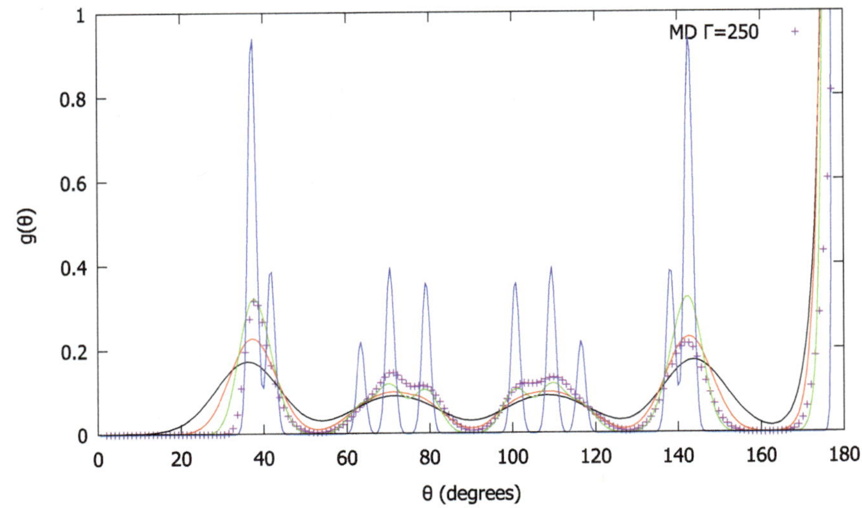

Fig. 9 Solid lines: evolution of the thermally broadened angular correlations from the Thomson problem as the parameter α increases. Peak height increases with larger α. Crosses: comparative results from MD at $\Gamma = 250$.

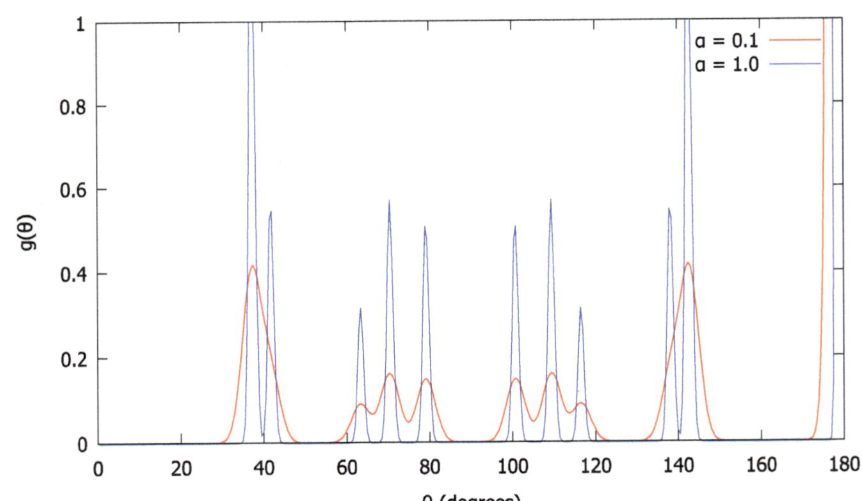

Fig. 10 Angular correlations from the thermally broadened Thomson system for two values of the fitting parameter α. Here, there are $N = 32$ charges in the system. The structure follows the same behavior as MD simulations for 32 particles in a shell.

In principle, the value of α in the ordered phase, for given $\Gamma, \bar{N}$, should be obtained from minimizing the above free energy functional using the assumed Gaussian density profile. In the fluid phase, the value of α would be large, representing a uniform profile. At very large Γ, approaching the ground state, the value of α would approach zero. However, the assumption of the Thomson sites should first come from solutions of (12). This could be quite difficult since ground-state studies from simulations suggest there are many metastable configurations as well.

4. Quantum mechanics

The above classical description of confined charges at strong coupling has exploited the methods of density functional theory, liquid state theory, MD, and MC. At low temperatures, at or below the Fermi temperature, quantum effects become important and many of these classical methods do not apply directly. An accurate quantum theory at finite temperatures, strong coupling, and confinement is still a challenging problem. Numerical methods such as quantum Monte Carlo are applicable at zero temperature but become less controlled as the temperature increases, particularly for fermions. A quite different approach is to develop an exact mapping of the quantum equilibrium structure to an effective classical problem. This has been done recently, allowing application of the above classical methods to quantum systems.[12] Structure and correlation calculated in this way for the OCP have shown good accuracy in comparison with recent quantum MC simulations.[13,14]

More recently, the effective classical representation of a system of quantum charges in a harmonic trap has been explored.[15,16] The primary differences from the classical description above are modifications of the Coulomb potential and the trap potential to accommodate quantum effects of diffraction and exchange symmetry. The exchange symmetry implies an additional pair interaction among charges even in the absence of Coulomb interactions. It is repulsive or attractive depending on the quantum statistics (Fermions or Bosons). The other important quantum effects is removal of the short distance Coulomb singularity. A new parameter appears, in addition to Γ and $\bar{N}$, the temperature relative to the Fermi temperature ($t = k_BT/e_F$ where e_F is the ideal gas Fermi energy per particle). For $t \gg 1$, the above description of shell structure and correlations is recovered for strong coupling. At smaller t, the effects of exchange degeneracy are incorporated by imposing the exact noninteracting density profile for the ideal gas. This is nontrivial since the classical representation of the quantum ideal gas has effective interactions. The additional effects of exchange and diffraction are included via the direct correlation function with modified Coulomb interactions. Fig. 11 shows a self-similar change in the classical two-shell structure being compressed due to quantum effects on the harmonic

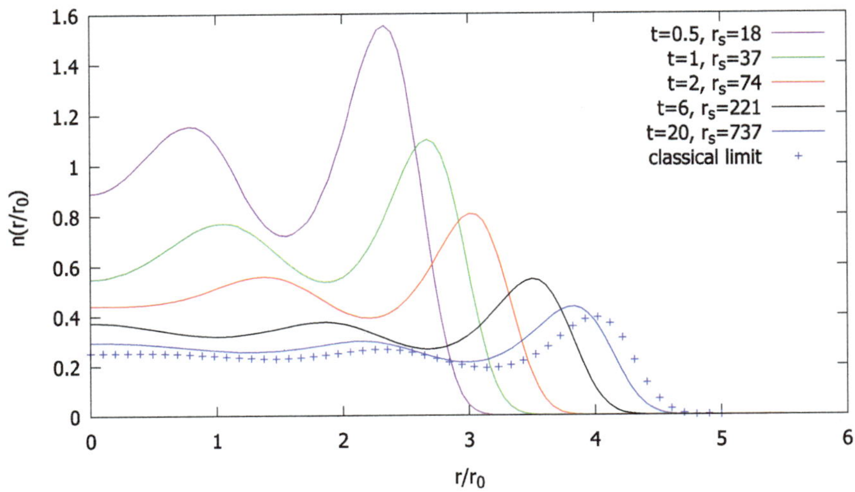

Fig. 11 Quantum effects for a trapped system of 100 particles. Here, $\Gamma = 20$ for all while the temperature parameter varies from $t = 0.5$ to $t = 20$.

potential. There is much more to be done with this classical description of a quantum system in the low t domain. At much smaller N, connection to other studies of quantum dots and ultracold gases should be useful. Other properties such as spin polarization, coherent control of trap properties, charge dependence, and others acceptable to direct observation can be addressed. A different direction for application of the results here is obtained by the replacement of the harmonic trap with a Coulomb potential to calculate the electron distribution about an ion. This is a solved problem of quantum chemistry, but its extension to a random configuration of ions is of intense current interest for warm, dense-matter applications, e.g., the electron density in the presence of an ion configuration. Such densities are required to compute the forces in quantum MD simulations for the ions in warm, dense matter at finite temperatures where traditional density functional methods fail (e.g., the traditional Kohn–Sham self-consistent equations for temperatures near the Fermi temperature). Here, those self-consistent equations are replaced with the classical integral equations of AHNC. This advantage has been stressed by Dharma-wardana.[24]

5. Discussion

The extreme conditions of long-range Coulomb charges, confined at strong coupling and finite temperatures, lead to complex structures: radial shell structure and broken symmetry angular ordering within the shells.

At the classical conditions, a density functional representation with strong coupling correlations from the OCP is able to capture quantitatively the transition from simple uniform filling at weak coupling to the formation of "atomic" shell structure. At infinite coupling, the ground state (ordered state) is closely related to the Thomson problem (sharp shell radii, no interactions between shells), extended here to finite temperature (e.g., like Debye–Waller broadening). The classical fluid phase with uniform angular distribution is now well studied by MC, MD, and theory—the features discussed in Ref. 9 are given quantitatively as a function of Γ, $\bar{N}$, e.g., number of shells, occupancy, amplitude, and location. The classical ordered phase also is well studied in the ground state, but less so at finite temperatures. The onset of localization within shells seems not to be sharp but rather gradual, as is the formation of shells, as a function of Γ (see however Refs. 25, 26 for some quantitative studies). The Thomson sites associated with the occupation number for a given shell provide a good reference for this localization, as confirmed by ground-state minimum energy models. The latter models, and simulation, indicated that there are metastable configurations with similar energy so the minimization requires care. It is possible that a more controlled limit is obtained from the limit of finite temperature studies as described here.

The quantum case is well studied at the ground state in the context of quantum dots, nanodevices, and related systems. The case of higher temperatures and transition to classical behavior is more limited, both from theory and simulation. The classical map method described here is particularly well suited for this domain, but has not been explored very much. Some of the advantages for states of warm, dense matter have been outlined in Ref. 24. Another interesting question is the line for Wigner crystallization (see Fig. 1), whose location is known only at $T = 0$ and in the high-temperature classical limit.

Acknowledgments

Much of the work summarized here was done in collaboration with the Bonitz group at Christian-Albrects University. The work of J.W. and J.D. was supported by US DOE Grant DE-SC0002139.

References

1. Thomson, J. J. On the Structure of the Atom. *Philos. Mag.* **1904**, 7, 237.
2. Pohl, T.; Pattard, T.; Rost, J. M. Coulomb Crystallization in Expanding Laser-Cooled Neutral Plasmas. *Phys. Rev. Lett.* **2004**, *92*, 155003.

3. Arp, O.; Block, D.; Piel, A.; Melzer, A. Dust Coulomb Balls: Three-Dimensional Plasma Crystals. *Phys. Rev. Lett.* **2004**, *93*, 165004; Arp, O.; Block, D.; Bonitz, M.; Fehske, H.; Golubnychiy, V.; Kosse, S.; Ludwig, P.; Melzer, A.; Piel, A. 3D Coulomb Balls: Experiment and Simulation. *J. Phys. Conf. Ser.* **2005**, *11*, 234.
4. Ludwig, P.; Kosse, S.; Bonitz, M. Structure of Spherical Three-Dimensional Coulomb Crystals. *Phys. Rev. E* **2005**, *71*, 046403.
5. Bonitz, M.; Block, D.; Arp, O.; Golubnychiy, V.; Baumgartner, H.; Ludwig, P.; Piel, A.; Filinov, A. Structural Properties of Screened Coulomb Balls. *Phys. Rev. Lett.* **2006**, *96*, 075001.
6. Golubnychiy, V.; Baumgartner, H.; Bonitz, M.; Filinov, A.; Fehske, H. Screened Coulomb Balls–Structural Properties and Melting Behaviour. *J. Phys. A Math. Gen.* **2006**, *39*, 4527.
7. Baumgartner, H.; Kählert, H.; Golubnychiy, V.; Henning, C.; Käding, S.; Melzer, A.; Bonitz, M. Structural and Dynamical Properties of Yukawa Balls. Contrib. *Plasma Phys.* **2007**, *47*, 281; Baumgartner, H.; Asmus, D.; Golubnychiy, V.; Ludwig, P.; Kählert, H.; Bonitz, M. Ground States of Finite Spherical Yukawa Crystals. *New J. Phys.* **2008**, *10*, 093019.
8. Cioslowski, J.; Grzebielucha, E. Parameter-Free Shell Model of Spherical Coulomb Crystals. *Phys. Rev. E* **2008**, *78*, 026416.
9. Wrighton, J.; Dufty, J. W.; Kählert, H.; Bonitz, M. Charge Correlations for Charges in a Harmonic Trap. *Phys. Rev. E* **2009**, *80*, 038912.
10. Wrighton, J.; Dufty, J. W.; Bonitz, M.; Kählert, H. Shell Structure of Confined Charges at Strong Coupling. *Contrib. Plasma Phys.* **2010**, *50*, 26.
11. Wrighton, J.; Kählert, H.; Ott, T.; Ludwig, P.; Thomsen, H.; Dufty, J.; Bonitz, M. Charge Correlations in a Harmonic Trap. *Contrib. Plasma Phys.* **2012**, *52*, 45.
12. Dufty, J. W.; Dutta, S. Classical Representation of a Quantum System at Equilibrium: Theory, Phys. Rev. E 87, 032101 (2013). *Contrib. Plasma Phys.* **2012**, *52*, 100.
13. Dutta, S.; Dufty, J. Classical Representation of a Quantum System at Equilibrium: Applications. *Phys. Rev. E* **2013**, *87*, 032102.
14. Dutta, S.; Dufty, J. Uniform Electron Gas at Warm, Dense Matter Conditions. *Euro. Phys. Lett.* **2013**, *102*, 67005.
15. Wrighton, J.; Dufty, J.; Dutta, S. Finite Temperature Quantum Effects in Many-Body Systems by Classical Methods. In Advances in Quantum Chemistry, Vol. 71; Elsevier: NY, 2015.
16. Wrighton, J.; Dufty, J.; Dutta, S. Finite-Temperature Quantum Effects on Confined Charges. *Phys. Rev. E* **2016**, *94*, 053208.
17. Dufty, J. W. Density Functional Theory for Electron Gas and for Jellium. *Langmuir* **2017**, *33* (42), 11570.
18. Silvestri, L. G.; Stanek, L. J.; Dharuman, G.; Choi, Y.; Murillo, M. S. Sarkas: A Fast Pure-Python Molecular Dynamics Suite for Plasma Physics. *Comput. Phys. Commun.* **2022**, *272*, 108245.
19. Sarkas., 2022. https://github.com/murillo-group/sarkas.
20. Hansen, J.-P.; MacDonald, I. *Theory of Simple Liquids*; Academic Press: San Diego, CA, 1990.
21. Ng, K. C. Hypernetted Chain Solutions for the Classical One-Component Plasma up to $\Gamma = 7000$. *J. Chem. Phys.* **1974**, *61*, 2680.
22. Thompson, A. P.; Aktulga, H. M.; Berger, R.; Bolintineanu, D. S.; Brown, W. M.; Crozier, P. S.; in 't Veld, P. J.; Kohlmeyer, A.; Moore, S. G.; Nguyen, T. D.; Shan, R.; Stevens, M. J.; Tranchida, J.; Trott, C.; Plimpton, S. J. LAMMPS–A Flexible Simulation Tool for Particle-Based Materials Modeling at the Atomic, Meso, and Continuum Scales. *Comput. Phys. Commun.* 2022, *271*, 10817.
23. LAMMPS., 2022. http://lammps.org.

24. Dharma-wardana, M. W. C. A Review of Studies on Strongly-Coupled Coulomb Systems Since the Rise of DFT and SCCS-1977. *Contrib. Plasma Phys.* **2015**, *55*, 85; Dharma-wardana, M. W. C. Current Issues in Finite-T Density-Functional Theory and Warm-Correlated Matter. *Computation* **2016**, *4*, 16.
25. Kählert, H.; Bonitz, M. How Spherical Plasma Crystals Form. *Phys. Rev. Lett.* **2010**, *104*, 015001.
26. Galván-Moya, J. E.; Altantzis, T.; Nelissen, K.; Peeters, F.; Grzelczak, M.; Liz-Marzán, L. M.; Bals, S.; Van Tendeloo, G. Self-Organization of Highly Symmetric Nanoassemblies: A Matter of Competition. *ACS Nano* **2014**, *8*, 3869.

CHAPTER FIVE

New insights on nonlinear solvatochromism in binary mixture of solvents

Evanildo G. Lacerda Jr., Sylvio Canuto*, and Kaline Coutinho
Instituto de Física, Universidade de São Paulo, São Paulo, SP, Brazil
*Corresponding author: e-mail address: canuto@if.usp.br

Contents

Abstract

In the last few decades we have seen enormous theoretical developments to incorporate solvent effects in molecular properties and spectroscopy. The vast majorities of these methods and applications have considered homogeneous solvents. Binary mixtures of solvents are also a common situation in laboratories and deserve more attention. This is a more complex situation and new characteristics arise. The two solvent components are not expected to contribute equally and consequently there is a preferential solvation of one of the two solvents. This leads to nonlinear solvation shift of the maximum of the UV–visible absorption spectrum as a function of the concentration of the mixture. In this investigation we have focused in the 1-methyl-8-oxyquinolinium betaine (QB) molecule in methanol-water mixtures as well as in acetone-water mixtures, in several concentrations. The sequential Quantum Mechanics/Molecular Mechanic method was used to obtain solute-solvent configurations composed of the chromophore and the

Advances in Quantum Chemistry, Volume 85
ISSN 0065-3276
https://doi.org/10.1016/bs.aiq.2022.04.004
Copyright © 2022 Elsevier Inc.
All rights reserved.

surrounded binary solvent molecules used subsequently to calculate the absorption spectrum and the solvatochromic shift as a function of water concentration. A simple explanation of the nonlinear behavior of the solvatochromism is put forward here based on two local solvation properties: the fraction of water mass in the first solvation shell and the fraction of solute-water hydrogen bonds in the mixture with respect to the pure water solution, which give an excellent description of the experiment data. Therefore, we conclude that these two properties related with the preferential solvation are the main aspects to describe the nonlinear solvatochromism.

1. Introduction

It is well known that solvents affect the physical-chemistry properties, spectroscopy and reactivity of molecules. In the last few decades great progress has been obtained to incorporate the solvent effects in electronic properties using theoretical models and methods.[1–4] As a natural step, these developments have been mostly devoted to homogeneous solvents. But many experimental measurements of molecular properties are conducted in binary mixtures of solvents.[5–10] Hence, the theoretical study of electronic properties of a solute (*X*) interacting with binary mixtures is important. However, because of the complexity involved, most studies consider continuum models in quantum mechanics with an adjustable dielectric constant.[5,6] Although this approach has certainly some merits the detailed consideration of the solute-solvent interaction is lacking and it is important. The situation is indeed more complex when the solute-solvent interaction involves two types of solvents (*S* and *W*) and new solvent(*S*)-solvent(*W*) interactions arise creating a competition between the three component of the system (*X*, *S* and *W*).[11] Additionally, in these simplified continuum models the polarity of the medium is adjusted with the concentration of the mixture[5,6] neglecting the possibility of a preferential solvation. In other words, a simple average parameter cannot consider that the solvation may be different for one solvent to the other.

One previous theoretical study[12] shows the effect of the preferential solvation of eumelanin monomers, dimers and tetramers in binary mixtures of methanol and water with different concentrations in the UV–visible absorption spectrum. These are preferably solvation by methanol, but a preferential micro-solvation (hydrogen bonds) of water for most species considered. It is clear that in a binary mixture of solvents, the solute can present more affinity for one of the solvents. This is termed as preferential solvation, i.e., the difference between the local and bulk concentration.

The nonlinear solvatochromism caused by binary mixtures has been thoroughly analyzed experimentally by the group of El Seoud[5,13–18] that has advanced some possible rationalization and published several important experimental data. Preferential solvation affects, of course, also other properties, such as dipole moment, excitation wavelengths or NMR chemical shifts, among others. The Fig. 1 (top) illustrates for the 1-methyl-8-oxyquinolinium betaine (QB, see Fig. 2 for chemical structure), how binary mixtures can affect electronic property.[14–18] The wavelength, E, of the first band maximum of the UV–visible spectra of the QB is plotted against the molar fraction of water, X_W, in several organic cosolvent. For $X_W = 0$, the solution has only the organic solvent and the wavelength of the first band maximum is so-called E_S; for $X_W = 1$, the solution has only the aqueous solvent and it is so-called E_W and for a water concentration X_W, it is $E(X_W)$. Therefore, the solvatochromic shift of the mixture is defined as the difference between the first band maximum wavelength of the mixture with a water concentration X_W and the pure organic solvent (S),

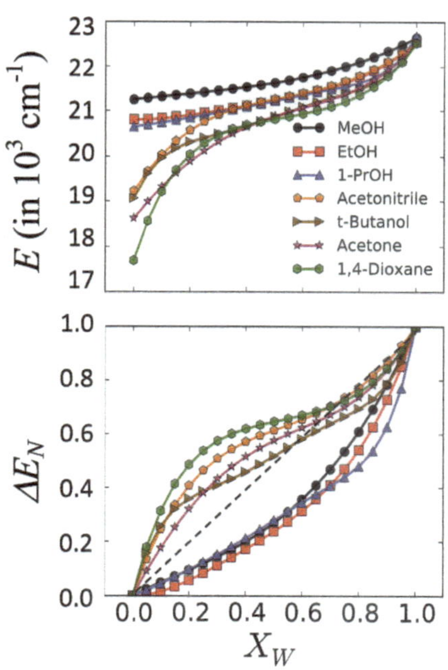

Fig. 1 (Top) Experimental values for the wavelength of the first maximum UV-visible absorption band, E (10^3 cm^{-1}), of the QB in several binary mixtures of organic solvent and water as a function of the water concentration X_W in the mixture.[10–14] (bottom) The normalized solvatochromic shift, ΔE_N (see Eq. 1 for definition). The dashed line corresponds to the linear behavior of ΔE_N with respect to X_W.

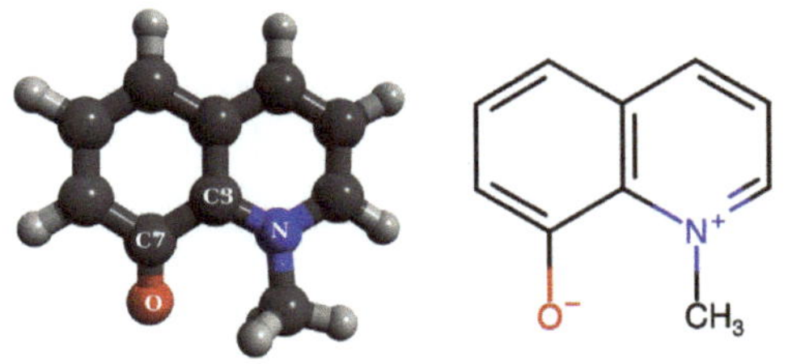

Fig. 2 (Right) The chemical structure and (left) optimized structure of 1-methyl-8-oxyquinolinium betaine (QB) and some atomic labels.

$\Delta E(X_W) = E(X_W) - E_S$. For some cosolvents, such as acetonitrile, t-butanol, acetone and 1,4-dioxane, a small increase of the water concentration X_W from 0.0 to 0.2 causes a considerable increase in E (large solvatochromism, $\Delta E(0.2) = 2260\ \text{cm}^{-1}$ or 0.28 eV or 63 nm for 1,4-dioxane). While for other cosolvents, such as methanol (MeOH), ethanol (EtOH) and 1-propanol (1-PrOH), cause small increase in E (small solvatochromism, $\Delta E(0.2) = 160\ \text{cm}^{-1}$ or 0.02 eV or 3 nm for MeOH). This is a manifestation of the preferential solvation and it deserves a closer look at the molecular level.

To better compare the effect of different organic cosolvent in the solvatochromism, the normalized solvatochromic scale ΔE_N is used as defines in Eq. (1), with respect to the water concentration X_W.

$$\Delta E_N(X_W) = [E(X_W) - E_S]/(\mathrm{E_W} - \mathrm{E_S}) \tag{1}$$

where E_W–E_S is the total solvatochromic shift of the reference molecule measured in aqueous solution W and in a given organic solvent S. Independent on the solvent S, this normalized scale ranges from $0 \leq \Delta E_N \leq 1$ (if $X_W=0 \Rightarrow E(X_W)=E_S \Rightarrow \Delta E_N=0$ and if $X_W=1 \Rightarrow E(X_W)=E_W \Rightarrow \Delta E_N=1$). Therefore, ΔE_N represents a percentage of the spectral shift due to the presence of the water concentration. Fig. 1 (bottom) represents this normalized scale for the QB solute. Note that ΔE_N does not show a linear behavior with X_W. For the protic solvents MeOH, EtOH and 1-PrOH, increasing the water concentration X_W presents a shift ΔE_N that is undervalued compared to the linear behavior (dashed line). Well different from the other four solvents considered where the shift ΔE_N is overvalued. This illustrates in a very clear way that there is a preferential solvation and suggests an important role of the solute-solvent hydrogen bonds. The detailed understanding of the origin of this preference is the subject of the present investigation. As expected preferential solvation affects molecular properties in general such as absorption, infrared and NMR spectroscopy.[5–10,13–20] On the theoretical side, there are some previous studies of the electronic properties of molecules in binaries mixtures[12,21–24] and even ternary mixtures.[25] Frutos-Puerto et al[24] investigated p-nitroaniline in cyclohexane–triethyamine mixtures by a mean-field sequential quantum mechanics/molecular mechanics (S-QM/MM) scheme. They indicated that the nonlinear effects in the absorption spectrum the solute are related to the hydrogen bond formations between the solute and the triethylamine molecules.

In this work, we apply the sequential S-QM/MM[26,27] to study the solvent effects of acetone-water (AW) and methanol-water (MW) mixtures on the first absorption band of the molecular probe QB (Fig. 2). This molecule belongs to a class of push-pull molecule that has two groups, electron-acceptor (carbonyl group) and electron-donor (amide group), separated by a group with a conjugated and delocalized π orbital (naphthalene group). Therefore, its electronic structure is very sensitive to the solvent environment. We show that the normalized solvatochromic ΔE_N is very well modeled by a combination of two local solvation properties: the mass fraction of water in the first solvation ($m_W{}^{1st}$) and the fraction of solute-water hydrogen bonds in the mixture with respect to the pure water solution ($f_W{}^{HB}$). Our results allow a discussion of the dominant cause of the preferential solvation in the solvatochromic shift.

2. Methods and details

Molecular Dynamics (MD) simulations and quantum mechanical (QM) calculations were performed to study the binary mixture effects on the QB polarization and electronic absorption wavelength with distinct water concentration X_W. Our theoretical approach was composed by three stages: first the solute was polarized considering each concentration X_W; then, the several solute-solvent configurations were generated and the solvation properties obtained; finally, the QM calculations were performed using the 200 solvent-solvent configurations to calculate the averaged electronic excitations.

To describe well the solute-solvent interactions in classical molecular mechanics simulations, it is important to consider the solute polarization, as reported in previous studies.[28–30] We used successfully an iterative polarization procedure[28] based on the S-MM/QM, which describes the solute polarization due to the solvent calculating the atomic charges of the solute surrounded by the electrostatic embedding of the solvent molecules. Therefore, in a mixture, it is crucial to describe the solvation shells as accurate as possible since these will influence the solute polarization and the shifts in the absorption spectrum.

After the polarization, MD simulations were performed using the solute polarized atomic charges. Thus, we calculated structural properties, such as molar and mass fraction of water in the first solvation shell and number of solute-solvent hydrogen bonds. To calculate the first electronic excitation wavelength of the QB, 200 statistically uncorrelated configurations were selected from the MD simulations and used in the subsequent QM calculations. It follows a description of the computational details.

2.1 Force field parameters for the solute and solvent

As a starting point the QB geometry was optimized using the density functional theory (DFT) method B3LYP[31] and the Dunning basis set cc-pVDZ.[32] For the solute, the bond lengths and bond angles of the optimized geometry were used as equilibrium parameters of the bonded potentials, the force constants and the Lennard-Jones parameters were taken from the OPLS force field[33] and the atomic charges for the Coulomb potential were obtained using the polarization procedure described in next section. This combination of OPLS parameters with polarized atomic charges were used before successfully,[29,30] showing even better description of the solute-solvent interactions when compared with the OPLS charges.[34] For the solvent molecules, methanol and acetone, all parameters where taken from the OPLS force field.[33] For water, the TIP4P model[35] was used.

2.2 Solute polarization

To define the partial atomic charges of the solute we used an iterative polarization procedure based on the S-MM/QM previous proposed.[28] The procedure consists of the following steps: (1) initially QM calculation was performed to obtain the dipole moment and the set of atomic charges for the isolated solute, μ_{vac} and $\{q_i\}_{vac}$; (2) using the set of charges obtained in the previous step, MD simulation of 50 ps was performed; (3) using the last MD configuration of the previous step, QM calculation was performed to obtain a new dipole moment and set of atomic charges for the solute embedded in the electrostatic field (point charges) of the solvent molecules $\mu_{i\text{-}pol}$

and $\{q_i\}_{i\text{-}pol}$; (4) repeat steps (2) and (3) for 100 times until reaches a total simulation time of 5 ns. In other words, the 5 ns MD simulation, used in the solute polarization process, started with the solute atomic charges in vacuum $\{q_i\}_{vac}$ and was interrupted each 50 ps to update the solute atomic charges $\{q_i\}_{i\text{-}pol}$ obtained with a QM calculation of the solute surrounded by the electrostatic embedding of the solvent. Typically, there is a fast growth in the values of dipole moment $\mu_{i\text{-}pol}$ compared with μ_{vac} in less than 10 steps. Then, $\mu_{i\text{-}pol}$ reaches an equilibrium trend, which the values fluctuate around an average. Therefore, the final set of the polarized solute atomic charges $\{q_i\}_{pol}$ was obtained as an average over the last 80 sets of $\{q_i\}_{i\text{-}pol}$ for each atom. We calculated the atomics charges from the QM electrostatic potentials using a grid-based method (CHELPG)[36] at the B3LYP/cc-pVDZ level. All QM calculations were carried out with Gaussian 09.[37]

2.3 Molecular dynamics simulations

All MD simulations were performed in the *NPT* ensemble, at room condition (T=25 °C and P=1 atm), using one solute QB molecule and 1000 solvent molecules. For the methanol and acetone aqueous mixtures, the molar fractions of water simulated were X_W=0.0, 0.2, 0.4, 0.5, 0.6, 0.8 and 1.0. The Berendsen thermostat[38] and barostat[39] were employed with coupling constants of 0.1 ps for the temperature, and 1.0 ps for the pressure. All bond length between hydrogen atoms and other atoms were kept rigid using the linear constraint solver (LINCS) method.[40] Molecular interactions for distances smaller than 1.5 nm were calculated and long-range corrections were applied for distances beyond this cut-off radius, using the particle mesh Ewald (PME) method[41,42] for Coulomb potential and the potential shift for the van der Waals potential. The leap-frog algorithm was used to integrate the equations of motion using a time step of 0.5 fs. After an equilibration stage for 2 ns, the following 5 ns was used to generate a set of atomic charges $\{q_i\}_{pol}$ that describes an average solute polarization due to solvent environment, then with the polarized solute the following 20 ns was used to generate 200 thousand solute-solvent configurations saved for statistical analysis. All the MD simulations were carried out with the GROMACS program.[43,44]

2.4 Solvation analysis

We analyzed the composition, i.e., the number of water molecules (N_W) and organic cosolvent molecules (N_M for methanol and N_A for acetone), of the first and second solvation shells using the minimum distance distribution function (MDDF) with rectangular normalization of $9 \times 10 \times 5$ Å.[45,46] The first and second solvation shells were described by the first and second well-defined peaks of the MDDF, respectively. Therefore, the number of molecules in the first and second solvation shells (N^{1st} and N^{2nd}) were obtained summing the histograms, that generate the MDDF, from zero to the minimum r^{1st} and from r^{1st} to r^{2nd}, respectively. Additionally, the fraction of water molecules in the first and second solvation shells ($X_W{}^{1st}$ and $X_W{}^{2nd}$) were also computed as.

$$X_W{}^{1st}(X_W) = N_W{}^{1st}(X_W)/N(X_W) \tag{2}$$

and using equivalent definition for the second solvation shell. Considering that the water molecules and the organic cosolvent may have different size, volume and mass, it would be interest to analyze any solvation property that consider also this information. Therefore, we analyzed also the fraction of water mass in the first solvation shell, as.

$$m_W{}^{1st}(X_W) = \left[m_W\, N_W{}^{1st}\left(X_W{}^{1st}\right)\right] / \left[m_W\, N_W{}^{1st}\left(X_W{}^{1st}\right) + m_S\, N_S{}^{1st}\left(X_W{}^{1st}\right)\right] \tag{3}$$

where m_W is the water mass, m_S is the organic cosolvent mass and $S=M$ for methanol or A for acetone for each mixture with different water concentration X_W.

The number of hydrogen bonds HB between the solute and the solvent (N^{HB}) were also analyzed. They were calculated using the standard geometrical criterion given by the distance between the acceptor-donor atoms $R_{AD} \leq 3.3$ Å and the angle formed between the acceptor-donor-hydrogen atoms $\theta_{ADH} \leq 40°$. Additionally, the fraction of water forming HB with QB were also computed with respect to the total number of HB as.

$$X_W{}^{HB}(X_W) = N_W{}^{HB}(X_W)/N^{HB}(X_W) \tag{4}$$

and with respect to the pure water solution as

$$f_W{}^{HB}(X_W) = N_W{}^{HB}(X_W)/N^{HB}(X_W = 1). \tag{5}$$

2.5 Calculation of the excitation wavelength

The first excitation wavelength of the QB were calculated with QM using the TD-DFT method, with the functional BHandHLYP[47] and the 6–311+G(d,p) basis set.[48] To describe the solvent effect in the QM calculations, a total of four solvent models were considered to have a systematic analysis of the contributions: (*i*) only the solute molecule with its geometry affected by the solvent generated by the MD simulations, named as QB model; (*ii*) the solute and the solvent molecules that form hydrogen bonds, named as QB+HB model; (*iii*) the solute surrounded by all solvent molecules treated as atomic point charges describing an electrostatic embedding, named as QB+EE model; and (*iv*) the solute and the solvent molecules that form hydrogen bonds surrounded by the electrostatic embedding, named as QB+HB+EE model. For each mixture with different X_W and each model, 200 configurations were used in the QM calculations to obtain the first excitation wavelength. The maximum band wavelength was achieved from the band profile generated by convoluting Lorentzians for each configuration with individual width of 10 cm^{-1}.

For comparison, the solvent effect was also calculated using the simple polarizable continuum model (PCM)[49] with dielectric constant ε value for the mixture composition taken from experimental data.[50,51]

3. Results

3.1 Solute polarization

Before starting the solute-mixture MD simulation, we performed also pure mixture simulations using the same procedures and thermodynamics condition to validate the combination of the force field choice. We calculated the average mass density of the simulated systems and compared with the experimental data[18] as a function of the water concentration X_W. This comparison is show in Fig. 3 revealing a very good agreement between the values. Therefore, the combination of OPLS force field for acetone and methanol with the TIP4P model for water is a reliable choice to describe well the macroscopic density behavior of these mixtures. The density of methanol-water mixture follows almost a linear behavior with the X_W variation. However, for the acetone-water mixtures the density is undervalued compared to the linear behavior, showing differences in the interactions of the cosolvent methanol and acetone with water.

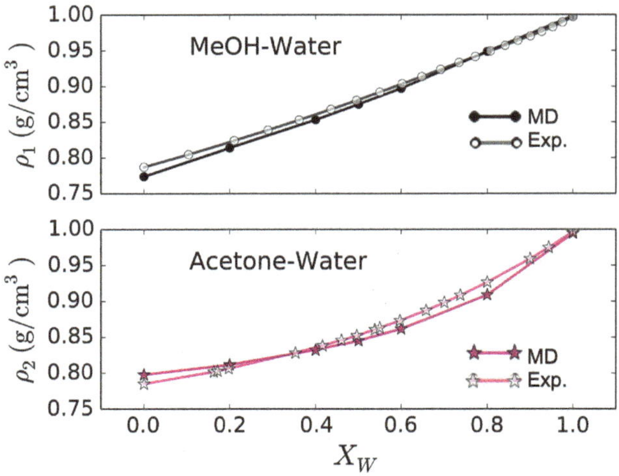

Fig. 3 Densities (in g/cm^3) for the methanol-water mixture (ρ_1) and acetone-water mixture (ρ_2) as a function of the water concentration X_W: (open symbols) experimental data[8] and (solid symbols) values obtained from the MD simulations.

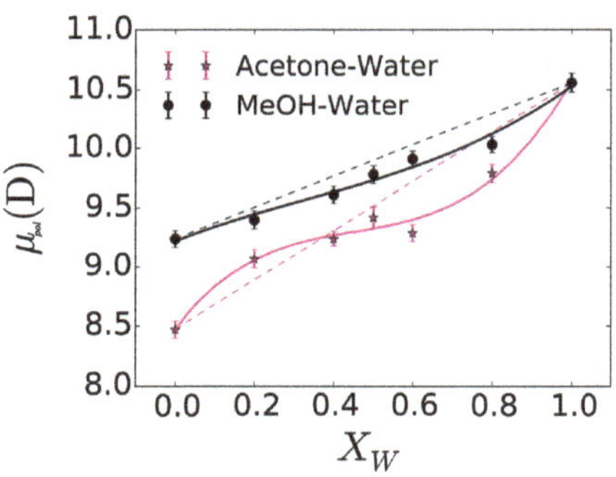

Fig. 4 Calculated dipole moment of polarized QB, μ_{pol} (in Debye), as a function of the water concentration X_W. The error bars are the estimated standard errors. The solid lines are the polynomial fits of the calculated values and the dashed lines are the linear behavior only shown for comparison.

The QB molecule was electronic polarized in every molar fraction of water for methanol-water and acetone-water mixtures. In Fig. 4, it is shown the calculated dipole moment of the polarized QB, μ_{pol}, as a function of the composition of the mixture X_W. In methanol-water mixtures, μ_{pol} increases 13% from pure methanol (μ_{pol}=9.2 D for X_W=0.0) to pure water (μ_{pol}=10.6 D for X_W=1.0), while in acetone-water mixtures increases 20% (μ_{pol}=8.5 D for X_W=0.0). In neither case, the increase of the μ_{pol} is linear with X_W, as indicated by the dashed lines in comparison with calculated data. The polynomial fits of the calculated μ_{pol} are also shown in Fig. 4, which describes the calculated tendency of the μ_{pol} for a given molar fraction X_W. For the methanol-water mixtures the calculated tendency is always below the linear behavior, indicating higher effective contribution of the methanol molecules. For acetone-water mixtures the behavior is more complex, where μ_{pol} is above the linear behavior only for small concentration of water (small X_W) but changing for higher concentrations of water (large X_W). To better understand these variations of μ_{pol} in both mixtures, we also analyzed the averaged atomic charges $\{q_i\}_{pol}$. In Fig. 5, these charges for some selected atoms are shown with their respective error bar. The O atom becomes more negative with the increase of water concentration in both mixtures and the variation is larger for acetone-water than for methanol-water mixture that is in agreement with the dipole increase respectively in both mixtures. However, the charges in the N, C3 and C7 atom remains essentially constant. This is interesting because it is not normally expected,[3,4] the charge transfer to the Oxygen atom comes from the ring but not from these specific atoms such as the Nitrogen.

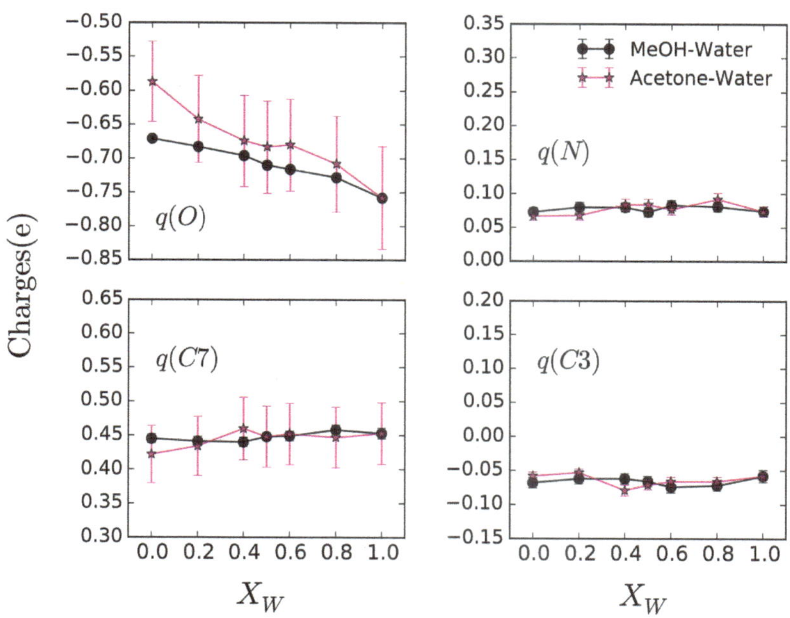

Fig. 5 Atomic charges (in *e*) for the atoms O, N, C3 e C7 of polarized QB molecule as a function of the concentration X_W, calculated with CHELPG method using B3LYP/cc-pVDZ level (see Section 2.2). Atomic labels are defined in Fig. 2. The error bars are also shown.

Note that in the present approach we focus in the solute polarization improving the description of the solute-solvent interactions. This affects especially the solute-solvent hydrogen bonds and solvation shells, as will be discussed in the next section and was discussed previously in detail.[45] For the solvent, an average polarization is implicitly included in the classical force field and it is essential to describe properly the solvent properties, such as density and solvation shells. We have shown for single solvent that including solute polarization is enough to well describe the solvent effects in the electronic properties of the solute.[28–30,34,45,46] Now we extend it also for binary mixture. However, for more complex systems, such as ionic liquids, a more detailed description of the solvent polarization is necessary and the use of polarizable force field[52] may be recommended.

3.2 Solvation analysis

Using the 200 thousand configurations generated in each MD simulation of polarized QB solute in mixtures with different X_W, we calculated the number of hydrogen bonds formed between the solute and solvent molecules using the geometrical criterion. To identify all HB, even the non-conventional HB, for the QB solute we consider the Nitrogen and the Oxygen atoms as acceptor atoms of HB and all HC bonds as donor groups. For methanol and water molecules, the Oxygen atom was considered as acceptor of HB and HO bonds as donor groups and for acetone the Oxygen atom was considered as acceptor and all HC bonds as donor groups. After the analyses, only conventional HB were identified between the acceptor Oxygen atom of the QB solute and the donor HO bonds of the methanol and water solvents. None HB between QB and acetone were identified, as expected. In Table 1, we present the results of this HB analysis. For methanol-water mixture, both solvents formed HB with QB solute, $N^{HB} = 1.62$ HB for pure methanol solution ($X_W = 0$) and 3.02 HB for pure water solution ($X_W = 1$). Comparing the fraction of water forming HB, ${X_W}^{HB}(X_W) = {N_W}^{HB}(X_W) / N^{HB}(X_W)$, with the total fraction of water in the mixture, X_W, is possible to see that ${X_W}^{HB}$ is always larger than X_W. For example, with 20% of water in the mixture, $X_W = 0.2$, there is 35% of water and 65% of methanol forming HB with QB. Then, we can conclude that although there is a competition between methanol and water forming HB with QB, the water molecules are preferred. For the acetone-water mixture, as the acetone molecules do not form HB with QB, the QB-water HB is formed even with a very small water concentration. As example, for 20%

Table 1 The total number of solute-solvent hydrogen bonds (N^{HB}) and its composition of methanol, acetone and water (N_M, N_A and N_W) for methanol-water and acetone-water mixtures with different water concentration X_W, obtained from geometrical criteria (see Section 2.4).

	Methanol-water					Acetone-water				
X_W	N^{HB}	$N_M{}^{HB}$	$N_W{}^{HB}$	$X_W{}^{HB}$	$f_W{}^{HB}$	N^{HB}	$N_A{}^{HB}$	$N_W{}^{HB}$	$X_W{}^{HB}$	$f_W{}^{HB}$
0.0	1.62	1.62	0.00	**0.00**	**0.00**	0.00	0.00	0.00	**0.00**	**0.00**
0.2	1.85	1.20	0.65	**0.35**	**0.22**	1.32	0.00	1.32	**1.00**	**0.44**
0.4	2.08	0.97	1.11	**0.53**	**0.37**	1.72	0.00	1.72	**1.00**	**0.57**
0.5	2.17	0.66	1.51	**0.70**	**0.50**	1.91	0.00	1.91	**1.00**	**0.63**
0.6	2.36	0.63	1.73	**0.73**	**0.57**	1.99	0.00	1.99	**1.00**	**0.66**
0.8	2.54	0.32	2.22	**0.87**	**0.74**	2.37	0.00	2.37	**1.00**	**0.78**
1.0	3.02	0.00	3.02	**1.00**	**1.00**	3.02	0.00	3.02	**1.00**	**1.00**

of water in the mixture, X_W=0.2, there are 1.32 QB-water HB formed. In this case, the fraction of water forming HB is always 100% independent of X_W. But considering fraction of water forming HB with respect to the pure water solution, 44% of the available water HB is already formed for 20% of water in the mixture. This reveals the strong affinity of the QB solute to water HB formation.

We also analyzed the first and second solvation shells and its composition. The solvation shells are obtained from the MDDF and these are shown in the Fig. 6 for QB in pure organic solution (X_W=0) and pure water solution (X_W=1) and in the 1:1 mixture (X_W=0.5). In both cases, for pure water and methanol solutions the first shape peak characterizes the solute-solvent HB in approximately 1.9 Å around the QB. For pure acetone solution, this peak does not exist, as expected. In both mixtures, the first solvation shell is defined by the first well-defined intense peak that ends around 4.0 Å and the second shell around 8.0 Å. Summing the histogram, used to generate the MDDF, until the minima of the first and the second solvation shells we obtained the total number of solvent molecules (N) and the separate number of water molecules (N_W) and organic cosolvent molecules (N_S) that compose those two shells. In Table 2, these values are shown for the methanol-water mixture for different water concentration X_W and fractions of water molecules $X_W{}^{1st}$ and water mass $m_W{}^{1st}$ in the first solvation shell are show in Table 3 and in Fig. 7. In the 1:1 mixture, there are approximately 16 methanol and 12 water molecules in the first solvation shell and 45 and 41,

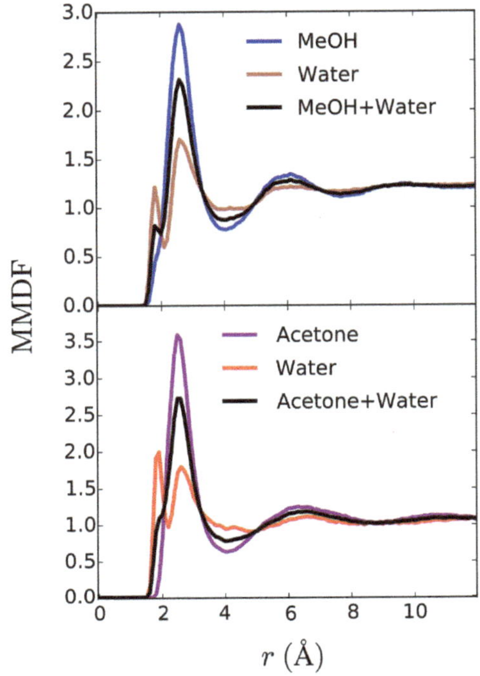

Fig. 6 Minimum distance distribution function, MDDF, of the solvent molecules around the QB for the case of X_W=0.5, for both methanol-water (top) and acetone-water (bottom) mixtures.

Table 2 The size (*r* in Å), the total number of solvent molecules (*N*) in the first and second solvation shells of methanol-water mixture with different water concentration X_W and its composition of methanol and water (N_M and N_W), obtained from the histograms that generate the minimum distance distribution function, MDDF (see Fig. 6).

	Methanol-water (1st solvation shell)				Methanol-water (2nd solvation shell)			
X_W	r_{1st}	N^{1st}	N_M^{1st}	N_W^{1st}	r_{2nd}	N^{2nd}	N_M^{2nd}	N_W^{2nd}
0.0	4.1	23.3	23.3	0.0	7.8	61.1	61.1	0.0
0.2	4.0	24.7	20.5	4.2	7.8	70.1	57.0	13.1
0.4	4.0	26.9	17.9	9.0	7.9	79.4	49.5	29.9
0.5	4.0	28.1	16.1	12.0	7.8	85.2	44.7	40.5
0.6	4.0	29.4	14.3	15.1	7.8	91.1	39.2	51.9
0.8	4.1	33.9	9.7	24.2	7.8	105.4	23.0	82.4
1.0	4.0	37.6	0.0	37.6	7.5	113.4	0.0	113.4

Table 3 The fraction of water molecules (X_W^{1st}) and the fraction of water mass (m_W^{1st}) in the first solvation shell for methanol-water and acetone-water mixture with different water concentration X_W (see Section 2.4 for definitions).

	Methanol-water		Acetone-water	
X_W	X_W^{1st}	m_W^{1st}	X_W^{1st}	m_W^{1st}
0.0	0.00	0.00	0.00	0.00
0.2	0.17	0.08	0.19	0.09
0.4	0.33	0.18	0.38	0.21
0.5	0.43	0.25	0.48	0.29
0.6	0.51	0.31	0.57	0.37
0.8	0.71	0.52	0.78	0.61
1.0	1.00	1.00	1.00	1.00

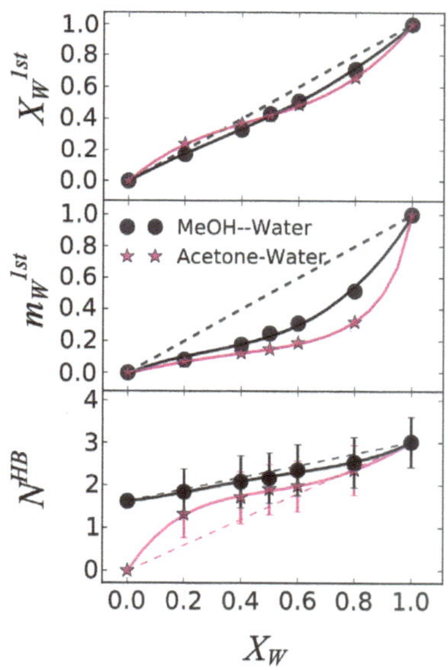

Fig. 7 The molar fraction of water X_W^{1st} (top) and the local mass fraction of water m_W^{1st} (middle), in the first solvation shell (see Section 2.4) and N^{HB} (bottom) is the number of solute–solvent HB as a function of X_W. The solid lines are polynomial fit of the calculated values and the dashed lines are references for the linear behavior for X_W^{1st} and N^{HB} and the bulk behavior for water mass fraction, i.e., $m_W(X_W) = [m_W N_W(X_W)] / [m_W N_W(X_W) + m_S N_S(X_W)]$.

respectively in the second solvation shell, showing a preferential solvation of methanol. The fraction of water molecules is 43% that is smaller than the 50% of the mixture, a reduction of ~85%. In the second solvation shell X_W^{2nd} is 48% that is slightly smaller than 50% of the mixture, a reduction of ~95%.

Therefore, major contribution of the preferential solvation appears in the first solvation shell, while the second is quite similar to the bulk concentration. Considering that methanol molecules are larger and heavier than water molecules, another interesting property is the fraction of water mass in the first solvation shell, ${m_W}^{1st}$, that shows a much smaller water contribution of 25% compared with 50% of the mixture. These results of the first solvation shell reveal the hydrophobic character of the QB which prefer interact with methanol than water molecules, as expected due to the naphthalene ring. For the acetone-water mixture, the fraction of water molecules and mass in the first solvation shell is larger than for methanol-water mixture, which may be considered nonsense at first sight considering that acetone is a less polar solvent and should be preferred to water. However, analyzing the HB also shown in Fig. 7 and Table 1, we identified a much larger contribution of water compared with the linear behavior. Therefore, QB-water HB promote an excess of expected water contribution in the first solvation shell.

3.3 Excitation wavelength

We now make analysis of the calculated absorption maximum wavelength E of the QB molecule in both mixtures with different water concentration X_W. These values are presented in the Fig. 8 for the different solvent models considered and explained in Section 2.5. Each point of this graphic is an average over 200 MD configurations. For the QB model, in which only the solute with flexibly geometry obtained from the MD simulation in each X_W is included in the QM calculations, the E values seem to behave independently of X_W. This means that the effect of the solvent in the geometry is

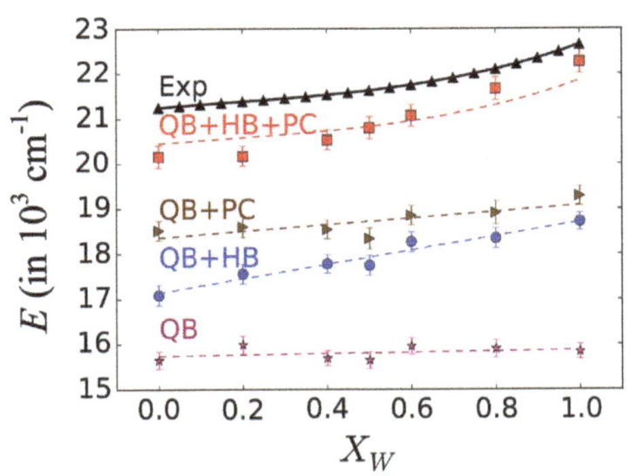

Fig. 8 Calculated excitation wavelength E of the absorption maximum of the QB molecule in methanol-water mixture using different theoretical models (see Section 2.5) with different concentrations X_W compared with experimental data.[14]

minor. The QB + HB and the QB + EE models show both linear increase with X_W with larger slope for QB + HB, indicating that E is more sensitivity to HB than the electrostatic embedding EE. The QB + HB + EE model shows a very good qualitative behavior compared to the experimental values.[14] The difference of the calculated values of E compared with the experimental data are 800 cm^{-1} or 0.1 eV or 18 nm. In literature, this is considered a reasonable good agreement considering the precision of many functional used in the QM calculation. Therefore, we conclude that in explicit inclusion of the solute and the hydrogen bonded solvent molecules with the electrostatic embedding of the remain solvent molecules is a good model the describe the solvent effect of the mixture in the UV–visible absorption solvatochromic shift.

3.4 Energy-solvation relationship

To understand the effect of the solvation properties with the normalized solvatochromism $\Delta E_N(X_W)$, we analyzed correlation between calculated solvation properties with $\Delta E_N(X_W)$. So, we identified that a two-descriptor model can be built to describe the $\Delta E_N(X_W)$ using the fraction of water mass in the first solvation shell $m_W{}^{1st}(X_W)$ and the fraction of water involved in solute-solvent HB with respect with the pure water solution $f_W{}^{HB}(X_W)$ as.

$$\Delta E_N(X_W) = \mathrm{A}\,\mathrm{m_W}{}^{1\mathrm{st}}(\mathrm{X_W}) + \mathrm{B}\,\mathrm{f_W}{}^{\mathrm{HB}}(\mathrm{X_W}) \tag{6}$$

where A and B are fitted coefficients with $A + B = 1$, $m_W{}^{1st}(X_W)$ was defined in Eq. (3) and $f_W{}^{HB}(X_W)$ was defined in Eq. (5). Then, A and B are interpreted as the relative weight of these two properties, or descriptors. The best fit of the $\Delta E_N(X_W)$ experimental data with Eq. (6), gave $A = 0.88$ and $B = 0.12$ for methanol-water mixture with high correlation coefficient of $R^2 = 0.996$ and random mean square error (RMSE) of only 0.035. Therefore, the fraction of water mass in the first solvation shell has 88% of contribution and 12% comes from fraction of water involved in solute-solvent HB with respect with the pure water solution, showing that the first solvation shell concentration is the most important descriptor to explain the behavior of E_N for protic solvents mixed with water. For acetone-water mixture, the best fit gave $A = 0.16$ and $B = 0.84$ with high correlation coefficient of $R^2 = 0.990$ and RMSE $= 0.031$. Therefore, the fraction of water mass in the first solvation shell has 16% of contribution and 84% comes from fraction of water involved in solute-solvent HB with respect with the pure water solution, showing that HB are the most important descriptor to explain the behavior of E_N for non-protic solvents mixed with water.

In the case of protic solvent mixed with water, such as methanol, ethanol and propanol, there are solute-solvent HB formed for any concentration of water (from 1.62 to 3.02 HB for X_W varying from 0.0 to 1.0 in methanol-water mixture, see Table 1), while for aprotic solvent mixed with water, such as acetonitrile, acetone and dioxane, with $X_W = 0$ there is none HB. In the case of butanol-water mixture the ability of butanol to form HB is small due to its large size, then it has the behavior of non-protic solvent. Therefore, the effect of the solvent in the solute polarization is much smaller (increase of 13% from $X_W = 0.0$ to 1.0 in methanol-water) compared with non-protic solvent mixed with water (increase of 20% from $X_W = 0.0$ to 1.0 in acetone-water). This result shows a strong effect of HB in the solute polarization and consequently in the solvatochromic shift when a small amount of water is added to the non-protic solvent which changes the situation from none to some solute-solvent HB. Fig. 9 shows the very good description of these best fit compared with the experimental leading great credence to the two-descriptor model obtained from the preferential

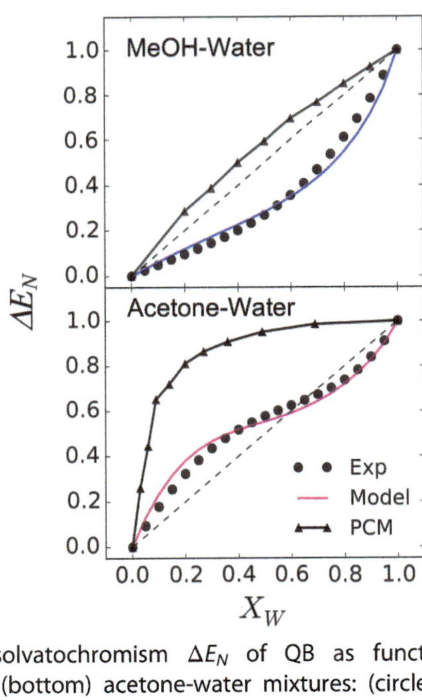

Fig. 9 Normalized solvatochromism ΔE_N of QB as function of X_W for: (top) methanol-water and (bottom) acetone-water mixtures: (circle) experimental data,[14] (triangle) calculated values of QB with solvent described with PCM, and (colored solid lines) best fit with the two-descriptor model of Eq. (6) (see Section 3.4).

solvation. For comparison, Fig. 9 also show the values obtained using PCM calculations with dielectric constants taken from experiments[50,51] varying 32.3 to 78.3 for methanol-water mixture and from 20.8 to 78.3 for acetone-water mixture. As it can be seen, this continuum model cannot describe the experimental trend, due to an over estimation of the water contribution showing overvalued shifts compared to the linear behavior for both mixtures. This provides evidence of the limitations of continuum solvation models for a correct description of liquid mixtures.

4. Conclusions

The theoretical description of solvent effects in molecular properties and spectroscopy is a topic of great interest and increasing developments. In more recent years, attention has been placed on the description of binary mixtures of solvents. Compared to simple homogeneous liquids the use of binary solvents, as expected, adds some additional complexities. From experiments, it has been seen that the solvatochromic shift due to the solvent exhibits some peculiarities. This solvatochromism has been analyzed experimentally and the observation of a nonlinear behavior has been emphasized and still challenges a simple explanation. El Seoud and collaborators[17] have made an interesting suggestion that a third species composed of the two solvent aggregates would give an important contribution. We look for this third species but we could not find any interpretation in this direction. In this investigation, we have focused in the 1-methyl-8-oxyquinolinium betaine, QB, molecule in methanol-water mixtures as well as acetone-water mixtures, in several concentrations. The sequential QM/MM method was used to calculate the electronic excitation of QB solute in the mixtures. MD simulations were performed to obtain solute-solvent configurations composed of the polarized chromophore and the surrounded binary solvent molecules. We were not able to find this third component of the solvent with aggregation and have focused on the importance of the local solvation properties, corresponding to the interactions of the two solvents in the vicinities of the chromophore. A simple explanation is put forward here based on two local solvation properties, the fraction of water mass in the first solvation shell and fraction of water involved in solute-solvent hydrogen bonds with respect with the pure water solution. As shown here this simple model gives a very good correlation between this two-descriptor model and experiment, leading credence to the explanation given here. Additionally, our results show limitations of the continuum solvation models for describing the solvent effects of liquid mixtures.

Acknowledgments

This work was partially supported by the Brasilein agencies: FAPESP, CNPQ and CAPES. All authors thank the National Institute of Science and Technology of Complex Fluids (INCT-FCx) with the CNPq grant no. 141260/2017-3 and FAPESP grant no. 2014/50983-3, and the BioMol project with CAPES grant no. 23038.004630/2014-35. K.C. also tanks FAPESP for the grants no. 2021/09016-3 and 2017/11631-2.

We dedicate this contribution to Prof. John Sabin.

References

1. Canuto, S., Ed. *Solvation Effects on Molecules and Biomolecules. Computational Methods and Applications*; Springer, 2008. ISBN:978-1-4020-8269-6; e-ISBN: 978-1-4020-8270-2.
2. Memnnucci, B.; Cammi, R., Eds. *Continuum Solvation Models in Chemical Physics: From Theory to Applications*; J. Wiley, 2008.
3. Canuto, S.; Brändas, E.; Sabin, J. Combining Quantum Mechanics and Molecular Mechanics. Some Recent Progresses in QM/MM Methods. *Special Issue of the Advances in Quantum Chemistry*, vol. 59; 2010.
4. Marenich, A. V.; Cramer, C. J.; Truhlar, D. G. Universal Solvation Model Based on Solute Electron Density and a Continuum Model of the Solvent Defined by the Bulk Dielectric Constant and Atomic Surface Tensions. *J. Phys. Chem. B* **2009**, *113*, 6378–6396. https://doi.org/10.1021/jp810292n.
5. El Seoud, O. A. Understanding Solvation. *Pure Appl. Chem.* **2009**, *81*, 697–707. https://doi.org/10.1351/PAC-CON-08-09-27.
6. Bagno, A.; Campulla, M.; Pirana, M.; Scorrano, G.; Stiz, S. Preferential Solvation of Organic Species in Binary Solvent Mixtures Probed by Intermolecular 1H NOESY NMR Spectroscopy. *Chem. – Eur. J.* **1999**, *5*, 1291–1300. https://doi.org/10.1002/(SICI)1521-3765(19990401)5:4<1291::AID-CHEM1291>3.0.CO;2-M.
7. Chatterjee, P.; Laha, A. K.; Bagchi, S. Preferential Solvation in Mixed Binary Solvents - Ultraviolet-Visible Spectroscopy of N- Alkylpyridinium Iodides in Mixed-Solvents Containing Cyclic Ethers. *J. Chem. Soc., Faraday Trans.* **1992**, *88*, 1675–1678. https://doi.org/10.1039/FT9928801675.
8. Navarro, A. M.; García, B.; Hoyuelos, F. J.; Penãcoba, I. A.; Leal, J. M. Preferential Solvation in Alkan-1-Ol/Alkylbenzoate Binary Mixtures by Solvatochromic Probes. *J Phys Chem B* **2011**, *115*, 10259. https://doi.org/10.1021/jp202019x.
9. Amenta, V.; Cook, J. L.; Hunter, C. A.; Low, C. M. R.; Vinter, J. G. Influence of Solvent Polarity on Preferential Solvation of Molecular Recognition Probes in Solvent Mixtures. *J. Phys. Chem. B* **2012**, *116*, 14433–14440. https://doi.org/10.1021/jp310379h.
10. Santina, L. R. R.; Santos, S. C.d.; Novo, D. L. R.; Bianchini, D.; Gerola, A. P.; Braga, G.; Caetano, W.; Moreira, L. M.; Bastos, E. L.; Romani, A. P.; Oliveira, H. P. M.d. Study between Solvatochromism and Steady-State and Time-Resolved Fluorescence Measurements of the Methylene Blue in Binary Mixtures. *Dyes Pigm.* **2015**, *119*, 12–21. https://doi.org/10.1016/j.dyepig.2015.03.004.
11. Marcus, Y. On the Preferential Solvation of Drugs and PAHs in Binary Solvent Mixtures. *J. Molec. Liquids* **2008**, *140* (1–3), 61–67. https://doi.org/10.1016/j.molliq.2008.01.005.
12. Oliveira, L. B. A.; Fonseca, T. L.; Cabral, B. J. C.; Coutinho, K.; Canuto, S. Preferential Solvation and Optical Properties of Eumelanin Building Blocks in Binary Mixture of Methanol and Water. *J. Chem. Phys.* **2021**, *155*. https://doi.org/10.1063/5.0065723, 174504.

13. El Seoud, O. A. Solvation in Pure and Mixed Solvents: Some Recent Developments. *Pure Appl. Chem.* **2007**, *79*, 1135. https://doi.org/10.1351/pac200779061135.
14. Tada, E. B.; Novaki, L. P.; El Seoud, O. A. Solvatochromism in Pure and Binary Solvent Mixtures: Effects of the Molecular Structure of the Zwitterionic Probe. *J. Phys. Org. Chem.* **2000**, *13*, 679. https://doi.org/10.1002/1099-1395(200011)13:11<679::aid-poc299>3.0.co;2-r.
15. Antonious, M. S.; Tada, E. B.; El Seoud, O. A. Thermo-Solvatochromism in Aqueous Alcohols: Effects of the Molecular Structures of the Alcohol and the Solvatochromic Probe. *J. Phys. Org. Chem.* **2002**, *15*, 403. https://doi.org/10.1002/poc.510.
16. Tada, E. B.; Silva, P. L.; El Seoud, O. A. Thermo-Solvatochromism of Betaine Dyes in Aqueous Alcohols: Explicit Consideration of the Water-Alcohol Complex. *J. Phys. Org. Chem.* **2003**, *16*, 691. https://doi.org/10.1002/poc.632.
17. Tada, E. B.; Silva, P. L.; Tavares, C.; El Seoud, O. A. Thermo-Solvatochromism of Zwitterionic Probes in Aqueous Aliphatic Alcohols and in Aqueous 2-Alkoxyethanols: Relevance to the Enthalpies of Activation of Chemical Reactions. *J. Phys. Org. Chem.* **2005**, *18*, 398. https://doi.org/10.1002/poc.887.
18. Bastos, E. L.; Silva, P. L.; El Seoud, O. A. Thermosolvatochromism of Betaine Dyes Revisited: Theoretical Calculations of the Concentrations of Alcohol-Water Hydrogen-Bonded Species and Application to Solvation in Aqueous Alcohols. *J Phys Chem A* **2006**, *110*, 10287. https://doi.org/10.1021/jp062250t.
19. Chatterjee, P.; Bagchi, S. Preferential Solvation of a Dipolar Solute in Mixed Binary Solvent: A Study of UV-Visible Spectroscopy. *J. Phys. Chem.* **1991**, *95*, 3311. https://doi.org/10.1021/j100161a064.
20. González, G.; Clavijo, E. Solvent Effects on the Infrared Spectra of Nitro-N-Methylanilines: Intra- and Inter-Molecular Interactions and Molecular Configurations. *J Chem Soc Perkin Trans* **1985**, *2, vol. 1*, 1751. https://doi.org/10.1039/p29850001751.
21. Saielli, G.; Scorrano, G.; Bagno, A.; Wakisaka, A. Solvation of Tetraalkylammonium Chlorides in Acetonitrile-Water Mixtures: Mass Spectrometry and Molecular Dynamics Simulations. *ChemPhysChem* **2005**, *6*, 1307. https://doi.org/10.1002/cphc.200500050.
22. Saielli, G.; Bagno, A. Preferential Solvation of Glucose and Talose in Water-Acetonitrile Mixtures: A Molecular Dynamics Simulation Study. *Phys. Chem. Chem. Phys.* **2010**, *12*, 2981. https://doi.org/10.1039/b922550a.
23. Pratihar, S.; Chandra, A. A First Principles Molecular Dynamics Study of Lithium Atom Solvation in Binary Liquid Mixture of Water and Ammonia: Structural, Electronic, and Dynamical Properties. *J Chem Phys* **2011**, *134*. https://doi.org/10.1063/1.3511702, 024519.
24. Frutos-Puerto, S.; Aguilar, M. A.; Galván, I. F. Theoretical Study of the Preferential Solvation Effect on the Solvatochromic Shifts of Para-Nitroaniline. *J Phys Chem B* **2013**, *117*, 2466. https://doi.org/10.1021/jp310964k.
25. Ray, N.; Bagchi, S. Fluorimetric Study of Solvation in Ternary Solvent Mixtures. Ketocyanine Dye in Ethanol+Benzene+Water and Ethanol+Benzene+Acetone. *J. Mol. Liq.* **2004**, *111*, 19. https://doi.org/10.1016/j.molliq.2003.09.018.
26. Coutinho, K.; Canuto, S. Solvent Effects from a Sequential Monte Carlo-Quantum Mechanical Approach. *Adv. Quantum Chem.* **1997**, *28*, 89. https://doi.org/10.1016/S0065-3276(08)60209-9.
27. Coutinho, K.; De Oliveira, M. J.; Canuto, S. Sampling Configurations in Monte Carlo Simulations for Quantum Mechanical Studies of Solvent Effects. *Int. J. Quantum Chem.* **1998**, *66*, 249. https://doi.org/10.1002/(SICI)1097-461X(1998)66:3<249::AID-QUA6>3.0.CO;2-V.

28. Georg, H.; Coutinho, K.; Canuto, S. Converged Electronic Polarization of Acetone in Liquid Water and the Role in the $n\pi$ Transition. *Chem. Phys. Lett.* **2006**, *429*, 119. https://doi.org/10.1016/j.cplett.2006.08.047.
29. Coutinho, K.; Georg, H.; Fonseca, T.; Ludwig, V.; Canuto, S. An Efficient Statistically Converged Average Configuration for Solvent Effects. *Chem. Phys. Lett.* **2007**, *437*, 148. https://doi.org/10.1016/j.cplett.2007.02.012.
30. Vequi-Suplicy, C. C.; Orozco-Gonzalez, Y.; Lamy, M. T.; Canuto, S.; Coutinho, K. A New Interpretation of the Absorption and the Dual Fluorescence of Prodan in Solution. *J. Chem. Phys.* **2020**, *153*, 244104. https://doi.org/10.1063/5.0025013.
31. Becke, A. D. Density-Functional Thermochemistry. III. The Role of Exact Exchange. *J Chem Phys* **1993**, *98*, 5648. https://doi.org/10.1063/1.464913.
32. Dunning, T. H. Gaussian Basis Sets for Use in Correlated Molecular Calculations. I. the Atoms Boron through Neon and Hydrogen. *J Chem Phys* **1989**, *90*, 1007. https://doi.org/10.1063/1.456153.
33. Jorgensen, W. L.; Maxwell, D. S.; Tirado-rives, J. Development and Testing of the OPLS all-Atom Force Field on Conformational Energetics and Properties of Organic Liquids. *J. Am. Chem. Soc.* **1996**, *118*, 11225. https://doi.org/10.1021/ja9621760.
34. Barreto, R. C.; Coutinho, K.; Georg, H. C.; Canuto, S. Combined Monte Carlo and Quantum Mechanics Study of the Solvatochromism of Phenol in Water. The Origin of the Blue Shift of the Lowest P–P* Transition. *Phys. Chem. Chem. Phys.* **2009**, *11*, 1388. https://doi.org/10.1039/b816912h.
35. Jorgensen, W. L.; Chandrasekhar, J.; Madura, J. D.; Impey, R. W.; Klein, M. L. Comparison of Simple Potential Functions for Simulating Liquid Water. *J Chem Phys* **1983**, *79*, 926. https://doi.org/10.1063/1.445869.
36. Breneman, C. M.; Wiberg, K. B. Determining Atom-Centered Monopoles from Molecular Electrostatic Potentials. The Need for High Sampling Density in Formamide Conformational Analysis. *J. Comput. Chem.* **1990**, *11*, 361. https://doi.org/10.1002/jcc.540110311.
37. Frisch, M. J.; Trucks, G. W.; Schlegel, H. B.; Scuseria, G. E.; Robb, M. A.; et al. *Gaussian 09, Revision D.01*; 2009.
38. Berendsen, H. J. C.; Postma, J. P. M.; van Gunsteren, W. F.; DiNola, A.; Haak, J. R. Molecular Dynamics with Coupling to an External Bath. *J. Chem. Phys.* **1984**, *81*, 3684. https://doi.org/10.1063/1.448118.
39. Bussi, G.; Donadio, D.; Parrinello, M. Canonical Sampling through Velocity Rescaling. *J. Chem. Phys.* **2007**, *126*, 014101. https://doi.org/10.1063/1.2408420.
40. Hess, B.; Bekker, H.; Berendsen, H. J. C.; Fraaije, J. G. E. M. LINCS: A Linear Constraint Solver for Molecular Simulations. *J. Comput. Chem.* **1997**, *18*, 1463. https://doi.org/10.1002/(SICI)1096-987X(199709)18:12<1463::AID-JCC4>3.0.CO;2-H.
41. Darden, T.; York, D.; Pedersen, L. Particle Mesh Ewald: An Nlog(N) Method for Ewald Sums in Large Systems. *J. Chem. Phys.* **1993**, *98*, 10089. https://doi.org/10.1063/1.464397.
42. Essmann, U.; Perera, L.; Berkowitz, M. L.; Darden, T.; Lee, H.; et al. A Smooth Particle Mesh Ewald Method. *J Chem Phys* **1995**, *103*, 8577. https://doi.org/10.1063/1.470117.
43. Hess, B.; Uppsala, S.; Lindahl, E. GROMACS 4 : Algorithms for Highly Efficient, Load-Balanced, and Scalable Molecular Simulation. *J. Chem. Theory Comput.* **2008**, *vol. 4*, 435. https://doi.org/10.1021/ct700301q.
44. Van Der Spoel, D.; Lindahl, E.; Hess, B.; Groenhof, G.; Mark, A. E.; et al. GROMACS: Fast, Flexible, and Free. *J. Comput. Chem.* **2005**, *26*, 1701. https://doi.org/10.1002/jcc.20291.
45. Georg, H. C.; Coutinho, K.; Canuto, S. Solvent Effects on the UV-Visible Absorption Spectrum of Benzophenone in Water: A Combined Monte Carlo Quantum Mechanics Study Including Solute Polarization. *J Chem Phys* **2007**, *126*, 034507. https://doi.org/10.1063/1.2426346.

46. Rocha, W. R.; Martins, V. M.; Coutinho, K.; Canuto, S. Solvent Effects on the Electronic Absorption Spectrum of Formamide Studied by a Sequential Monte Carlo/Quantum Mechanical Approach. *Theor. Chim. Acta* **2002**, *108*, 31. https://doi.org/10.1007/s00214-002-0353-y.
47. Becke, A. D. A New Mixing of Hartree-Fock and Local Density-Functional Theories. *J. Chem. Phys.* **1993**, *98*, 1372. https://doi.org/10.1063/1.464304.
48. Petersson, G. A.; Al-Laham, M. A. A Complete Basis Set Model Chemistry. II. Open-Shell Systems and the Total Energies of the First-Row Atoms. *J. Chem. Phys.* **1991**, *94*, 6081. https://doi.org/10.1063/1.460447.
49. Tomasi, J.; Mennucci, B.; Cancès, E. The IEF Version of the PCM Solvation Method: An Overview of a New Method Addressed to Study Molecular Solutes at the QM Ab Initio Level. *J. Mol. Structure (Theochem)* **1999**, *464*, 211. https://doi.org/10.1016/S0166-1280(98)00553-3.
50. Smith, R., Jr.; Lee, S.; Komori, H.; Arai, K. Relative Permittivity and Dielectric Relaxation in Aqueous Alcohol Solutions. *Fluid Phase Equilib.* **1998**, *144*, 315. https://doi.org/10.1016/S0378-3812(97)00275-6.
51. Kumbharkhane, A. C.; Helambe, S. N.; Lokhande, M. P.; Doraiswamy, S.; Mehrotra, S. C. Structural Study of Aqueous Solutions of Tetrahydrofuran and Acetone Mixtures Using Dielectric Relaxation Technique. *Pramana* **1996**, *46*, 91. https://doi.org/10.1007/BF02848226.
52. Borodin, O. Polarizable Force Field Development and Molecular Dynamics Simulations of Ionic Liquids. *J. Phys. Chem. B* **2009**, *113* (33), 11463. https://doi.org/10.1021/jp905220k.

CHAPTER SIX

Mean total and orbital excitation energies of atomic ions in two approaches of the Thomas–Fermi theory

Remigio Cabrera-Trujillo[a,b] and Salvador A. Cruz[c,*]
[a]Instituto de Ciencias Físicas, Universidad Nacional Autónoma de México, Cuernavaca, Morelos, Mexico
[b]Department of Physics, University of Gothenburg, Gothenburg, Sweden
[c]Departamento de Física, Universidad Autónoma Metropolitana-Iztapalapa, Mexico City, Distrito Federal, Mexico
*Corresponding author: e-mail address: cruz@xanum.uam.mx

Contents

Abstract

The mean excitation energy is a nontrivial material property that depends on the excitation spectra of atoms and molecules. It is a crucial input for the evaluation of the energy deposition of swift charged particles. In Bethe's theory of the stopping cross section, it is determined from the dipole oscillator strength, while in the Lindhard and Scharff theory, it is given in terms of the material response through the Local Plasma Approximation. This work is dedicated to Jack Sabin's motto: a simple model is worth exploring before a sophisticated theory. Following this idea, here we regain old theories such as the Thomas–Fermi model and the original Local Plasma Approximation to study the mean excitation energy of neutral atoms and their cations. We implement two approaches, the Thomas–Fermi theory with Amaldi corrections for ionic systems and complement our work with a more sophisticated DFT-like Thomas–Fermi–Dirac–

Advances in Quantum Chemistry, Volume 85
ISSN 0065-3276
https://doi.org/10.1016/bs.aiq.2022.03.003
Copyright © 2022 Elsevier Inc.
All rights reserved.

Weizsäcker approximation. Using the obtained total mean excitation energies for C, F, Ne, Si, Cl, Sc, Mn, and Zn and their cations, the single-electron mean excitation energy is determined allowing for a reasonable description of orbital and shell contributions as compared with more sophisticated theories.

1. Introduction

The concept of mean excitation energy as the ability of a medium to absorb energy from any kind of incident radiation (either photons or charged particles) found its origins in the pioneering work by Niels Bohr in his classical theory of penetration of swift charged particles in matter[1–3] whereby the absorbed energy was envisaged as a sum of contributions from harmonically bound electrons forming the target structure. Shortly after the advent of modern quantum mechanics, Hans Bethe developed the first formal quantum treatment for the inelastic energy loss process between a bare charged projectile and a target atom.[4] In this theory, the mechanisms of energy absorption by the target system depend on the allowed electronic transitions, consistent with the quantum-mechanical description of its electronic structure. According to Bethe's original work, the rate of energy-loss per scatterer (stopping cross section, S_e) for a swift bare ion of charge Z_1 and velocity v is given as[4]

$$S_e(v) = \frac{4\pi e^4}{mv^2}\, Z_1^2 Z_2 \,\ln\left(\frac{2mv^2}{I_0}\right), \tag{1}$$

where Z_2 is the number of electrons composing an atomic target and m and e the electron mass and charge, respectively.

The energy transfer to the internal degrees of freedom of the target system is accounted for by the excitation function I_0 in Eq. (1), which is defined as the mean excitation energy and expressed as[5]

$$Z_2 \ln I_0 = \sum_n f_{n0} \ln\left(E_n - E_0\right). \tag{2}$$

The sum involves all discrete and continuum states and f_{n0} is the dipole oscillator strength (DOS) for the transition from the ground state energy E_0 to a state of energy E_n:

$$f_{n0} = \frac{2}{3}(E_n - E_0)\left|\left\langle n\left|\sum_{i=1}^{Z_2} r_i\right|0\right\rangle\right|^2. \tag{3}$$

We note at this stage that the structure of Eqs. (2) and (3) indicates a clear dependence on the internal states of the perturbed system and hence the proper assignment of I_0 as the mean excitation energy. It is worth mentioning here that this definition is not only related with ion-atom collisions but also to processes involving induced dipole optical transitions from different sources, such as photon or electron beams[5] or atoms under extreme conditions (Ref. 6 and references there in).

The formal evaluation of I_0 from Eqs. (2) and (3) requires of proper knowledge of the energy spectrum and eigenfunctions corresponding to different electronic states in a many-electron atom. In his pioneering work, Bethe used hydrogenic wave functions, which was the best available approach at the time. Since then, several formal treatments to evaluate I_0 in terms of the DOS have been developed (see, e.g., Refs. 5,7–10 for an account) with successful agreement with experiment. These calculations require of well-defined ab initio methods to construct accurate atomic (molecular) wave functions, thus making the DOS computation more complicated but rendering accurate reference values for other approximate treatments.

Looking at I_0 as a measure of the response of the medium to an external perturbing agent, an useful and physically appealing alternative model to account for I_0, known as the Local Plasma Approximation (LPA), was developed by Lindhard and Scharff in the celebrated electron gas model.[11,12] In this model, considering the atom as an inhomogeneous electron gas with local electron density $\rho(\mathbf{r})$, its dynamical response is characterized by local plasma oscillations of frequency

$$\omega_p(\mathbf{r}) = \sqrt{4\pi e^2 \rho(\mathbf{r})/m} \tag{4}$$

so that absorption spectra result from independent absorption frequencies from each atomic volume element. Using dielectric response theory, according to Lindhard, the mean excitation energy, I_{N_e}, for an atom with N_e electrons and nuclear charge Z is given by

$$N_e \ln I_{N_e} = \int \rho(\mathbf{r}) \ln\left(\gamma \omega_p(\mathbf{r})\right) d^3\mathbf{r} \tag{5}$$

with $\omega_p(\mathbf{r})$ is given by Eq. (4) and γ is an empirical parameter introduced to account for differences in single-particle excitations between light atoms ($\gamma = 1$), where polarization effects are not so important, and heavier atoms ($\gamma = \sqrt{2}$).[13]

Since its development, the LPA was proved to have reasonable success in estimates of the mean excitation energy of atoms and molecules.[14] A major drawback of the LPA seems to be its failure to describe the shell structure of the oscillator strength spectra and the apparent inability to derive it from the spectral distribution of oscillator strengths.[15] An important step forward in this direction was set by Sabin and Oddershede[7,16] by proposing a decomposition of the total mean excitation energy in terms of orbital mean excitation energies, I_k, such that

$$\ln I_{N_e} = \frac{1}{N_e} \sum_k w_k \ln I_k \tag{6}$$

with w_k the orbital weights $w_k = (n_k + n_k f_k)/2$ given in terms of the orbital occupation numbers n_k and total oscillator strengths $n_k f_k$ of all optical transitions from level k into the unoccupied states (both discrete and in the continuum).[16] The orbital weight factors were derived from corresponding Hartree–Fock orbital oscillator strengths, which require of a proper ab initio treatment. These ideas were further exploited by Meltzer, Sabin, and Trickey[17–19] to develop the Orbital Local Plasma Approximation (OLPA), which consists in adapting the original LPA (Eq. 5) to an orbital scheme by defining the orbital mean excitation energy as

$$\ln I_k = \frac{1}{n_k} \int \rho_k(\mathbf{r}) \ln \left[\gamma_k \hbar \sqrt{\frac{4\pi e^2 \rho_k(\mathbf{r})}{m}} \right] d^3r, \tag{7}$$

with $\rho_k(\mathbf{r})$ the local electronic density of orbital k and γ_k the associated scaling parameter as proposed in the LPA theory. By using Eq. (6) with the orbital weight factors w_k substituted by the orbital occupation numbers n_k, the total mean excitation energy could be partitioned into the orbital contributions, I_k, well-based on the LPA theory. This scheme proved to be in general good agreement with the DOS treatment by Sabin and Oddershede[7,16] and constitutes a useful means to incorporate shell-structure effects within the LPA theory.

Note that the DOS sum in Eq. (6) as well as its OLPA implementation with I_k given by Eq. (7) assign the same orbital mean excitation energy to all electrons occupying a given k-orbital. However, recent ab initio calculations by Sauer et al.[9,20–22] based on the Random Phase Approximation (RPA) to study the total mean excitation energy I_{0,N_e} of atomic ions with fixed nuclear charge Z and electron populations varying from $N_e = 1$ to Z indicate the possibility to estimate I_{0,N_e} from knowledge of I_{0,N_e-1} and I_{0,N_e+1}.[22]

This simple rule points to a relation of total mean excitation energies among atoms with different electronic populations but the same nuclear charge Z. Moreover, since neighboring cations differ by only one electron, a further refinement of the theory was set forward by Cabrera-Trujillo et al.[8] who proposed a way to calculate the single-electron orbital mean excitation energy $I_{0,1}$ as deduced from the total mean excitation energy I_{0,N_e} for an ionic system with N_e electrons and that from the system with $(N_e - 1)$ electrons as

$$N_e \ln I_{0,N_e} = (N_e - 1) \ln I_{0,N_e-1} + \ln I_{0,1}, \tag{8}$$

This expression has the advantage in assigning each electron its own mean excitation energy as it is added to the system, in contrast with previous work based on the DOS sum in Eq. (6) as well as the OLPA which, as mentioned before, assigns the same orbital mean excitation energy to all electrons in that orbital.

Eq. (8) is remarkably simple and amenable for the exploration of the use of approximate approaches designed to evaluate the total mean excitation energy of an atomic/ionic system of nuclear charge Z composed by N_e and $(N_e - 1)$ electrons and estimate the one-electron mean excitation energy. This procedure would shed light on the ability of those approximate treatments to account for shell contributions to the mean excitation energy, even if no shell structure is incorporated in their assumptions. This is a challenging idea which, we believe Jack Sabin would like to explore according to his inquiring motto.

The aim of this work is to draw a simple theoretical treatment for a system with N_e electrons using the original LPA as given by Eqs. (4) and (5) whereby the electron density is constructed through the Thomas–Fermi theory with Amaldi (TFA) corrections[23] for a ionic system and using its Tietz[24] approximate analytical solution. Accordingly, a closed expression is obtained for the total mean excitation energy for a ionic system with N_e electrons and nuclear charge Z. Using this scheme, the one-electron mean excitation energies are obtained through Eq. (8) and total and shell mean excitation energies are estimated and compared with the corresponding accurate RPA results for the selected set of atomic systems: C, F, Ne, Si, Cl. Sc, Mn, and Zn as well as all their cations. For completeness and for the sake of comparison, the orbital mean excitation energies for neutral atoms obtained through the HF-DOS[7] sums according to Eq. (6) are also considered. On the other hand, an alternative subsidiary calculation using the Thomas–Fermi–Dirac–Weizsäcker (TFDW) density functional formalism[25,26] together with the original LPA theory (Eqs. 4, 5, and 8) is performed for the same set of systems and compared with the TFA, RPA, and HF-DOS results. Interestingly, a general overall qualitative and quantitative agreement is observed among all calculations, considering the crudeness of the approximate treatments as compared with the accurate ones. The results of this work point to the adequacy of Eq. (8) for the use of reasonably physically based approximate models to estimate orbital and total mean excitation energies.

The layout of this work is the following. In Section 2 the main arguments leading to the evaluation of the total mean excitation energy of an atomic system with nuclear charge Z and N_e electrons are presented in the frame of the Thomas–Fermi theory with Amaldi corrections and the LPA. Also, on the same footing, a subsidiary calculation based on the Thomas–Fermi–Dirac–Weizsäcker density functional theory is included in this section for comparison purposes. In Section 3, the results of this study for the total, one-electron, as well as shell-wise mean excitation energies are presented and discussed when compared with available corresponding accurate calculations. Finally, the conclusions of this work are given in Section 4.

2. Theory

2.1 Thomas–Fermi theory with Amaldi corrections

Let us consider an ionic atom with nuclear charge Z and $N_e \leq Z$ electrons. According to Fermi and Amaldi,[23] the screening function $\varphi(x)$ for such a system must follow the celebrated Thomas–Fermi equation

$$\frac{d^2\varphi(x)}{dx^2} = \frac{\varphi^{3/2}(x)}{x^{1/2}}, \tag{9}$$

where $\varphi(x)$ is related to the effective Coulomb potential $V(r)$ due to the nucleus and electronic cloud and is given as

$$V(r) = \frac{Ze}{r}\varphi(x) = \frac{Ze}{\mu^* x}\varphi(x), \tag{10}$$

where, according to Fermi and Amaldi, the following convenient definitions hold for an atom with nuclear charge Z and N_e active electrons:

$$\begin{aligned}\mu^* &= \frac{1}{4}\left(\frac{9\pi^2}{2Z}\right)^{1/3}\left(\frac{N_e}{N_e-1}\right)^{2/3} a_0,\\ x &= r/\mu^*,\\ a_0 &= \hbar^2/me^2\end{aligned} \tag{11}$$

being the Bohr radius. Furthermore, in consistency with the original Thomas–Fermi theory, the screening function $\varphi(x)$ in Eq. (9) for positive ions must satisfy the following boundary conditions:

$$\varphi(0) = 1; \quad \varphi(x_0) = 0; \quad -x_0 \frac{d\varphi(x)}{dx}\bigg|_{x_0} = q + \frac{1}{Z}, \tag{12}$$

where $q = (Z - N_e)/Z$ and $x_0 = r_0/\mu^*$ is the ionic atom radius. Within this scheme, the electron density is cast in terms of $\varphi(x)$ and x as

$$\rho(x) = \frac{Z}{4\pi\mu^*3}\left(\frac{N_e}{N_e-1}\right)\left(\frac{\varphi(x)}{x}\right)^{3/2}. \tag{13}$$

This expression, together with the LPA definition given by Eqs. (4) and (5) is now used to calculate the total mean excitation energy of an ionic atom with nuclear charge Z and N_e electrons, which is given by

$$\ln I_{N_e} = \frac{1}{2}\ln\left(\frac{ZN_e}{\mu^*3(N_e-1)}\right) + \frac{Z}{2(N_e-1)}\int_0^{x_0} x\frac{d^2\varphi}{dx^2}\ln\left(\frac{1}{x}\frac{d^2\varphi}{dx^2}\right)dx. \tag{14}$$

An approximate analytical solution $\varphi(x)$ for the Thomas–Fermi equation (9) satisfying the boundary conditions (12) has been given by Tietz[24] as

$$\varphi(x) = \frac{1}{(1+ax)^2}\left[1 - \left(\frac{1+ax}{1+ax_0}\right)^5\right] \tag{15}$$

with $a = 37/65$. With this, the mean excitation energy as indicated by Eq. (14) becomes

$$I_{N_e} = \frac{8\sqrt{2}}{3\pi} Z\sqrt{\frac{(N_e-1)}{N_e}} f(x_0)^{Z/2(N_e-1)} \tag{16}$$

where $f(x) = \exp(g(x))$ and

$$g(y_0) = \int_0^{y_0} \frac{6y}{(1+y)^4}\left[1 - \left(\frac{1+y}{1+y_0}\right)^5\right] \ln\left[\frac{6a^3}{y(1+y)^4}\left[1 - \left(\frac{1+y}{1+y_0}\right)^5\right]\right] dy \tag{17}$$

with $y_0 = ax_0$. This expression can be evaluated analytically, but it results in a long polynomial expression and hyper-confluent functions. We carry out numerically the integration and fit it to a more simple analytical expression

$$f(y) = \alpha + \frac{1}{\beta + \gamma y + \delta y^2} \tag{18}$$

with $\alpha = 0.04211$, $\beta = 1.06481$, $\gamma = -0.42977$, and $\delta = 0.58515$.

The atomic radius is determined from Tietz analytical expression by taking into account that for $q < 0.0325$, Tietz solution gives a good description of the TF ionic atom while for $q > 0.0325$ we use a fitted expression to the Lee and Wu[27] exact numerical solution, resulting in

$$y_0 = \begin{cases} -\frac{3}{2}\left(1 - \sqrt{1 + \frac{20}{9(q + 1/Z)}}\right), & q < 0.0325 \\ a(A - \sqrt{A^2 + B\ln(q + 1/Z)}), & q > 0.0325 \end{cases} \tag{19}$$

with $A = 29.7123$ and $B = 226.2770$.

The above described strategy to construct an accurate analytical representation of both $f(y_0)$ and y_0 as given by Eqs. (18) and (19), respectively, proofs to be quite satisfactory as shown next. Fig. 1 displays a comparison between the exact numerical solution by Lee and Wu[27] (open circles) and our analytical expression (Eq. 19) (solid line) for the radius of an ion with charge fraction $q = (Z - N_e)/Z$. An excellent agreement is observed between both calculations. In Fig. 2, the function $f(y_0)$ corresponding to the exponentiation of the numerical integration indicated by Eq. (17) (solid line) is compared with the fitted analytical expression (Eq. 18) (dashed line). An excellent agreement is also obtained. These results provide a reliable representation of the basic functions needed to evaluate the mean excitation energy of an atomic ion of nuclear charge Z and N_e electrons as prescribed by Eq. (16) within the TFA approach.

As mentioned in Section 1, we may now apply the TFA approach for the calculation of I_{0,N_e} and I_{0,N_e-1} in Eq. (8) and evaluate the single-electron orbital mean excitation energy $I_{0,\,1}$ for different atomic systems and their cations and compare its predictions with the accurate RPA calculations.

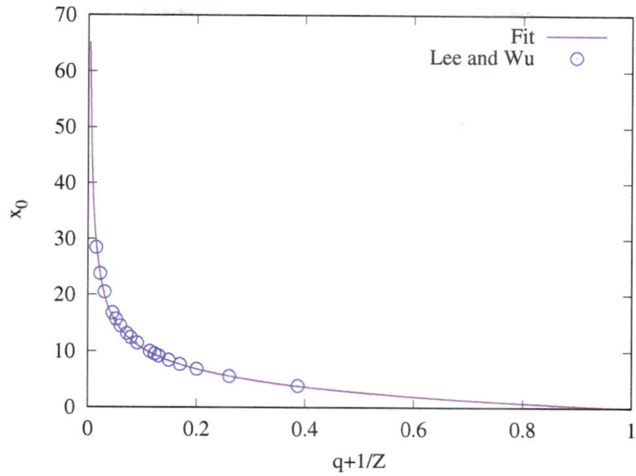

Fig. 1 Ion atomic radii, x_0, as a function of the ion charge fraction, $q = (Z - N_e)/Z$ as obtained by Eq. (19) within the Thomas–Fermi model of an ion with Amaldi corrections and using Tietz approximated solution. The *open circle symbols* are the exact numerical results by Lee and Wu[27] and the *solid line* is Eq. (19). Note the excellent agreement to our fitted simple analytical expression. See text for discussion.

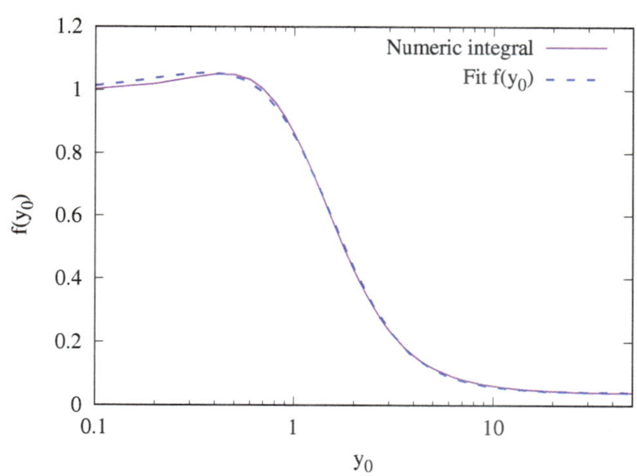

Fig. 2 Comparison of the numeric and fitted solutions to the exponentiation of Eq. (17). The *dashed line* is our analytical approximation (Eq. 18) while the *solid line* is the numerical integration of Eq. (17). See text for discussion.

2.2 Thomas–Fermi–Dirac–Weizsäcker density functional approach

Before proceeding any further, we find it appropriate to briefly describe a complementary, subsidiary calculation based on the orbital implementation of the Thomas–Fermi–Dirac–Weizsäcker (TFDW) density functional. The details of this formulation are given elsewhere.[25,26,28] Here, we highlight the relevant ideas concerning the main ingredients necessary to calculate I_{0,N_e} and I_{0,N_e-1} according to the original LPA theory (Eqs. 4 and 5) and obtain the single-electron orbital mean excitation energy through Eq. (8), as done for the TFA treatment.

The TFDW energy-density functional is based on the original Thomas–Fermi electron gas theory, including the Weizsäcker inhomogeneity correction $T_W(\rho)$ in the kinetic energy and the Dirac exchange term $X_{ex}(\rho)$ in the Coulomb electron–electron interaction.[26] Accordingly, the TFDW density functional for an atom with nuclear charge Z and N_e electrons may be cast in general as

$$E_{TFDW}(\rho,\lambda) = T_0(\rho) + \lambda T_W(\rho) + V_{en}(\rho) + V_{ee}(\rho) + X_{ex}(\rho) \tag{20}$$

with the different terms defined as

$$\begin{aligned} T_0(\rho) &= \frac{3}{10}\left(3\pi^2\right)^{2/3} \int \rho^{5/3} d\tau \\ T_W(\rho) &= \frac{1}{8}\int \frac{\nabla\rho \cdot \nabla\rho}{\rho} d\tau \\ V_{en}(\rho) &= -Z\int \frac{\rho}{r} d\tau \\ V_{ee}(\rho) &= \frac{1}{2}\int \rho\varphi d\tau \\ X_{ex}(\rho) &= -\frac{3}{4}(3/\pi)^{1/3} \int \rho^{4/3} d\tau, \end{aligned} \tag{21}$$

where ρ is the total electron density, φ the electrostatic potential due to charge density ρ and λ a scaling parameter for the Weizsäcker term (see further below).

In the orbital implementation of the TFDW scheme[25,28] the total electronic density is given as the sum of orbital densities:

$$\rho(r) = \sum_k^{n_{occ}} \rho_k(r,\xi_k) \tag{22}$$

with ρ_k given in terms of Slater-type ansatz orbital radial densities defined as

$$\rho_k(r,\xi_k) = N_k r^{2n_k-2} e^{-2\xi_k r}. \tag{23}$$

Here n_k is the principal quantum number associated to orbital k, ξ_k a variational parameter, and N_k a normalization factor so that

$$\int \rho_k(r,\xi_k) d\tau = \omega_k \tag{24}$$

gives the electron population ω_k of orbital k. Consequently

$$\int \rho d\tau = N_e \tag{25}$$

provides the total number of electrons N_e.

Once the energy functional, given by Eq. (20), is constructed, its variational optimization in terms of the orbital parameters ξ_k is carried out such that

$$\frac{\partial E}{\partial \xi_k} = 0, \qquad k = 1,\ldots,n_{occ}. \tag{26}$$

The factor λ before the Weizsäcker term in Eq. (20) has been adjusted in an *ad hoc* manner so that the optimized energy corresponds to three decimal digits with corresponding Hartree–Fock calculations[29] for free atoms and ions. In this manner, the orbital and total electron densities are obtained showing a good correspondence with HF densities with the shell features moderately reproduced.[28]

This methodology has been previously applied to the calculation of mean orbital excitation energies within the OLPA approach according to Eqs. (6) and (7).[30] Here we shall use the total electron density given by Eqs. (22) and (23) to obtain the LPA total mean excitation energy with no resource to Eq. (6). This will bring into the same footing—as the TFA—for the evaluation of the single-electron orbital mean excitation energy for proper comparison.

3. Results

3.1 Total mean excitation energies

Table 1 displays the total mean excitation energy I_{0,N_e} for a system with N_e electrons as obtained through the TFA approach (Eq. 16). Also shown in this

Table 1 Total mean excitation energy (I_{0,N_e}) in the LPA approximation as a function of the number of electrons (N_e) for C, F, Ne, Cl, Si, Sc, Mn, and Zn and their cations obtained through the TFA and TFDW approaches.

	C		F		Ne		Si		Cl		Sc		Mn		Zn	
N_e	$I_{0,N}^{TFA}$	$I_{0,N}^{TFDW}$	$I_{0,N}^{TFA}$	$I_{0,N}^{TFDW}$	$I_{0,N}^{TFA}$	$I_{0,N}^{TFDW}$	$I_{0,N}^{TFA}$	$I_{0,N}^{TFDW}$	$I_{0,N}^{TFA}$	$I_{0,N}^{TFDW}$	$I_{0,N}^{TFA}$	$I_{0,N}^{TFDW}$	$I_{0,N}^{TFA}$	$I_{0,N}^{TFDW}$	$I_{0,N}^{TFA}$	$I_{0,N}^{TFDW}$
1		182.95		333.26		389.67		642.7		858.34		1176.49		1526.42		2004.51
2	212.02	235.74	335.42	443.07	374.53	521.29	522.42	873.63	625.18	1174.98	752.30	1620.53	868.88	2111.43	1000.87	2783.04
3	180.23	138.68	337.81	274.25	389.19	325.62	587.16	558.02	728.63	757.56	909.32	1053.09	1081.92	1379.37	1287.29	1826.40
4	129.17	104.17	286.74	215.74	342.65	258.21	564.84	450.94	726.27	616.88	934.17	863.28	1134.65	1135.64	1376.10	1509.25
5	92.88	80.91	228.25	182.20	281.95	221.98	510.45	392.06	683.19	540.81	908.32	762.16	1126.72	1007.28	1391.06	1343.96
6	69.68	57.30	179.82	152.70	226.93	188.94	445.41	352.92	621.67	491.08	857.02	697.16	1087.82	925.84	1368.56	1240.38
7	–	–	144.21	125.46	183.84	158.54	381.71	309.99	553.84	441.12	792.55	638.30	1031.17	857.83	1323.90	1159.86
8	–	–	118.25	104.94	151.71	134.86	325.55	273.99	487.46	396.39	722.67	580.59	964.49	786.72	1265.30	1070.59
9	–	–	97.64	88.43	127.27	116.01	278.89	245.61	427.03	360.58	652.83	534.40	893.15	729.21	1198.06	999.15
10	–	–	–	–	106.89	100.39	241.15	222.04	374.52	330.90	586.70	496.28	821.14	682.23	1126.05	940.59
11	–	–	–	–	–	–	210.69	187.68	330.12	290.74	526.43	446.77	751.40	621.97	1052.31	865.26
12	–	–	–	–	–	–	185.61	160.18	293.01	256.98	472.95	403.69	685.91	568.63	979.22	798.15
13	–	–	–	–	–	–	164.02	137.39	261.93	228.37	426.30	366.60	625.84	522.41	908.59	739.48
14	–	–	–	–	–	–	143.57	117.10	235.58	204.14	385.96	334.89	571.68	482.66	841.72	688.89
15	–	–	–	–	–	–	–	–	212.62	183.23	351.13	307.55	523.41	448.36	779.41	645.13
16	–	–	–	–	–	–	–	–	191.72	161.51	320.93	283.55	480.67	418.29	722.09	607.02

17	–	–	–	–	–	–	–	–	170.73	141.72	294.46	259.97	442.94	391.69	669.84	573.33
18	–	–	–	–	–	–	–	–	–	–	270.86	237.20	409.59	365.13	622.52	543.22
19	–	–	–	–	–	–	–	–	–	–	249.24	209.80	379.96	336.53	579.83	511.78
20	–	–	–	–	–	–	–	–	–	–	228.53	184.28	353.41	308.33	541.38	480.92
21	–	–	–	–	–	–	–	–	–	–	206.55	161.56	329.33	281.58	506.71	447.23
22	–	–	–	–	–	–	–	–	–	–	–	–	307.10	256.94	475.36	417.04
23	–	–	–	–	–	–	–	–	–	–	–	–	286.03	234.54	446.88	389.28
24	–	–	–	–	–	–	–	–	–	–	–	–	265.12	213.82	420.83	363.07
25	–	–	–	–	–	–	–	–	–	–	–	–	241.96	191.53	396.78	339.64
26	–	–	–	–	–	–	–	–	–	–	–	–	–	–	374.28	318.14
27	–	–	–	–	–	–	–	–	–	–	–	–	–	–	352.87	298.31
28	–	–	–	–	–	–	–	–	–	–	–	–	–	–	331.95	279.99
29	–	–	–	–	–	–	–	–	–	–	–	–	–	–	310.53	258.60
30	–	–	–	–	–	–	–	–	–	–	–	–	–	–	285.67	237.71

All values given in eV.

table are the corresponding TFDW values obtained within the LPA theory (Eqs. 4 and 5) using the total electronic densities given by Eqs. (22) and (23) with orbital parameters defined after optimization of the TFDW energy functional for each system as indicated by Eq. (26). In all cases, the optimal Weizsäcker λ-factors were set so that the total energy coincides to three decimal digits with corresponding Roothaan–Hartree–Fock (RHF) calculations by Clementi and Roetti.[29] The values of these parameters are given in Table 2 for the neutral atoms and their cations treated here.

As may be gathered from the results shown in Table 1, an overall reasonable quantitative agreement between the TFA and TFDW calculations is observed pointing to the adequacy of the TFA analytical expression (Eq. 16) to describe the total mean excitation energy for atoms with $N_e \geq 2$. Note that the TFA approach may not be applied for $N_e = 1$ in view of the limitation imposed by Eqs. (11) and (13). However, we do not expect the Thomas–Fermi model to work well for systems with very few electrons. This is also evident in the TFDW results for $N_e = 1$ where a sudden drop in the mean excitation energy is observed. Fig. 3 shows graphically the TFA (dashed lines) and TFDW (solid lines) numerical results to gain a better idea of their relative qualitative and quantitative behavior. As may be gathered from this figure, the TFA calculations remain systematically above the TFDW ones following a similar qualitative trend for $N_e \geq Z/2$. For $N_e \leq Z/5$ (highly ionized system) a turnover of the TFA values is observed in contrast with the TFDW case. A possible reason for this behavior is the fact that the original Thomas–Fermi theory does not include the Weiszäcker inhomogeneity correction to the kinetic energy functional. Note that the TFDW approach also shows a sudden drop for $N_e = 1$ and, as indicated in Table 2, this case requires of a much higher λ value for the Weizsäcker term in the energy evaluation. On the other hand, the TFDW curves show in general a non monotonic behavior in terms of N_e with some moderate kinks at $N_e = 10$, 18, which reflect the closed shell configuration of Ne and Ar. This effect may be due to the inherent construction of the total electronic density in terms of different representations of the orbital densities in this method.

Having discussed the comparison between the predicted TFA and TFDW total mean excitation energies for the atomic systems considered and their cations, we now compare the TFA results with the accurate ab initio RPA results.[8,9,16,20,21] Fig. 4 shows the RPA curves (solid lines) where a clear shell structure is observed. The TFA curves (dashed lines) lie systematically below the corresponding RPA curves but show a uniform tendency to the RPA values as the number of electrons increases with a

Table 2 Optimal Weizsäcker λ-factors for the neutral atoms and their cations treated in this work.

N_e	C	F	Ne	Si	Cl	Sc	Mn	Zn
1	0.388750	0.399700	0.401880	0.407540	0.410050	0.412270	0.413780	0.415105
2	0.066095	0.071693	0.072778	0.075520	0.076710	0.077754	0.078457	0.079070
3	0.114835	0.133324	0.137010	0.146472	0.150633	0.154320	0.156819	0.159005
4	0.128807	0.157880	0.163830	0.179259	0.186123	0.192230	0.196393	0.200036
5	0.129856	0.164925	0.172110	0.190720	0.198940	0.206272	0.211255	0.215615
6	0.124790	0.158215	0.165660	0.185445	0.194270	0.202100	0.207407	0.212038
7	–	0.151380	0.153340	0.172767	0.178750	0.186012	0.191000	0.195390
8	–	0.139120	0.143400	0.155400	0.161035	0.166177	0.169728	0.172867
9	–	0.131220	0.133090	0.138260	0.140740	0.143035	0.144645	0.146080
10	–	–	0.123310	0.120733	0.119457	0.118345	0.117610	0.116993
11	–	–	–	0.123685	0.124789	0.126287	0.127562	0.128832
12	–	–	–	0.124790	0.127589	0.131189	0.134180	0.137123
13	–	–	–	0.125725	0.130019	0.135335	0.139700	0.143982
14	–	–	–	0.125450	0.130785	0.137414	0.142901	0.148313
15	–	–	–	–	0.130257	0.137803	0.144160	0.150490
16	–	–	–	–	0.129495	0.137479	0.144413	0.151391

Continued

Table 2 Optimal Weizsäcker λ-factors for the neutral atoms and their cations treated in this work.—cont'd

N_e	C	F	Ne	Si	Cl	Sc	Mn	Zn
17	–	–	–	–	0.128310	0.136045	0.143278	0.150702
18	–	–	–	–	–	0.133962	0.141025	0.148650
19	–	–	–	–	–	0.134680	0.141406	0.148747
20	–	–	–	–	–	0.135255	0.141423	0.148160
21	–	–	–	–	–	0.135347	0.141375	0.147339
22	–	–	–	–	–	–	0.141370	0.146336
23	–	–	–	–	–	–	0.141360	0.145212
24	–	–	–	–	–	–	0.141530	0.144545
25	–	–	–	–	–	–	0.141485	0.143710
26	–	–	–	–	–	–	–	0.143230
27	–	–	–	–	–	–	–	0.142678
28	–	–	–	–	–	–	–	0.142103
29	–	–	–	–	–	–	–	0.142000
30	–	–	–	–	–	–	–	0.141706

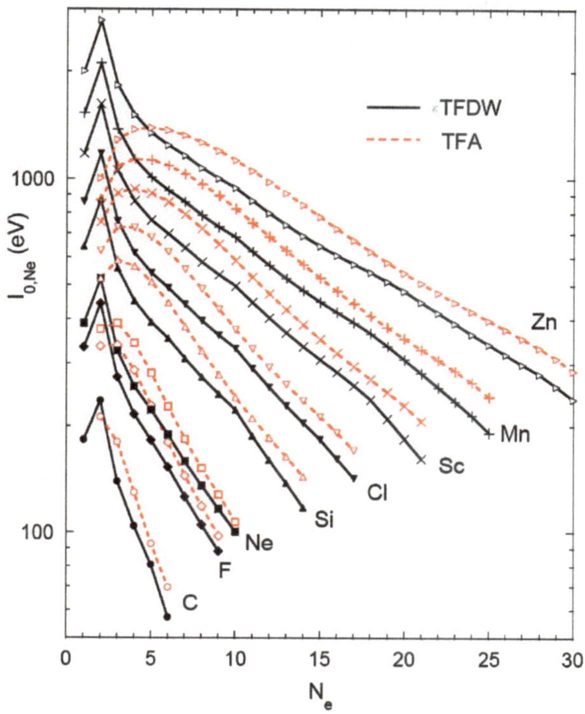

Fig. 3 Total mean excitation energy, (I_{0,N_e}) in the LPA approximation as a function of the number of electrons (N_e) for C, F, Ne, Cl, Si, Sc, Mn, and Zn and their cations. *Dashed lines with symbols*: values obtained within the TFA model through Eq. (16) and the electronic density given by Eq. (13). *Solid lines with symbols*: results from the TFDW treatment with the optimized electronic density given by Eqs. (22) and (23).

surprisingly good correspondence for the neutral systems. In fact, the TFA values show a better quantitative correspondence with the RPA ones than the TFDW values, as may be gathered from Fig. 3.

3.2 One-electron mean excitation energies

We now proceed to evaluate the one-electron mean excitation energies $I_{0,i}$ by using Eq. (8) and the results given in Table 1 for the TFA and TFDW total mean excitation energies I_{0,N_e}, I_{0,N_e-1} of the different atoms and their cations. Since we may assign a mean excitation energy to each electron as it is added to the system, according to the ground state orbital electronic configuration of each atomic system and following the *aufbau* principle, the mean excitation energy of electron "*i*" occupying a subshell (*nl*) is now denoted by $I_{nl,i}$. Accordingly, in Table 3 we show the one-electron mean excitation

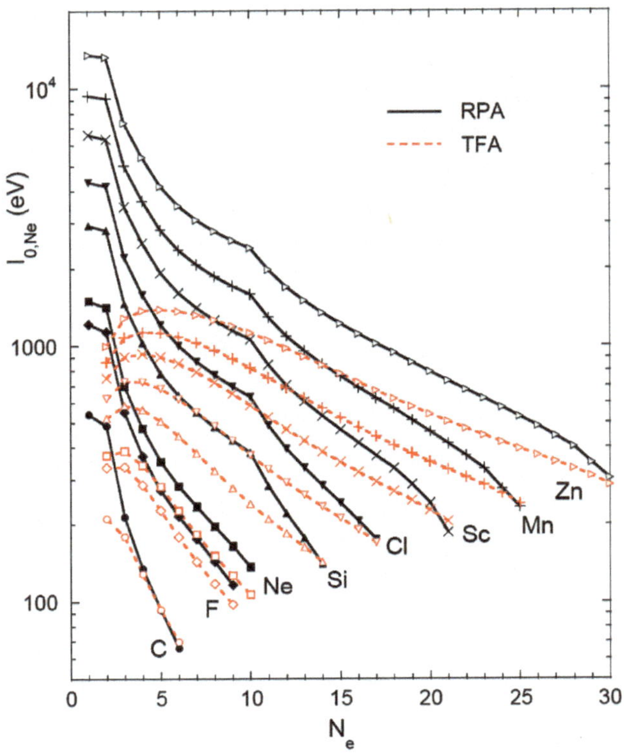

Fig. 4 Comparison of the TFA results for the total mean excitation energy (*dashed lines with symbols*) shown in Fig. 3 and corresponding accurate RPA calculations (*solid lines with symbols*) by Sauer et al.[9,20] See text.

energies calculated within the TFA and TFDW schemes where the contribution from electrons in different subshells may be appreciated. At this stage, we note that accurate RPA calculations of $I_{nl,i}$ following Eq. (8) have been recently reported for the same systems,[8,9] thus providing an important reference for proper comparison. Fig. 5 shows the RPA (solid lines) and TFA (dashed lines) behavior of $I_{nl,i}$ as each electron is added to the system. A clear shell effect is observed for the RPA values, as is also indicated in Refs. 8 and 9. Interestingly, the one-electron TFA values constitute a good average behavior of the RPA results both qualitatively and quantitatively. This is in spite of the quantitative differences observed among the RPA and TFA total mean excitation energies as discussed before. The same comparison is shown in Fig. 6 between the RPA $I_{nl,i}$ values (solid lines) and the TFDW ones (dashed lines). In this case, the TFDW results show also a good average value but also with some shell-like features.

Table 3 One-electron mean excitation energies ($I_{nl,\,i}$) calculated from Eq. (8) and using the total mean excitation energies for each system with nuclear charge Z and N_e electrons within the TFA and TFDW methods in the LPA approximation.

		C		F		Ne		Si		Cl		Sc		Mn		Zn	
N_e	$(nl)_i$	$I_{nl,i}^{TFA}$	$I_{nl,i}^{TFDW}$	$I_{nl,i}^{TFA}$	$I_{nl,i}^{TFDW}$	$I_{nl,i}^{TFA}$	$I_{nl,i}^{TFDW}$	$I_{nl,i}^{TFA}$	$I_{nl,i}^{TFDW}$	$I_{nl,i}^{TFA}$	$I_{nl,i}^{TFDW}$	$I_{nl,i}^{TFA}$	$I_{nl,i}^{TFDW}$	$I_{nl,i}^{TFA}$	$I_{nl,i}^{TFDW}$	$I_{nl,i}^{TFA}$	$I_{nl,i}^{TFDW}$
1	$1s_1$	–	182.95	–	333.26	–	389.67	–	642.7	–	858.34	–	1176.49	–	1526.42	–	2004.51
2	$1s_2$	212.02	303.76	335.42	589.06	374.53	697.37	424.65	1187.54	455.36	1608.43	481.05	2232.16	494.59	2920.65	18.37	3863.94
3	$2s_1$	130.23	47.99	342.65	105.07	420.26	127.05	741.72	227.66	989.70	314.91	1328.51	444.715	1677.50	588.69	78.29	786.59
4	$2s_2$	47.55	44.15	175.37	105.02	233.82	128.75	502.83	237.97	719.24	333.08	1012.88	475.57	1308.78	633.75	61.80	851.64
5	$2p_1$	24.83	29.44	91.63	92.69	129.27	121.25	340.46	224.02	534.97	319.46	811.87	463.04	1095.53	623.43	53.40	845.06
6	$2p_2$	16.57	10.21	54.58	63.14	76.65	84.41	225.31	208.59	387.81	303.17	640.86	446.44	912.57	607.39	46.37	830.61
7	$2p_3$	–	–	38.36	38.59	51.98	55.34	151.21	142.36	276.93	231.73	495.71	375.99	748.11	542.74	39.89	775.37
8	$2p_4$	–	–	29.46	30.06	39.53	43.46	106.86	115.46	199.43	187.54	378.75	299.07	604.05	429.29	33.88	611.15
9	$2p_5$	–	–	21.10	22.48	31.21	34.79	80.89	102.41	148.12	169.06	289.52	275.32	482.98	397.29	28.46	575.03
10	$2p_6$	–	–	–	–	22.24	27.32	65.16	89.56	114.99	152.74	224.36	254.96	385.36	374.66	23.70	546.17
11	$3s_1$	–	–	–	–	–	–	54.60	34.94	93.46	79.72	178.08	156.19	309.31	246.69	19.65	375.50
12	$3s_2$	–	–	–	–	–	–	46.03	28.04	78.91	66.11	145.57	132.33	251.56	212.08	16.31	328.40
13	$3p_1$	–	–	–	–	–	–	37.19	21.78	68.23	55.40	122.59	115.33	208.37	188.88	13.60	295.82
14	$3p_2$	–	–	–	–	–	–	25.42	14.67	59.33	47.50	105.98	103.31	176.23	172.51	11.45	274.19
15	$3p_3$	–	–	–	–	–	–	–	–	50.63	40.36	93.44	93.35	152.21	159.74	9.76	257.40
16	$3p_4$	–	–	–	–	–	–	–	–	40.59	24.34	83.28	83.82	133.97	147.65	8.44	243.52

Continued

Table 3 One-electron mean excitation energies ($I_{nl,i}$) calculated from Eq. (8) and using the total mean excitation energies for each system with nuclear charge Z and N_e electrons within the TFA and TFDW methods in the LPA approximation.—cont'd

		C		F		Ne		Si		Cl		Sc		Mn		Zn	
N_e	$(nl)_i$	$I_{nl,i}^{TFA}$	$I_{nl,i}^{TFDW}$	$I_{nl,i}^{TFA}$	$I_{nl,i}^{TFDW}$	$I_{nl,i}^{TFA}$	$I_{nl,i}^{TFDW}$	$I_{nl,i}^{TFA}$	$I_{nl,i}^{TFDW}$	$I_{nl,i}^{TFA}$	$I_{nl,i}^{TFDW}$	$I_{nl,i}^{TFA}$	$I_{nl,i}^{TFDW}$	$I_{nl,i}^{TFA}$	$I_{nl,i}^{TFDW}$	$I_{nl,i}^{TFA}$	$I_{nl,i}^{TFDW}$
17	$3p_5$	–	–	–	–	–	–	–	–	26.71	17.50	74.29	64.81	119.76	136.89	7.40	229.95
18	$3p_6$	–	–	–	–	–	–	–	–	–	–	65.45	49.93	108.23	110.67	6.59	217.11
19	$3d_1$	–	–	–	–	–	–	–	–	–	–	55.76	23.03	98.33	77.51	5.94	174.99
20	$3d_2$	–	–	–	–	–	–	–	–	–	–	$(4s_1)$ 43.96	15.67	89.26	58.46	5.40	147.53
21	$3d_3$	–	–	–	–	–	–	–	–	–	–	$(4s_2)$ 27.33	11.62	80.30	45.85	4.96	104.64
22	$3d_4$	–	–	–	–	–	–	–	–	–	–	–	–	70.77	37.55	4.57	96.11
23	$3d_5$	–	–	–	–	–	–	–	–	–	–	–	–	59.85	31.53	4.22	85.53
24	$3d_6$	–	–	–	–	–	–	–	–	–	–	–	–	$(4s_1)$ 46.27	25.48	3.89	73.07
25	$3d_7$	–	–	–	–	–	–	–	–	–	–	–	–	$(4s_2)$ 26.97	13.64	3.55	68.50
26	$3d_8$	–	–	–	–	–	–	–	–	–	–	–	–	–	–	3.20	62.03
27	$3d_9$	–	–	–	–	–	–	–	–	–	–	–	–	–	–	2.80	55.97
28	$3d_{10}$	–	–	–	–	–	–	–	–	–	–	–	–	–	–	2.34	50.58
29	$4s_1$	–	–	–	–	–	–	–	–	–	–	–	–	–	–	1.76	27.94
30	$4s_2$	–	–	–	–	–	–	–	–	–	–	–	–	–	–	0.93	20.66

Each electron occupying an orbital $(nl)_i$ gets assigned its own mean excitation energy. Note the specific distinction of the $4s$ orbitals for the transition elements. All values given in eV.

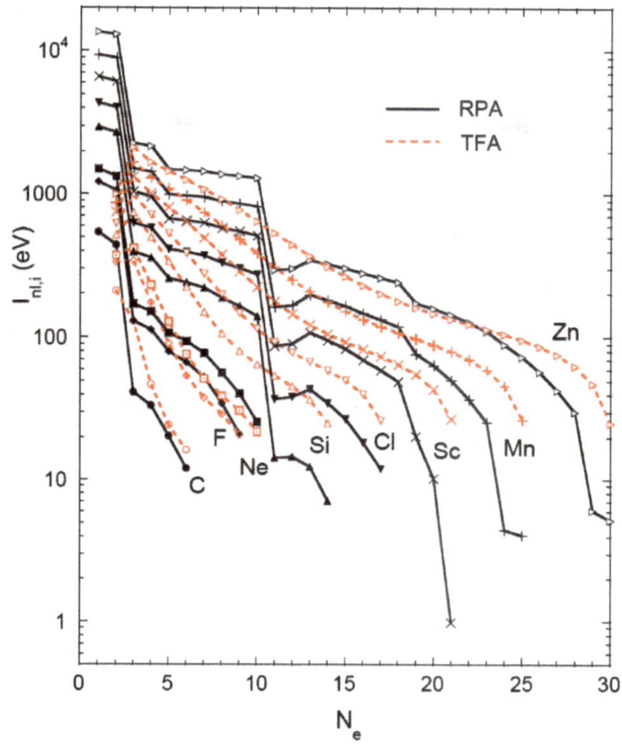

Fig. 5 One-electron mean excitation energies ($I_{nl,\,i}$) as electrons are added to an atomic system with nuclear charge Z up to its neutral electronic configuration (see text for discussion). *Dashed lines*: TFA results from Table 3. *Solid lines*: corresponding RPA results from Sauer et al.[9,20]

3.3 Shell-wise mean excitation energies

The possibility to assign each electron its own orbital mean excitation energy $I_{nl,i}$, as indicated in Table 3, allows us to group the contributions from electrons in the same atomic shell so that the shell-wise mean excitation energy I_n is defined by:

$$\ln I_n = \frac{1}{\nu_n} \sum_{l=0}^{n-1} \sum_{i=1}^{\nu_{nl}} \ln I_{nl,i}, \tag{27}$$

where n is the principal quantum number defining the usual assignment of K, L, M, ... shells, ν_{nl} is the electron occupation number in orbital nl and ν_n the total number of electrons in shell n such that

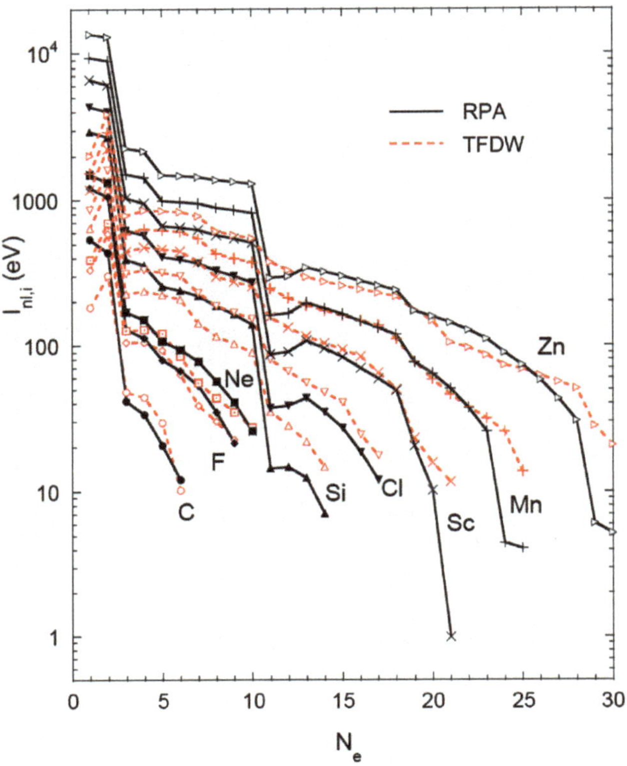

Fig. 6 Same as in Fig. 5 for the TFDW results (*dashed lines*) as compared with the RPA calculations (*solid lines*).

$$\nu_n = \sum_{l=0}^{n-1} \nu_{nl}, \tag{28}$$

Note that Eq. (27) corresponds to the logarithmic average of single-electron contributions, in contrast with previous similar assumptions[30,31] where the same orbital mean excitation energy is assigned to all electrons in that orbital. Also, although the same reasoning as that by Sauer et al.[9] leading to Eq. (27) is applied here. In that work the subshell contributions are reported (i.e., fixed *nl* values) whereas here the shell values correspond to all subshell contributions for a given value of *n*.

Table 4 displays the calculated shell mean excitation energies according to Eqs. (27) and (28) considering the TFA and TFDW results given in Table 3. Also, following the same procedure, and for the sake of comparison, the RPA results for single-electron mean excitation energies of first and

Table 4 Shell mean excitation energies of some first and second row atoms and transition elements obtained according to Eqs. (27) and (28) through the TFA and TFDW approaches in this work and available RPA data.[9,20]

Shell	***K***			***L***			***M***			***N***		
Atom	RPA	TFA	TFDW	RPA	TFA	TFDW	RPA	TFA	TFDW	RPA	TFA	TFDW
C	486.08	212.02	235.74	24.26	39.95	28.25	–	–	–	–	–	–
F	1132.79	335.42	443.07	60.79	68.63	55.80	–	–	–	–	–	–
Ne	–	374.53	521.29	–	78.13	66.50	–	–	–	–	–	–
Si	2810.95	522.42	873.63	231.38	198.77	157.65	11.64	39.26	23.65	–	–	–
Cl	4178.11	625.18	1174.98	393.97	329.49	241.05	27.90	55.58	42.20	–	–	–
Sc	6357.42	752.30	1620.53	675.63	551.34	369.85	67.17	96.27	80.60	3.17	34.67	13.50
Ti	7058.47	–	–	755.01	–	–	73.86	–	–	3.46	–	–
V	7724.08	–	–	840.61	–	–	81.45	–	–	3.74	–	–
Cr	8419.66	–	–	930.68	–	–	77.84	–	–	2.4	–	–
Mn	9145.22	868.88	2111.43	1025.24	809.62	514.36	98.80	127.08	103.16	4.26	35.33	18.64
Fe	9838.74	–	–	1126.04	–	–	107.62	–	–	4.54	–	–
Co	10686.30	–	–	1227.78	–	–	117.20	–	–	4.82	–	–
Ni	11501.80	–	–	1335.78	–	–	127.35	–	–	5.07	–	–
Cu	12298.70	–	–	1449.69	–	–	123.37	–	–	3.33	–	–
Zn	13222.80	1000.87	2783.04	1565.22	1159.72	717.17	149.35	168.40	142.82	5.61	34.92	24.03

All values given in eV.

second row atoms as well as the transition elements reported in Ref. 9 have been considered to obtain the corresponding shell contributions included in this table.

Fig. 7 shows the mean excitation energy contributions from the *K*, *L*, *M*, and *N* shells as a function of atomic number as generated from the TFA (dashed lines), TFDW (dash-dot lines) and RPA (solid lines) approaches. We also include in this figure the HF-DOS values (dotted lines) reported by Oddershede and Sabin.[7] While reasonable quantitative and qualitative agreement is observed for the *L*, *M*, and *N* shell contributions between the RPA and OS calculations, large differences are observed for the *K* shell contribution. The reasons for this discrepancy are discussed by Sauer et al.[9] in terms of the differences between the RPA and the HF-DOS calculations.

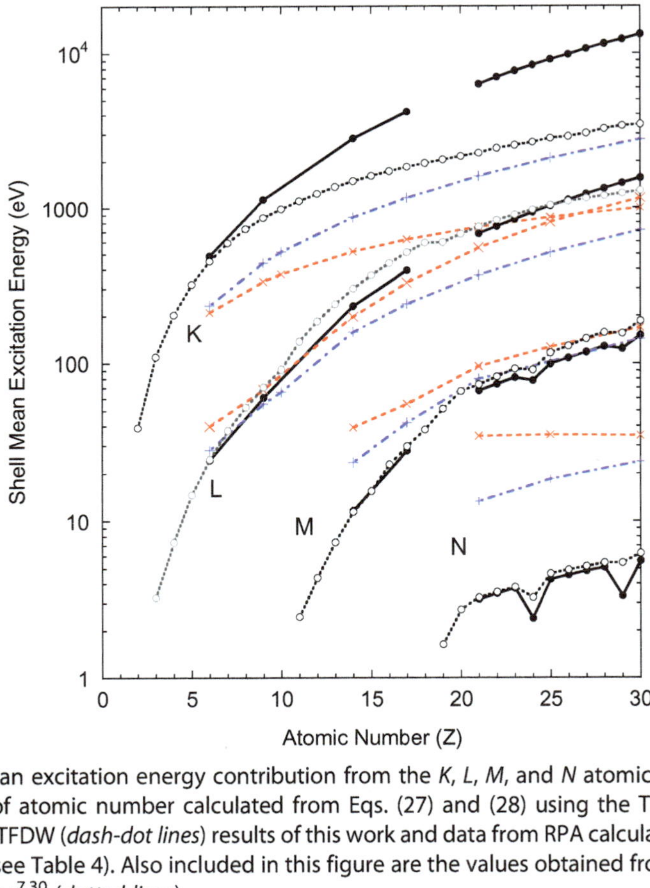

Fig. 7 Mean excitation energy contribution from the *K*, *L*, *M*, and *N* atomic shells as a function of atomic number calculated from Eqs. (27) and (28) using the TFA (*dashed lines*) and TFDW (*dash-dot lines*) results of this work and data from RPA calculations (*solid lines*)[9,20] (see Table 4). Also included in this figure are the values obtained from HF-DOS calculations[7,30] (*dotted lines*).

On the other hand, as may be gathered from Fig. 7, the calculated TFA and TFDW shell contributions show a reasonable behavior for the *L* and *M* shells in close correspondence with the other accurate calculations. In contrast, the TFA *K*-shell contribution remains lower than the TFDW and others as the atomic number increases, which reflects an underestimation of the mean excitation energy of systems with very few electrons as was shown in Figs. 3 and 4. In the *n*-shell case, the TFA and TFDW values for the systems treated in this work remain consistently larger than the accurate RPA and HF-DOS calculations. In spite of the observed quantitative differences, a clear distinction between different shell contributions is apparent from the TFA theory which does not incorporate any shell structure in its conception. A key aspect leading to this analysis within the TFA theory is the appropriate description of the total mean excitation energy of an atomic system with nuclear charge Z and N_e electrons given by Eq. (16) together with Eq. (8) which allows an estimate of single-electron mean excitation energies. The same may be said for the TFDW and the RPA approaches where Eq. (8) plays a central role.

4. Conclusions

Very powerful and sophisticate methods have been developed in recent years to estimate mean excitation energies of atomic and molecular systems. As a tribute to Jack Sabin's motto, in this work we have recovered old ideas and theories such as the original Thomas Fermi Theory with Amaldi corrections (TFA) for ionic systems and Lindhard's LPA, to account for the mean excitation energy of atoms and their cations. A practical analytical expression for the total mean excitation energy of an atomic system of nuclear charge Z and N_e electrons has been derived within the TFA theory. This, together with a previously derived—well founded—simple expression to account for one-electron mean excitation energies expressed in terms of the total mean excitation energy of neighboring cations, has allowed the use of the TFA theory to survey shell contributions to the mean excitation energy. The results of this study have been compared with ab initio RPA calculations and with subsidiary calculations based on the TFDW theory. It has been found that the TFA approach renders a good average behavior of the single-electron mean excitation energy as compared with the accurate RPA values and the TFDW ones. Interestingly, the TFA shell mean excitation energies obtained from the one-electron ones indicate a clear distinction between *K*, *L*, *M*, and *N* shell contributions although quantitatively different for the *K* and *N* shells when compared with the corresponding RPA calculations.

In summary, the use of old atomic theories together with modern concepts on atomic structure brings a consistent physical description of total and mean excitation energies as shown here. This, in Jack's jargon would mean "dos cervezas por favor."

Acknowledgments

Jack Sabin guided one of the authors during the first years of his academic carrier. Jack deeply considers that a simple model is more worth than a sophisticated theory. Guided by feeling and simple expression one can understand the behavior of atoms in a collision process. With this motto, we dedicate this contribution to one of his passions, the study of the mean excitation energy of atomic targets as required in energy deposition problems. R.C.-T. acknowledges support from grant UNAM-DGAPA-PAPIIT IN-111-820 and LANCAD-UNAM-DGTIC-228.

References

1. Bohr, N. On the Theory of the Decrease of Velocity of Moving Electrified Particles on Passing Through Matter. *Philos. Mag.* **1913**, *25*, 10.
2. Bohr, N. LX. On the Decrease of Velocity of Swiftly Moving Electrified Particles in Passing Through Matter. *Lond. Edinb. Dublin Philos. Mag. J. Sci.* **1915**, *30* (178), 581–612. https://doi.org/10.1080/14786441008635432.
3. Bohr, N. The Penetration of Atomic Particles Through Matter. *Mat. Fys. Medd. Dan. Vid. Selsk.* **1948**, *18* (8), 1–144.
4. Bethe, H. Zur Theorie des Durchgangs schneller Korpuskularstrahlen durch Materie. *Ann. Phys.* **1930**, *397* (3), 325–400. https://doi.org/10.1002/andp.19303970303.
5. Inokuti, M. Inelastic Collisions of Fast Charged Particles With Atoms and Molecules—The Bethe Theory Revisited. *Rev. Mod. Phys.* **1971**, *43* (3), 297–347. https://doi.org/10.1103/RevModPhys.43.297.
6. Cabrera-Trujillo, R.; Cruz, S. A. Confinement Approach to Pressure Effects on the Dipole and the Generalized Oscillator Strength of Atomic Hydrogen. *Phys. Rev. A* **2013**, *87* (1), 012502. https://doi.org/10.1103/PhysRevA.87.012502.
7. Oddershede, J.; Sabin, J. R. Orbital and Whole-Atom Proton Stopping Power and Shell Corrections for Atoms With $Z \leq 36$. *At. Data Nucl. Data Tables* **1984**, *31* (2), 275–297. https://doi.org/10.1016/0092-640X(84)90024-X.
8. Cabrera-Trujillo, R.; Sauer, S. P.; Sabin, J. R.; et al. Chapter Five–Bound and Continuum State Contributions to Dipole Oscillator Strength Sum Rules: Total and Orbital Mean Excitation Energies for Cations of C, F, Si, and Cl. In *Rufus Ritchie, A Gentleman and A Scholar*; Sabin, J. R., Oddershede, J., Eds.; Advances in Quantum Chemistry, 80; Academic Press, 2019; pp. 127–146.
9. Sauer, S. P. A.; Sabin, J. R.; Oddershede, J. Calculation of Mean Excitation Energies of 3d-Elements and Their Cations. *Mol. Phys.* **2021**, *119* (5), e1823508. https://doi.org/10.1080/00268976.2020.1823508.

10. Kamakura, S.; Sakamoto, N.; Ogawa, H.; Tsuchida, H.; Inokuti, M. Mean Excitation Energies for the Stopping Power of Atoms and Molecules Evaluated From Oscillator-Strength Spectra. *J. Appl. Phys.* **2006**, *100* (6), 064905. https://doi.org/10.1063/1.2345478.
11. Lindhard, J.; Scharff, M. Energy Loss in Matter by Fast Particles of Low Charge. *Kgl. Dan. Vidensk. Selsk. Mat. Fys. Medd.* **1953**, *27* (15), 1–31.
12. Lindhard, J.; Scharff, M. On the Properties of a Gas of Charged Particles. *Kgl. Dan. Vidensk. Selsk. Mat. Fys. Medd.* **1954**, *28* (8), 1–57.
13. Kamaratos, E. The Mean Excitation Energy for Stopping Power I, the Bragg Rule, and Chemical and Phase Effects. Application of a Statistical Treatment to the Determination of I for Chemically Bound Particles. *Chem. Rev.* **1984**, *84* (6), 561–576.
14. Cruz, S. A.; Soullard, J. Advances in the Core-and-Bond Formalism for Proton Stopping in Molecular Targets. In *Theory of the Interaction of Swift Ions with Matter. Part 2*; Advances in Quantum Chemistry, Vol. 46; Academic Press, 2004; pp. 195–239. https://doi.org/10.1016/S0065-3276(04)46008-0.
15. Johnson, R. E.; Inokuti, M. The Local-Plasma Approximation to the Oscillator-Strength Spectrum: How Good Is It and Why. *Comments At. Mol. Phys.* **1983**, *14*, 19–31.
16. Sabin, J. R.; Oddershede, J. Shell Corrections to Electronic Stopping Powers From Orbital Mean Excitation Energies. *Phys. Rev. A* **1982**, *26* (6), 3209–3219. https://doi.org/10.1103/PhysRevA.26.3209.
17. Meltzer, D. E.; Sabin, J. R.; Trickey, S. B. Calculation of Mean Excitation Energy and Stopping Cross Section in the Orbital Local Plasma Approximation. *Phys. Rev. A* **1990**, *41* (1), 220–232. https://doi.org/10.1103/PhysRevA.41.220.
18. Meltzer, D. E.; Sabin, J. R.; Trickey, S. B. Erratum: Calculation of Mean Excitation Energy and Stopping Cross Section in the Orbital Local Plasma Approximation. *Phys. Rev. A* **1990**, *42* (1), 666. https://doi.org/10.1103/PhysRevA.42.666.
19. Meltzer, D. E.; Sabin, J. R.; Trickey, S. B.; Wu, J. Z. Stopping of Swift Projectiles in Material Thin Films: Hydrogen. *Nucl. Instrum. Methods* **1993**, *B82*, 493.
20. Sauer, S. P. A.; Oddershede, J.; Sabin, J. R. Chapter Three—The Mean Excitation Energy of Atomic Ions. In *Concepts of Mathematical Physics in Chemistry: A Tribute to Frank E. Harris—Part A*; Sabin, J. R., Cabrera-Trujillo, R., Eds.; Advances in Quantum Chemistry, Vol. 71; Academic Press, 2015; pp. 29–40. https://doi.org/10.1016/bs.aiq.2015.02.001.
21. Sauer, S. P. A.; Sabin, J. R.; Oddershede, J. Z-Dependence of Mean Excitation Energies for Second and Third Row Atoms and Their Ions. *J. Chem. Phys.* **2018**, *148* (17), 174307. https://doi.org/10.1063/1.5027708.
22. Sauer, S. P. A.; Sabin, J. R.; Oddershede, J. Mean Excitation Energies of Singly Charged Atomic Anions With $Z \leq 18$. *J. Phys. B At. Mol. Opt. Phys.* **2019**, *52* (9), 095004. https://doi.org/10.1088/1361-6455/ab0e59.
23. Fermi, E.; Amaldi, E. *Le Orbite ∞ S Degli Elementi*, 6; Rom, 1934; p. 119.
24. Tietz, T. Approximate Solution of the Thomas-Fermi Equation With Amaldi-Fermi Boundary Condition. *J. Chem. Phys.* **1957**, *26* (2), 430. https://doi.org/10.1063/1.1743322.
25. Cruz, S. A. Thomas-Fermi-Dirac-Weizsäcker Density Functional Formalism Applied to the Study of Many-Electron Atom Confinement by Open and Closed Boundaries. In Advances in Quantum Chemistry, 57; Academic Press, 2009; pp. 255–283.
26. Parr, R. G.; Yang, W. *Density Functional Theory of Atoms and Molecules*; Oxford University Press: Oxford, 1989.
27. Lee, P. S.; Wu, T. Y. Statistical Potential of Atomic Ions. *Chin. J. Phys.* **1997**, *35*, 742.
28. Cruz, S. A.; Díaz-García, C.; Covarrubias, G. Statistical Atomic Models With Complete Neglect of Differential Overlap for the Study of Free and Confined Systems. *Int. J. Quantum Chem.* **2005**, *102* (5), 897–910. https://doi.org/10.1002/qua.20452.

29. Clementi, E.; Roetti, C. Roothaan-Hartree-Fock Atomic Wavefunctions. *At. Data Nucl. Data Tables* **1974**, *14*, 177.
30. Cabrera-Trujillo, R.; Cruz, S. A.; Soullard, J. From the Orbital Implementation of the Kinetic Theory to the Polarization Propagator Method in the Study of Energy Deposition Problems. In *Advances in Quantum Chemistry*; *Advances in Quantum Chemistry*, Vol. 48; Academic Press, 2005; pp. 335–367. https://doi.org/10.1016/S0065-3276(05)48018-1.
31. Chen, Y. F.; Kwei, C. M.; Tung, C. J. Analytical Representation of Atomic Shellwise Electron Densities and Applications. *J. Phys. B At. Mol. Opt. Phys.* **1993**, *26* (6), 1071–1080. https://doi.org/10.1088/0953-4075/26/6/008.

CHAPTER SEVEN

Recent progress in electron-propagator, extended-Koopmans-theorem and self-consistent-field approaches to the interpretation and prediction of electron binding energies

J.V. Ortiz*

Department of Chemistry and Biochemistry, Auburn University, Auburn, Alabama, United States
*Corresponding author: e-mail address: ortiz@auburn.edu

Contents

Abstract

Advances in three methodologies for the prediction and interpretation of molecular electron binding energies are reviewed. Electron-propagator self-energies in the diagonal, or quasiparticle, approximation have been widely used to interpret photo-electron spectra. A new generation of diagonal approximations with superior accuracy and computational efficiency has been developed and tested vs a data base whose quality approaches full-configuration-interaction calculations on vertical ionization energies of representative closed-shell molecules. The extended-Koopmans-theorem (EKT)

Advances in Quantum Chemistry, Volume 85
ISSN 0065-3276
https://doi.org/10.1016/bs.aiq.2022.03.004

Copyright © 2022 Elsevier Inc.
All rights reserved.

method often has been employed in the calculation of the lowest electron detachment energy where the initial state is represented by a correlated wavefunction or reduced-density matrices. The newly introduced complete-active-space EKT method produces electron detachment energies that approach exact results for a given active orbital space with the introduction basis functions that make much greater contributions to initial, uncharged states than to final, cationic states. Differences in self-consistent field total energies optimized for each state of interest define the ΔSCF method. Two recent theorems on the eigenvalue spectrum of the difference of two idempotent matrices have revealed how corresponding orbitals derived from SCF wavefunctions determine Dyson orbitals, probability factors, natural ionization orbitals with paired occupation numbers that describe relaxation effects and Fukui functions.

Measurements of electron binding energies and associated transition probabilities often benefit from interpretations based on ab initio electronic structure calculations. Confidence in these interpretations depends on the *accuracy* of the predicted data and on the *clarity* of the concepts that underly calculations. Advances in methods based on electron-propagator, extended-Koopmans-theorem and self-consistent-field theories for the prediction of electron binding energies are considered in this review. Each of the methods is associated with concepts that facilitate deeper understanding of electronic structure.

1. New diagonal self-energy approximations in electron-propagator theory

Ab initio electron propagator[1–3] calculations that employ diagonal self-energies in a Hartree-Fock (HF) orbital basis facilitate interpretations by producing Dyson orbitals[4] (DOs) that equal the product of the square root of a probability (i.e., intensity) factor times a canonical orbital.[5–8] This simplified DO sees an optical (relaxation-correlation) potential expressed by the self-energy matrix that corrects the uncorrelated, frozen-orbital approximation associated with Koopmans's identity.[9] For outer-valence ionization energies (IEs) of closed-shell molecules and molecular ions, neglect of the off-diagonal elements of the self-energy operator is usually an excellent approximation that also procures computational advantages.[10] Conceptual and computational economy has stimulated the widespread use of diagonal-self-energy methods of ab initio electron propagator theory (EPT).[11–13] The diagonal self-energy approximation also is made in G_0W_0 calculations that usually employ a Kohn-Sham (KS) orbital basis.[14–16] Numerical comparisons on the IEs of organic acceptor molecules have demonstrated that outer valence Green's function[12,13,17] (OVGF) and approximately renormalized partial third order[18–23] (P3+) EPT calculations are competitive with their G_0W_0 counterparts based on a variety of KS orbitals.[24]

A new generation of diagonal self-energy (or quasiparticle) approximations that retains the conceptual and computational advantages of its antecedents in EPT has recently achieved a higher level of accuracy.[25] All of these approximations require only diagonal elements of the self-energy matrix in a canonical HF basis. Arithmetic bottlenecks that exceed that of the fifth-power, four-index, electron-repulsion-integral (ERI) transformation are avoided. No intermediate matrices with more than three orbital indices are needed. For these reasons, the new methods remain as feasible as their OVGF or P3+ predecessors. *No empirical or adjustable parameters are introduced in the design of the self-energy operator or in the generation of molecular orbitals from basis functions.* The new methods are at least as useful as OVGF and P3+ in interpreting experiments, especially on species with *unprecedented* patterns of chemical bonding and electronic structure.

One of the chief differences between the OVGF and P3+ EPT methods and those of the G_0W_0 family pertains to renormalization (i.e., infinite-order summations of selected classes of self-energy terms or diagrams). In these two EPT methods, evaluation of diagrams of second and third order is followed by computationally trivial estimates of higher-order effects estimated with ratios of terms.[12,20] In this manner, the underestimates of diagonal second-order (D2) and overestimates of diagonal third-order (D3) IEs that typically persist with basis-set saturation are systematically mitigated.[22–24] Three OVGF renormalization procedures have been defined, but P3+ has only one version. (A parametrized algorithm for choosing among the OVGF alternatives has been widely applied,[26,27] but will not be examined here.) The G_0W_0 set of methods incorporates infinite-order summations of ring diagrams, usually with the neglect of exchange in the two-electron vertices.[28] Infinite-order rings (with exchange), ladders and mixed ring-ladder diagrams occur in the post-second-order, non-diagonal self-energies of the 1970s and 1980s.[12,29] The new EPT methods discussed below span both approaches. Only second-order and third-order diagrams are summed in some cases, whereas in others higher-order terms are evaluated explicitly. Although the latter approach requires more matrix multiplications, no fifth-power arithmetic bottleneck is ever exceeded. A version of G_0W_0 that employs HF orbitals and excitation energies and amplitudes at the Tamm-Dancoff[30,31] (single excitation) level is included in the present set of explicitly renormalized self-energies. This member of the G_0W_0 family is a simplification of the non-diagonal EPT method known as the two-particle-one-hole Tamm-Dancoff approximation (2ph-TDA).[32] It has recently been examined computationally[33] and will be denoted by HF-TDA G_0W_0.

Quasiparticle methods that involve only second-order terms have cubic arithmetic scaling and therefore require practically no extra effort once an ERI transformation has been performed. D2 underestimates of vertical IEs (VIEs) persist with improvements in basis sets and converge toward average errors of 0.5 eV.[22,23] A non-Dyson, second-order method[34] has a non-iterative instead of an iterative cubic bottleneck. Elimination of same-spin terms in Dyson and non-Dyson second-order methods also is examined computationally. (Adjustable parameters have been introduced in recent publications that follow the example of parametrized versions of second-order perturbation theory for total energies;[35–38] this kind of approach also is avoided herein.)

The new methods are designed for final states where the Koopmans description of a VIE is qualitatively valid. When probability factors (norms of DOs) are calculated to be lower than 0.85 (as in shake-up, or correlation, final states), quasiparticle calculations should give way to more flexible methods with renormalized, non-diagonal self-energies that treat simple and shake-up ionization operators on an equal basis. DOs of correlation states also may have large contributions from more than one canonical HF orbital. Efficient algorithms for finding the poles (electron binding energies) of the new, approximate electron propagators do not require diagonalizations. Only a small number (usually three) of energy iterations is needed to achieve convergence within 0.01 eV. In typical G_0W_0 calculations, just one iteration is performed at the value of the reference orbital energy.

Standard values of VIEs are needed to judge the quality of the new methods, but these data are seldom provided by experimental data, which necessarily reflect relaxation of nuclear positions from initial to final states and energy increments associated with vibrational levels. Relatively high-level ab initio results may involve coupled-cluster theory with single, double and perturbative triple substitution operators, or CCSD(T),[39] basis-set extrapolations and additivity schemes,[40] but error bars of 0.1 eV or greater are inevitably introduced. Another option could be provided by full configuration interaction (FCI) calculations in a fixed basis. Unfortunately, these results tend to be confined to just one state per irreducible representation for very small molecules. A better alternative may be provided by calculations in a fixed basis with a well-defined progression of approximate methods that is demonstrably converged toward the FCI limit. Such a base of data was created recently to examine the quality of orbital energies produced in

Kohn-Sham calculations. In this work, valence VIEs were calculated with a sequence of equation-of-motion, coupled-cluster methods[41] for ionization potentials (IPs). The last two levels of theory with full triple (IP-EOM-CCSDT) and full quadruple (IP-EOM-CCSDTQ) operators in the reference state were in very close (mean absolute deviation = 0.03 eV) agreement.[42] The latter methods therefore are acceptable substitutes for FCI in a cc-pVTZ[43] basis.

The theory that underlies old and new diagonal self-energy formulae is introduced in the next two sections. Numerical tests and a statistical analysis that reveal the strengths and weaknesses of each method precede recommendations for routine use.

1.1 Electron propagator theory

The electron propagator matrix has elements

$$\left\langle\left\langle a_p(t); a_q^\dagger(t')\right\rangle\right\rangle = -i\Theta(t-t')\left\langle \Psi_0^N \middle| a_p(t) a_q^\dagger(t') \middle| \Psi_0^N \right\rangle + i\Theta(t'-t)\left\langle \Psi_0^N \middle| a_q^\dagger(t') a_p(t) \middle| \Psi_0^N \right\rangle \quad (1)$$

with general spin-orbital indices p and q that refer to members of an orthonormal basis, u. The Heaviside step function (Θ) and the N-electron reference eigenstate, $\Psi_0{}^N$, occur in this definition. In its spectral form, the Fourier transform,

$$\left\langle\left\langle a_p; a_q^\dagger\right\rangle\right\rangle_E = \lim_{\eta\to 0^+}\left[\sum_R \frac{\langle \Psi_0^N | a_p | R\rangle \left\langle R \middle| a_q^\dagger \middle| \Psi_0^N\right\rangle}{E + E_0(N) - E_R(N+1) + i\eta} + \sum_S \frac{\left\langle \Psi_0^N \middle| a_q^\dagger \middle| S\right\rangle \langle S | a_p | \Psi_0^N\rangle}{E - E_0(N) + E_S(N-1) - i\eta}\right], \quad (2)$$

contains summations over final eigenstates with N + 1 or N − 1 electrons. This energy-dependent (i.e., E-dependent) function has poles in the R and S summations that correspond respectively to energies of electron attachment and detachment. Residues (numerators associated with vanishing denominators) are expressed in terms of Feynman-Dyson amplitudes,

$$f_p^R = \langle \Psi_0^N | a_p | R\rangle \quad (3)$$

$$g_p^S = \langle S | a_p | \Psi_0^N\rangle, \quad (4)$$

that may be used to construct Dyson spin-orbitals (DSOs) of electron attachment and detachment, where

$$\phi_R^{Dyson}(x_1) = \sqrt{N+1}\int \Psi_0^{N*}(x_2, x_3, x_4, \cdots, x_{N+1})\Psi_R(x_1, x_2, x_3, \cdots, x_{N+1})$$
$$dx_2 dx_3 dx_4 \cdots dx_{N+1} = \sum_R f_p^R u_p(x_1) \tag{5}$$

$$\begin{aligned}\phi_S^{Dyson}(x_1) &= \sqrt{N}\int \Psi_S^*(x_2, x_3, x_4, \cdots x_N)\Psi_0^N(x_1, x_2, x_3, \cdots x_N) dx_2 dx_3 dx_4 \cdots dx_N \\ &= \sum_S g_p^S u_p(x_1).\end{aligned} \tag{6}$$

The plus sign between the attachment and detachment summations in Eq. (2) is responsible for the anticommutator that appears in the definition of a super-operator metric:

$$[X|Y] = \langle \Psi_0^N | X^\dagger Y + Y X^\dagger | \Psi_0^N \rangle. \tag{7}$$

X and Y are members of a set of field-operators and their products wherein the number of creators exceeds the number of annihilators by one, e.g., $a_p^\dagger$ (hole and particle creators, or h and p), $a_p^\dagger a_q^\dagger a_r$ (2hp and 2ph), $a_p^\dagger a_q^\dagger a_r^\dagger a_s a_t$ (3h2p and 3p2h) and so on.[44] The identity and Hamiltonian super-operators,[45] defined by

$$\hat{I}X = X \tag{8}$$

$$\hat{H}X = HX - XH, \tag{9}$$

are used in the following expression of electron-propagator matrix elements:

$$\left\langle\left\langle a_p; a_q^\dagger \right\rangle\right\rangle_E = G_{pq}(E) = \left[a_p^\dagger \middle| \left(E\hat{I} - \hat{H}\right)^{-1} a_q^\dagger\right]. \tag{10}$$

To avoid expansion of the inverse occurring in the resolvent super-operator that appears in the electron propagator matrix,

$$\mathbf{G}(E) = \left[\mathbf{a}^\dagger \middle| \left(E\hat{I} - \hat{H}\right)^{-1} \mathbf{a}^\dagger\right], \tag{11}$$

an inner projection[46] with **w** operators is introduced according to

$$\mathbf{G}(E) = \left[\mathbf{a}^\dagger | \mathbf{w}\right]\left[\mathbf{w} | \left(E\hat{I} - \hat{H}\right)\mathbf{w}\right]^{-1}\left[\mathbf{w} | \mathbf{a}^\dagger\right]. \tag{12}$$

The **w** manifold may be partitioned into a set of simple, normalized field operators and an orthogonalized complement, **f**, such that $[\mathbf{a}^\dagger|\mathbf{f}] = \mathbf{0}$. The inverse form of the Dyson equation thereby emerges, where

$$\mathbf{G}^{-1}(E) = \left[\mathbf{a}^\dagger|\left(E\hat{I} - \hat{H}\right)\mathbf{a}^\dagger\right] - \left[\mathbf{a}^\dagger|\hat{H}\mathbf{f}\right]\left[\mathbf{f}|\left(E\hat{I} - \hat{H}\right)\mathbf{f}\right]^{-1}\left[\mathbf{f}|\hat{H}\mathbf{a}^\dagger\right]. \quad (13)$$

The first term that involves only the primary-space operators ($\mathbf{a}^\dagger$) has a familiar form that reads

$$\left[a_p^\dagger|\hat{H}a_q^\dagger\right] = h_{pq}^{(1)} + \sum_{rs}\langle pr||qs\rangle\rho_{rs} = F_{pq}, \quad (14)$$

where $\boldsymbol{\rho}$ is the one-electron reduced-density matrix (1-RDM) of Ψ_0^N and $\boldsymbol{F}$ is a *generalized* Fock matrix with the usual one-electron ($h^{(1)}$) and anti-symmetrized, two-electron (Dirac notation) matrix elements. In a spin-orbital basis that comprises eigenfunctions of a general, one-electron operator $h^{(1)}+v$ with eigenvalues e^v, an inverse zeroth-order electron-propagator matrix, defined by

$$\mathbf{G}_0^{-1}(E) = E\mathbf{1} - \mathbf{e}^v \quad (15)$$

has vanishing off-diagonal elements. The inverse Dyson equation now assumes a familiar form,

$$\mathbf{G}^{-1}(E) = \mathbf{G}_0^{-1}(E) - \boldsymbol{\Sigma}(E), \quad (16)$$

where the self-energy matrix, $\boldsymbol{\Sigma}(E)$, comprises constant (i.e., energy-independent) and energy-dependent terms:

$$\boldsymbol{\Sigma}(E) = \mathbf{F} - \mathbf{h}^{(1)} - \mathbf{v} + \left[\mathbf{a}^\dagger|\hat{H}\mathbf{f}\right]\left[\mathbf{f}|\left(E\hat{I} - \hat{H}\right)\mathbf{f}\right]^{-1}\left[\mathbf{f}|\hat{H}\mathbf{a}^\dagger\right]. \quad (17)$$

If one chooses v to be the Coulomb-exchange potential of a HF, self-consistent field, then

$$\boldsymbol{\Sigma}(E) = \boldsymbol{\Sigma}(\infty) + \left[\mathbf{a}^\dagger|\hat{H}\mathbf{f}\right]\left[\mathbf{f}|\left(E\hat{I} - \hat{H}\right)\mathbf{f}\right]^{-1}\left[\mathbf{f}|\hat{H}\mathbf{a}^\dagger\right], \quad (18)$$

where elements of the constant term in the self-energy matrix,

$$\Sigma_{pq}(\infty) = \sum_{rs}\langle pr||qs\rangle\left(\rho_{rs} - \rho_{rs}^{HF}\right), \quad (19)$$

depend only on the correlation contributions to the 1-RDM. One of the advantages of this uncorrelated choice of v is clear analysis of results in terms of final-state-orbital relaxation and differential correlation effects that are encountered only in $\mathbf{\Sigma}(\mathrm{E})$. Another advantage is that the sum of the first three terms in Eq. (17) vanishes through second order in the fluctuation potential. This sum corresponds to the energy-independent terms of Eq. (19) in higher orders. Orbital eigenfunctions obtained with alternative (e.g., Kohn-Sham) choices of v intended to capture reference-state exchange or correlation effects typically produce non-diagonal, energy-independent, first-order terms that involve the non-local, HF exchange operator anyway in Eq. (18).

For all choices of orbital bases, poles of $\mathbf{G}(E)$, or zeroes of $\mathbf{G}^{-1}(E)$, are equal to energies (E_X) where the quasiparticle eigenvalue equation is satisfied:

$$[\mathbf{F} + \mathbf{\Sigma}(E_X)]\mathbf{C}(E_X) = E_X\mathbf{C}(E_X). \tag{20}$$

The probability factor (also known as the pole strength), Π, equals the norm of the DSO[4] and may be calculated with

$$\Pi_X = \int \left|\varphi^{Dyson}(x)\right|^2 dx = \left(1 - \mathbf{C}^{\dagger}(E_X)\left[\frac{d\mathbf{\Sigma}(E)}{dE}\right]_{E=E_X}\mathbf{C}(E_X)\right)^{-1}. \tag{21}$$

Elements of $\mathbf{C}(E_X)$ determine the Feynman-Dyson amplitudes and DSOs:

$$\varphi_X^{Dyson}(x) = \Pi_X^{0.5}\sum_p C_p(E_X)u_p(x). \tag{22}$$

Methods that employ canonical, HF spin-orbitals and that ignore off-diagonal elements of the self-energy matrix have a variety of computational and conceptual advantages. In such cases, each canonical spin-orbital is associated with its own, independent set of electron binding energies that are obtained iteratively with the *diagonal quasiparticle equation*

$$e_p^{HF} + \Sigma_{pp}\left(E_{Xp}\right) = E_{Xp}, \tag{23}$$

where Koopmans results (e_p^{HF}) are improved by the relaxation and correlation terms that appear in diagonal elements of the self-energy matrix. Simplified probability factors read

$$\Pi_{Xp} = \left[1 - \left.\frac{d\Sigma_{pp}(E)}{dE}\right|_{E=E_{Xp}}\right]^{-1}. \tag{24}$$

The DSOs are equal to canonical, Hartree-Fock spin-orbitals times the square roots of the probability factors:

$$\varphi_{Xp}^{Dyson}(x) = \Pi_{Xp}^{0.5} u_p^{HF}(x). \tag{25}$$

For each canonical spin-orbital p, many values of E_{Xp} could be found in principle, but usually only one pole is of interest. A probability factor that exceeds 0.85 confirms the qualitative validity of the assumptions made in Koopmans's identity. The remaining electron binding energies may have some qualitative meaning, i.e., they may provide a crude description of correlation, or shake-up, final states. Such results are likely to require improvements from renormalized, non-diagonal self-energy approximations.

1.2 Self-energy approximations

Self-energy corrections to canonical, Hartree-Fock orbital energies appear in second and higher orders in the Møller-Plesset fluctuation potential.[47] Letting $|\Psi_o^N>$ be a HF wavefunction, restricting the secondary operator space to triple-field (2hp and 2ph, or $\mathbf{f}_3$) operator products and neglecting first-order couplings between the latter operators yields the non-diagonal, second-order (ND2) self-energy:

$$\boldsymbol{\Sigma}^{(2)}(E) = \left[\mathbf{a}^\dagger|\hat{H}\mathbf{f}_3\right]_1 \left[\mathbf{f}_3|(E\hat{I} - \hat{H})\mathbf{f}_3\right]_0^{-1} \left[\mathbf{f}_3|\hat{H}\mathbf{a}^\dagger\right]_1. \tag{26}$$

Subscripts to the right of the square brackets specify the highest order of the fluctuation potential that is kept in the corresponding matrix. (Zeroth-order couplings between $\mathbf{f}_3$ and $\mathbf{a}^\dagger$ operators vanish.) Diagonal second-order (D2) self-energy matrix elements have 2hp and 2ph terms:

$$\Sigma_{pp}^{D2}(E) = \frac{1}{2}\sum_a^{vir}\sum_{ij}^{occ}\frac{|\langle pa||ij\rangle|^2}{E + e_a - e_i - e_j} + \frac{1}{2}\sum_i^{occ}\sum_{ab}^{vir}\frac{|\langle pi||ab\rangle|^2}{E + e_i - e_a - e_b}. \tag{27}$$

Setting $E = e_p$ produces second-order, Rayleigh-Schrödinger perturbation theory (RSPT2) results for electron detachments or attachments.[48–50] Parametrized versions of D2 have assigned empirically adjusted weights to same-spin and opposite-spin terms in both summations.[35–38] A simpler, interpretable alternative is considered presently: neglect of the same-spin terms. (This approximation avoids the introduction of parameters adjusted with respect to a set of data.) The resulting opposite-spin, second-order method is denoted by os-D2. Another second-order approximation is founded on the non-Dyson (nD) approach, which neglects the electron affinity (EA) term in the electron-propagator matrix (see Eq. 2), has only one term in its super-operator metric therefore (compare with Eq. 7) and yields a diagonal second-order self-energy that is identical to D2 except that the second (2ph) term is energy-independent, with E set equal to e_p for electron detachments.[34] (The roles of the 2hp and 2ph terms are reversed in EA calculations.) These modifications yield the nD-D2 and os-nD-D2 methods. The nD methods have a minor computational advantage, for the OV^2 bottleneck (where O and V are dimensions of the occupied and virtual orbital spaces) of the 2ph term is no longer energy-dependent and therefore requires no iterations. In practice, the real bottleneck is the transformation to the HF basis of the ERIs; this operation may be reduced from fifth to fourth power dependence for a given p in D2 calculations.

Retention of first-order couplings between $\mathbf{f}_3$ operators yields the closely related 2ph-TDA[32] self-energy:

$$\mathbf{\Sigma}^{2ph-TDA}(E) = \left[\mathbf{a}^\dagger|\widehat{H}\mathbf{f}_3\right]_1 \left[\mathbf{f}_3|\left(E\widehat{I} - \widehat{H}\right)\mathbf{f}_3\right]_1^{-1} \left[\mathbf{f}_3|\widehat{H}\mathbf{a}^\dagger\right]_1. \tag{28}$$

No correlation corrections to the HF reference state are invoked. The inverted matrix generates ring and ladder terms of all orders with anti-symmetrized vertices, i.e., exchange terms are not discarded. The 2ph-TDA method provides a qualitatively correct account of primary and correlation states, but has not been employed with the diagonal, quasiparticle Eq. (23) in published calculations.

A member of the G_0W_0 set of methods that is based on HF spin-orbitals may be defined by neglecting off-diagonal elements of the self-energy matrix, ladder and ring-exchange terms in the $[\mathbf{f}_3|(E\widehat{I} - \widehat{H})\mathbf{f}_3]_1$ matrix and exchange terms in the $[\mathbf{a}^\dagger|\widehat{H}\mathbf{f}_3]_1$ and $[\mathbf{f}_3|\widehat{H}\mathbf{a}^\dagger]_1$ matrices. ERIs of the (single-bar, i.e., exchange-less) $<ab|ij>$ type suffice to calculate couplings between f_3 operators. In this method, the usual random-phase-approximation (RPA) excitation energies and transition amplitudes of G_0W_0 theory are replaced by their Tamm-Dancoff-approximation (TDA) counterparts. The resulting HF-TDA G_0W_0 self-energy contains a subset of terms that occur at the 2ph-TDA level.

The 3+ self-energy improves on 2ph-TDA by retaining all third-order contributions:

$$\mathbf{\Sigma}^{3+}(E) = \mathbf{\Sigma}_3(\infty) + \left[\mathbf{a}^\dagger|\widehat{H}\mathbf{f}_3\right]_2 \left[\mathbf{f}_3|\left(E\widehat{I} - \widehat{H}\right)\mathbf{f}_3\right]_1^{-1} \left[\mathbf{f}_3|\widehat{H}\mathbf{a}^\dagger\right]_2. \tag{29}$$

Higher (i.e., two and three) orders indicated in the subscripts involve first-order double or second-order single replacements with respect to $|HF>$

and < HF | in Eq. (7). As in 2ph-TDA, terms beyond third order are generated by the inverted matrix. The 3+ approximation is a member of the family of ADC(3) (third-order algebraic diagrammatic construction) methods; it is sometimes called strict ADC(3).[29] Other members of this family replace $\Sigma_3(\infty)$ with its fourth-order counterpart or with renormalized schemes[51] in which the trace of the 1-RDM used to evaluate the energy-independent term does not necessarily conserve the number of electrons. (The nD variant of ADC(3)[52] has recently been implemented efficiently.[53–55]) As with 2ph-TDA, the full (non-diagonal) 3+ self-energy has always been evaluated in practice; there is no published record of diagonal ADC(3) calculations.

By expanding the inverse matrix of 3+ in terms of its zeroth and first order components and neglecting terms beyond third order, it is possible identify the terms of the third order self-energy,

$$\begin{aligned}\mathbf{\Sigma}^{(3)}(E) = \mathbf{\Sigma}^{(2)}(E) + \mathbf{\Sigma}_3(\infty) + \left[\mathbf{a}^\dagger|\widehat{H}\mathbf{f}_3\right]_1\left[\mathbf{f}_3|(E\widehat{I}-\widehat{H})\mathbf{f}_3\right]_0^{-1}\left[\mathbf{f}_3|\widehat{V}\mathbf{f}_3\right]_1\\ \left[\mathbf{f}_3|(E\widehat{I}-\widehat{H})\mathbf{f}_3\right]_0^{-1}\left[\mathbf{f}_3|\widehat{H}\mathbf{a}^\dagger\right]_1 +\\ \left[\mathbf{a}^\dagger|\widehat{H}\mathbf{f}_3\right]_2\left[\mathbf{f}_3|(E\widehat{I}-\widehat{H})\mathbf{f}_3\right]_0^{-1}\left[\mathbf{f}_3|\widehat{H}\mathbf{a}^\dagger\right]_1\\ + \left[\mathbf{a}^\dagger|\widehat{H}\mathbf{f}_3\right]_1\left[\mathbf{f}_3|(E\widehat{I}-\widehat{H})\mathbf{f}_3\right]_0^{-1}\left[\mathbf{f}_3|\widehat{H}\mathbf{a}^\dagger\right]_2, \quad (30)\end{aligned}$$

where $\widehat{V}$ is the fluctuation-potential super-operator. The diagonal, third-order self-energy (D3) evaluated at $E=e_p$ provides the third-order, Rayleigh-Schrödinger perturbation theory (RSPT3) correction to Koopmans results.[48–50] Evaluations of the D3 self-energy precede estimates obtained with the A, B and C versions of OVGF.[12] Elements of the D3 self-energy matrix may be classified with the following expression,

$$\Sigma^{(3)} = S_{2hp} + S_{2ph} + C_D + C_S + R_{2hp} + R_{2ph} + P_{2hp} + P_{2ph}, \quad (31)$$

where the two *second-order* contributions that occur in Eq. (27) are represented by S_{2hp} and S_{2ph}, the *constant* terms arising from first-order double and second-order single replacements in the reference state are denoted by C_D and C_S, R_{2hp} and R_{2ph} are the *ring and ladder* terms and P_{2hp} and P_{2ph} are the energy-dependent terms that involve *perturbative* first-order double replacements. R_{2hp} and R_{2ph} correspond to the term with two energy-dependent denominators (i.e., the third term of Eq. 30) in $\mathbf{\Sigma}^{(3)}(E)$. The P_{2hp} and P_{2ph} terms have contributions from the last two matrix products in $\mathbf{\Sigma}^{(3)}(E)$ which are Hermitian conjugates of each other. (Energy dependence in the S, R and P terms has been suppressed in this notation.) Formulae for each of these terms have been published in several reviews. The three OVGF self-energies read as follows:

$$\Sigma^{OVGF-A} = \Sigma^{(2)} + \frac{S_{2hp} + S_{2ph}}{S_{2hp} + S_{2ph} - P_{2hp} - P_{2ph}} \left(C_D + C_S + R_{2hp} + R_{2ph} + P_{2hp} + P_{2ph}\right) \quad (32)$$

$$\Sigma^{OVGF-B} = \Sigma^{(2)} + C_D + C_S + \frac{S_{2hp}}{S_{2hp} - P_{2hp}}\left(R_{2hp} + P_{2hp}\right) + \frac{S_{2ph}}{S_{2ph} - P_{2ph}}\left(R_{2ph} + P_{2ph}\right) \quad (33)$$

$$\Sigma^{OVGF-C} = \Sigma^{(2)} + \left(C_D + C_S + R_{2hp} + R_{2ph} + P_{2hp} + P_{2ph}\right)\times \frac{R_{2hp} + R_{2ph} + P_{2hp} + P_{2ph}}{\frac{S_{2hp}-P_{2hp}}{S_{2hp}}\left(R_{2hp} + P_{2hp}\right) + \frac{S_{2ph}-P_{2ph}}{S_{2ph}}\left(R_{2ph} + P_{2ph}\right)}. \quad (34)$$

Despite their approximations of higher-order self-energy terms, the accuracy of the computationally less demanding OVGF methods is competitive with that of the family of ADC(3) methods for the lowest ionization energies of closed-shell molecules.[22,23] OVGF methods require only trivial computational work in addition to a D3 calculation.

Whereas ADC(3) and OVGF may be improved by adding reference-state replacements of higher orders to $|\Psi_0^N>$ and $<\Psi_0^N|$ in Eq. (7) (e.g., second-order double and triple substitutions) and by increasing operator manifolds to $\mathbf{f}_5$ and beyond, an alternative approach considers the consequences of adopting intermediate normalization (which accompanies size-extensive total energies in many-body theories[56]) in the choice of super-operator metric. In the derivation of the P3 self-energy,[18] such a choice recovers MP2 total energies evaluated with the electron propagator, but with the appearance of non-Hermitian super-operator matrices that are subsequently Hermitized in practical calculations. Separate approximations are necessitated for IE and EA calculations.[8] To supersede this approach, intermediate normalization will be combined with Hermitization and retention of all resulting third-order terms.

The new class of approximate self-energies is based on an intermediately normalized, Hermitized super-operator metric,

$$[X|Y] = \frac{1}{2}\left(\left\langle HF\middle|\left(X^{\dagger}Y + YX^{\dagger}\right)\left(1 + T_2^{(1)} + T_1^{(2)}\right)\middle|HF\right\rangle + \left\langle HF\middle|\left(1 + T_2^{(1)} + T_1^{(2)}\right)^{\dagger}\left(X^{\dagger}Y + YX^{\dagger}\right)\middle|HF\right\rangle\right) \quad (35)$$

where X and Y may be $\mathbf{a}^{\dagger}$ or $\mathbf{f}_3$ operators, and first-order, double and second-order, single replacements appearing in

$$T_2^{(1)} = \frac{1}{4}\sum_{ij}^{occ}\sum_{ab}^{vir}\frac{\langle ij||ab\rangle}{e_i + e_j - e_a - e_b} a_i^\dagger a_j^\dagger a_a a_b \tag{36}$$

$$T_1^{(2)} = \sum_{i}^{occ}\sum_{a}^{vir}\frac{1}{2(e_i - e_a)}\left(\sum_{jk}^{occ}\sum_{b}^{vir}\frac{\langle jk||ab\rangle\langle bi||jk\rangle}{e_j + e_k - e_a - e_b}\right. \tag{37}$$
$$\left. + \sum_{j}^{occ}\sum_{bc}^{vir}\frac{\langle ij||bc\rangle\langle bc||aj\rangle}{e_i + e_j - e_b - e_c}\right) a_i^\dagger a_a$$

occur *linearly* in the Hermitian super-operator matrix elements. (A non-symmetric super-operator metric that omitted single replacement operators was used to derive the P3 and non-diagonal, renormalized, second-order or NR2[7,8,57] self-energies; the IE and EA cases were derived independently.[8]) The RL3 self-energy, which retains all *third-order* terms that arise from this *linear* choice of metric and *ring-ladder renormalizations*, has matrix elements that read

$$\begin{aligned}\boldsymbol{\Sigma}^{RL3}(E) = \frac{1}{2}\Big(&\mathbf{C}_S + \left[\mathbf{a}^\dagger|\hat{H}\mathbf{f}_{2ph}\right]_2\left[\mathbf{f}_{2ph}|(E\hat{I} - \hat{H})\mathbf{f}_{2ph}\right]_1^{-1}\left[\mathbf{f}_{2ph}|\hat{H}\mathbf{a}^\dagger\right]_1\\
&+ \left[\mathbf{a}^\dagger|\hat{H}\mathbf{f}_{2ph}\right]_1\left[\mathbf{f}_{2ph}|(E\hat{I} - \hat{H})\mathbf{f}_{2ph}\right]_1^{-1}\left[\mathbf{f}_{2ph}|\hat{H}\mathbf{a}^\dagger\right]_2\\
&+ \left[\mathbf{a}^\dagger|\hat{H}\mathbf{f}_{2hp}\right]_1\left[\mathbf{f}_{2hp}|(E\hat{I} - \hat{H})\mathbf{f}_{2hp}\right]_1^{-1}\left[\mathbf{f}_{2hp}|\hat{H}\mathbf{a}^\dagger\right]_2\\
&+ \left[\mathbf{a}^\dagger|\hat{H}\mathbf{f}_{2hp}\right]_2\left[\mathbf{f}_{2hp}|(E\hat{I} - \hat{H})\mathbf{f}_{2hp}\right]_1^{-1}\left[\mathbf{f}_{2hp}|\hat{H}\mathbf{a}^\dagger\right]_1\Big),\end{aligned} \tag{38}$$

where $\mathbf{C}_S$ is the non-diagonal matrix generalization of C_S. To recover the third-order C_S term in Eq. (38), the $T_1^{(2)}$ operators in Eq. (35) are necessary. The C_D term is eliminated by the choice of Eq. (35), for it has a quadratic dependence on reference-state substitution amplitudes that often is a source of instability with respect to basis sets. Better stability and accuracy may require inclusion of the fourth-order, energy-independent self-energy. This improvement includes many computationally demanding terms (such as those that depend on $T_3^{(2)}$ and $T_2^{(1)}$ amplitudes) that are quadratic in reference-state substitution amplitudes. (See the $\Sigma(4+)$ self-energy proposed in nD-ADC(3) calculations.[58]) The high price of reduced efficiency and interpretability is unlikely to be proportionate to the benefits of increased accuracy. Omission of quadratic terms therefore provides a balanced compromise of efficiency, interpretability and (vide infra) accuracy. Ring and ladder terms that do not require reference-state substitution amplitudes are the same as those found in 2ph-TDA, 3+ and ADC(3), but terms that are linear in those amplitudes are now multiplied by a factor of one half because of Eq. (35). These P_{2hp} and P_{2ph} terms are usually the largest in the D3 self-energy and their double-weighting is often responsible for overestimates of IEs. Simplified versions of IP-EOM-CCSD equations that employ $T_2^{(1)}$ amplitudes instead of optimized CCSD T_2 elements produce the same coefficients for analogous terms.[59,60] This similarity was demonstrated in the original derivation of the P3 method,[18] which generates no analogues of C_D. Omission of quadratic terms also follows from an equivalent expression for $\hat{H}$ matrix elements in terms of a single-determinant reference state and the similarity-transformed Hamiltonian that pertains to unitary coupled-cluster theory.[61] A series of Hermitian model Hamiltonians which may be truncated after the linear term in the cluster *(T)* operator can be generated via a Baker-Campbell-Hausdorff expansion that involves commutators of the original Hamiltonian with T.

Expansion of the inverted matrices in the RL3 self-energy gives rise to fourth-order and higher-order terms. Couplings between 2ph operators that occur in third and higher orders require the most demanding arithmetic operations. Considerable numerical evidence indicates that these terms are of relatively little importance in calculations of electron removal energies for closed-shell molecules.[18] Neglecting them defines a renormalized, quasi-particle self-energy (RQ3) for the calculation of IEs that retains RL3's 2hp, third-order terms:

$$\begin{aligned}\mathbf{\Sigma}^{RQ3}(E) = &\frac{1}{2}\mathbf{C}_S + \left[\mathbf{a}^\dagger|\widehat{H}\mathbf{f}_{2ph}\right]_1\left[\mathbf{f}_{2ph}|\left(E\widehat{I}-\widehat{H}\right)\mathbf{f}_{2ph}\right]_0^{-1}\left[\mathbf{f}_{2ph}|\widehat{H}\mathbf{a}^\dagger\right]_1\\ &+\frac{1}{2}\left[\mathbf{a}^\dagger|\widehat{H}\mathbf{f}_{2hp}\right]_2\left[\mathbf{f}_{2hp}|\left(E\widehat{I}-\widehat{H}\right)\mathbf{f}_{2hp}\right]_1^{-1}\left[\mathbf{f}_{2hp}|\widehat{H}\mathbf{a}^\dagger\right]_1\\ &+\frac{1}{2}\left[\mathbf{a}^\dagger|\widehat{H}\mathbf{f}_{2hp}\right]_1\left[\mathbf{f}_{2hp}|\left(E\widehat{I}-\widehat{H}\right)\mathbf{f}_{2hp}\right]_1^{-1}\left[\mathbf{f}_{2hp}|\widehat{H}\mathbf{a}^\dagger\right]_2.\end{aligned} \tag{39}$$

The second term recovers the 2ph summation in Eq. (27). Omission of the constant term arising from $T_1^{(2)}$ in the reference state yields the RP3 self-energy:

$$\begin{aligned}\mathbf{\Sigma}^{RP3}(E) = &\left[\mathbf{a}^\dagger|\widehat{H}\mathbf{f}_{2ph}\right]_1\left[\mathbf{f}_{2ph}|\left(E\widehat{I}-\widehat{H}\right)\mathbf{f}_{2ph}\right]_0^{-1}\left[\mathbf{f}_{2ph}|\widehat{H}\mathbf{a}^\dagger\right]_1\\ &+\frac{1}{2}\left[\mathbf{a}^\dagger|\widehat{H}\mathbf{f}_{2hp}\right]_2\left[\mathbf{f}_{2hp}|\left(E\widehat{I}-\widehat{H}\right)\mathbf{f}_{2hp}\right]_1^{-1}\left[\mathbf{f}_{2hp}|\widehat{H}\mathbf{a}^\dagger\right]_1\\ &+\frac{1}{2}\left[\mathbf{a}^\dagger|\widehat{H}\mathbf{f}_{2hp}\right]_1\left[\mathbf{f}_{2hp}|\left(E\widehat{I}-\widehat{H}\right)\mathbf{f}_{2hp}\right]_1^{-1}\left[\mathbf{f}_{2hp}|\widehat{H}\mathbf{a}^\dagger\right]_2.\end{aligned} \tag{40}$$

RP3 is closely related to NR2 approximation for IEs; its diagonal third-order terms are the same as those of P3.

Omission of fourth and higher order terms associated with expansion of the inverted matrices in RL3, RQ3 and RP3 respectively defines the L3, Q3 and P3 self-energy matrices:

$$\Sigma^{L3} = S_{2hp} + S_{2ph} + \frac{1}{2}\left(C_S + P_{2hp} + P_{2ph}\right) + R_{2hp} + R_{2ph} \tag{41}$$

$$\Sigma^{Q3} = S_{2hp} + S_{2ph} + \frac{1}{2}\left(C_S + P_{2hp}\right) + R_{2hp} \tag{42}$$

$$\Sigma^{P3} = S_{2hp} + S_{2ph} + \frac{1}{2}P_{2hp} + R_{2hp}. \tag{43}$$

The new choice of super-operator metric results in factors of one half. Only P3 has been derived before.[18] The original derivation was based on a super-operator metric with first-order double but without second-order single replacements.

The approximately renormalized, partial third order self-energy[20] (P3+) for IEs may be expressed with S, R and P terms, where

$$\Sigma^{P3+} = \Sigma^{(2)} + \frac{S_{2hp}}{S_{2hp} - \frac{1}{2}P_{2hp}}\left(R_{2hp} + \frac{1}{2}P_{2hp}\right). \tag{44}$$

(P3+ is denominated the *approximately* renormalized partial third order method; it was previously denoted simply by renormalized partial third order.) The + sign pertains to approximately renormalized methods. R is reserved for methods in which higher-order terms are explicitly evaluated. Thus, the Q3+ approximation becomes

$$\Sigma^{Q3+} = \Sigma^{(2)} + \frac{1}{2}C_S + \frac{S_{2hp}}{S_{2hp} - \frac{1}{2}P_{2hp}}\left(R_{2hp} + \frac{1}{2}P_{2hp}\right). \tag{45}$$

A, B and C versions of L3+ may be defined by analogy to the three OVGF alternatives:

$$\Sigma^{L3+A} = \Sigma^{(2)} + \frac{S_{2hp} + S_{2ph}}{S_{2hp} + S_{2ph} - \frac{1}{2}\left(P_{2hp} + P_{2ph}\right)} \left[C_S + R_{2hp} + R_{2ph} + \frac{1}{2}\left(P_{2hp} + P_{2ph}\right)\right] \tag{46}$$

$$\Sigma^{L3+B} = \Sigma^{(2)} + C_S + \frac{S_{2hp}}{S_{2hp} - \frac{1}{2}P_{2hp}}\left(R_{2hp} + \frac{1}{2}P_{2hp}\right) + \frac{S_{2ph}}{S_{2ph} - \frac{1}{2}P_{2ph}}\left(R_{2ph} + \frac{1}{2}P_{2ph}\right) \tag{47}$$

$$\Sigma^{L3+C} = \Sigma^{(2)} + \left[C_S + R_{2hp} + R_{2ph} + \frac{1}{2}\left(P_{2hp} + P_{2ph}\right)\right] \times \frac{R_{2hp} + R_{2ph} + \frac{1}{2}\left(P_{2hp} + P_{2ph}\right)}{\frac{S_{2hp}-\frac{1}{2}P_{2hp}}{S_{2hp}}\left(R_{2hp} + \frac{1}{2}P_{2hp}\right) + \frac{S_{2ph}-\frac{1}{2}P_{2ph}}{S_{2ph}}\left(R_{2ph} + \frac{1}{2}P_{2ph}\right)}. \tag{48}$$

Pole search algorithms typically proceed as follows.

1. They employ HF orbital energies as initial guesses.
2. Derivatives of the diagonal self-energy matrix elements with respect to E are evaluated. (These data also are needed to determine probability factors).
3. Newton's method for finding the roots of a function of a single variable (see Eq. 23) is employed. Only three iterations are usually necessary for convergence to within 0.01 eV.

Pole searches with the approximately renormalized (i.e., OVGF, P3+, Q3+ and L3+) self-energies are initiated respectively at the D3, P3, Q3 and L3 poles; agreement to within 0.01 eV is usually obtained after only one Newton iteration. The three + (i.e., approximately renormalized) methods require almost no additional work after P3, Q3 or L3 calculations are finished.

1.3 Computational scaling and numerical results

For various self-energy approximations, arithmetic bottlenecks, the number of bottleneck steps and the memory scaling of the largest subset of ERIs in the canonical HF basis are displayed in Table 1. In determining arithmetic bottlenecks, the number of virtual orbitals (V) is assumed to be larger than the number of occupied orbitals (O). New methods are denoted by bold type. The D2 method requires only a quartic partial integral transformation step (instead of the usual fifth-power scaling) because all its required integrals for a given electron binding energy have a single fixed index, p. Second-order methods scale cubically (i.e., they are practically free after the integral transformation) and are aggregated in Table 1. The D3, OVGF, L3, L3+ and RL3 methods have OV^4 steps involving integrals with four virtual indices that are comparable in difficulty to the full integral transformation involving N_B atom-centered basis functions (i.e., $N_B^5 > V^5 > OV^4$). The P3, P3+, Q3, Q3+ and G_0W_0 methods have relatively fast O^2V^3 arithmetic bottlenecks. HF-TDA G_0W_0 has the additional advantage of not requiring integrals with three virtual indices. In several methods, the number of bottleneck steps is proportional to N_E, the number of energy iterations; in practice,

Table 1 Arithmetic scaling requirements for self-energy approximations[a].

Σ(E) Classification	**Methods[b]**	**Arithmetic bottleneck**	**Bottleneck Steps**	**Largest ERI[c] subset**
Second order, diagonal	D2, **os-D2**	OV^2	N_E	O^2V^2
	RSPT2, **nD-D2**, **os-nD-D2**	OV^2	1	O^2V^2
Post-second order, diagonal	RSPT3	OV^4	1	V^4
	D3, OVGF	OV^4	N_E	OV^4
	P3, P3+, **Q3, Q3+**	O^2V^3	1	OV^3
	L3, L3+	OV^4	N_E	V^4
	RP3, RQ3	O^2V^3	1	OV^3
	RL3	OV^4	N_IN_E	V^4
	HF-TDA G_0W_0	O^2V^3	N_IN_E	O^2V^2
Non-diagonal	ND2	OV^3	N_I	OV^3
	2ph-TDA	OV^4	N_I	V^4
	3+	O^2V^4	1	V^4
	ADC(3)	O^2V^4	N_I	V^4
	NR2	O^3V^3	1	OV^3
	BD-T1	O^2V^4	BD[d]	V^4

[a]O = # occupied orbitals, V = # virtual orbitals, N_E = energy iterations, N_I = inverse-expansion iterations.
[b]ERI = electron repulsion integrals.
[c]New methods in **bold type**.
[d]The O^3V^4 arithmetic bottleneck occurs in the BD, coupled-cluster reference calculation; the subsequent electron-propagator bottleneck is OV^4.

N_E is most often equal to three. In the RP3, RQ3, RL3 and HF-TDA G_0W_0 methods, expansion of an energy-dependent, inverse matrix is postulated to require N_I steps. N_I may increase when *E* approaches a zeroth order shakeup energy defined by differences of HF orbital energies. In RL3 and HF-TDA G_0W_0 calculations, the number of bottleneck steps may be estimated to be N_IN_E. For P3, P3+, Q3, Q3+, RP3 and RQ3, bottleneck contractions occur just once in the evaluation of the energy-independent matrix of second-order couplings between h and 2hp operators. Q3+ bottleneck steps are therefore less numerous than their counterparts in HF-TDA G_0W_0. Diagonal methods that include post-second-order terms are juxtaposed in Table 1.

A sixth-power, O^2V^4 arithmetic scaling factor is shared by the non-diagonal 3+, ADC(3) (with the Schirmer-Angonoa treatment of the constant self-energy terms[51]) and BD-T1[62–66] self-energy matrices; the same factor applies to the ground-state calculation that is necessary in IP-EOM-CCSD.[67] NR2's lower, O^3V^3 bottleneck arises from second-order couplings between h and 2hp operators that require no iterations. All sixth-power arithmetic bottlenecks stem from treatments of reference-state correlation. The iterative nature of the ADC(3) method's bottleneck is a consequence of its constant (energy-independent) self-energy terms. It is possible to avoid some energy-independent sixth-power steps in 3+ and ADC(3) by adding additional fifth-power contractions to iterative, matrix-vector multiplications that involve the h-2hp couplings mentioned above. Iterative diagonalization algorithms enable ND2 and 2ph-TDA respectively to have fourth-power and fifth-power arithmetic scaling factors.[68,69]

A standard data set of IP-EOM-CCSDT/cc-pVTZ results spans 33 closed-shell molecules with H, C, N, O and F nuclei and 170 valence VIEs; the accompanying IP-EOM-CCSDTQ subset has 11 molecules and 34 VIEs.[42] As the mean absolute error of IP-EOM-CCSDT with respect to IP-EOM-CCSDTQ in the latter set is only 0.03 eV, these data are excellent approximations to FCI results in the cc-pVTZ basis. Some of the final states correspond to inner-valence DOs whose probability factors (Π, see Eq. 21) are likely to approach the 0.85 criterion of applicability that customarily applies to diagonal self-energy methods.

In Table 2, mean absolute errors (MAEs), their standard deviations and maximum absolute errors with respect to the IP-EOM-CCSDT standards are displayed for the self-energy approximations that require evaluation of no terms above third order in the fluctuation potential. Cases where one of these diagonal methods has a probability factor below 0.85 are excluded from the error calculations. 116 VIEs therefore remain. Bold type denotes new methods in the second column and the best combinations of accuracy and efficiency in the third column.

The 1.6 eV MAE of Koopmans's theorem (KT) is reduced to 0.6 eV at the D2 level, but even greater improvements occur with os-D2 and os-nD-D2 (MAEs of 0.3 eV) without the introduction of adjustable parameters. Whereas D2 tends to underestimate VIEs even with saturated basis sets,[21–23] os-D2 and os-nD-D2 have mean signed errors that are positive

Table 2 Errors (eV) with respect to cc-pVTZ/IP-EOM-CCSDT standards[42] for 116 IEs.

Σ(E) Classification	**Method**[a]	**Mean absolute error**[b]	**Standard deviation**	**Maximum absolute error**
Uncorrelated	KT	1.595	0.857	3.780
Second order, diagonal	D2	0.601	0.349	1.389
	os-D2	0.301	0.223	1.838
	nD-D2	0.633	0.363	1.442
	os-nD-D2	**0.289**	0.220	1.817
Post-second order, diagonal	D3	0.724	0.462	2.324
	OVGF-A	0.242	0.213	1.739
	OVGF-B	0.367	0.265	1.900
	OVGF-C	0.382	0.271	1.933
	P3	0.267	0.202	1.623
	P3+	0.166	0.160	1.479
	Q3	0.246	0.204	1.624
	Q3+	**0.147**	0.155	1.480
	L3	0.227	0.201	1.648
	L3+A	**0.122**	0.154	1.418
	L3+B	**0.125**	0.158	1.505
	L3+C	**0.126**	0.158	1.504
Non-diagonal	ND2	0.632	0.468	3.081
	2ph-TDA	0.634	0.198	1.384
	3+	0.398	0.285	1.776
	ADC(3)	0.370	0.257	1.712
	NR2	0.113	0.135	1.166
	BD-T1	0.140	0.178	1.554
	IP-EOM-CCSD	0.174	0.165	1.593

[a]New methods in **bold type**.
[b]Optimal methods in **bold type**.

instead of negative. P3+ and Q3+ achieve additional gains in accuracy (i.e., MAEs of 0.17 and 0.15 eV) over the os-D2 and os-nD-D2 methods at the reasonable cost of non-iterative O^2V^3 steps and the employment of transformed integrals with three virtual indices. (Storage of the latter subset of ERIs can be avoided with a semi-direct algorithm.[70,71]) These added tasks in P3+ and Q3+ are faster than the preceding transformation of ERIs and yield more accurate results than D3 and the three OVGF variants. The three L3+ versions have similar computational requirements as the OVGF methods, clearly superior accuracy (MAE = 0.12–0.13 eV) and produce slight improvements over P3+ and Q3+ at the cost of higher arithmetic scaling and the need for integrals with four virtual indices. Additional 2ph terms in the L3+ methods diminish overestimates of VIEs with more accurate descriptions of initial-state correlation. The approximately renormalized P3+, Q3+ and L3+ methods lower errors by about 0.1 eV with respect to their P3, Q3 and L3 precursors with negligible additional effort. Insignificant differences in MAEs among the three L3+ alternatives suggest a slight preference in the interest of consistency for the B version, which has the same approximate renormalization scheme as Q3+ and P3+. ND2 is slightly worse than D2. The non-diagonal 2ph-TDA, 3+ and ADC(3) methods are less efficient and accurate than P3+, Q3+ and the three L3+ versions. NR2 and BD-T1 are the only competitive non-diagonal methods for this data set and attain better accuracy than IP-EOM-CCSD. Even the approximately renormalized, diagonal methods, P3+, Q3+ and L3+ are slightly better than IP-EOM-CCSD.

When probability factors are above 0.85, the os-nD-D2, Q3+ and L3+ methods offer the highest accuracy for their respective levels of computational effort, OV^2, O^2V^3 *and* OV^4. The non-diagonal NR2, BD-T1 and IP-EOM-CCSD methods remain useful when DOs that are not proportional to HF orbitals are needed (compare Eqs. 22 and 25). To facilitate comparisons with respect to efficiency and accuracy, Fig. 1 exhibits the errors of low-order and approximately renormalized diagonal self-energies according to their bottleneck arithmetic scaling factors. The best combinations of accuracy and efficiency (os-nD-D2, Q3+ and L3+B) are compared with older methods such as the three OVGF versions, ADC(3), P3+, NR2, BD-T1 and IP-EOM-CCSD.

The set of explicitly renormalized self-energies comprises HF-TDA G_0W_0, RP3, RQ3 and RL3, where the latter three methods respectively are (ring-ladder) renormalized extensions of P3, Q3 and L3. Cases where the expansion of the inverted matrices that occur in these four methods does not converge are excluded in addition to the cases with probability factors less than 0.85. 80 VIEs therefore remain in Table 3.

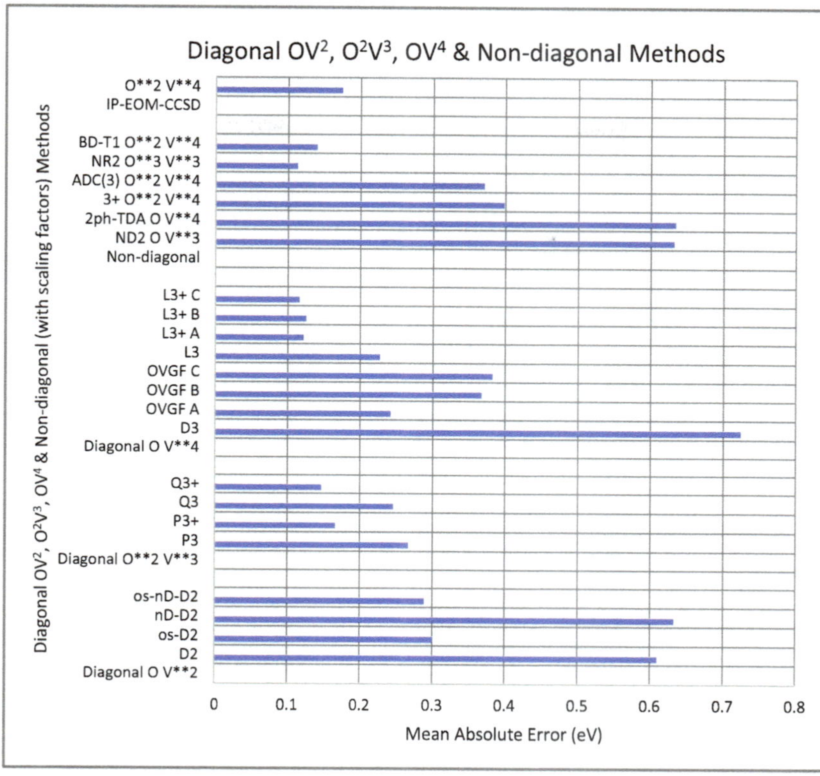

Fig. 1 Mean absolute errors[a] (eV) and arithmetic scaling factors[b] of low-order and approximately renormalized diagonal vs non-diagonal Self-Energies for 116 VIEs with $\Pi > 0.85$. [a]Errors are calculated with respect to cc-pVTZ/IP-EOM-CCSDT results.[42] [b]Powers of occupied (O) and virtual (V) orbitals are preceded by **.

Most errors in this more restricted data set are lower. The MAE of os-nD-D2 is 0.23 eV. HF-TDA G_0W_0 realizes slightly greater accuracy: MAE = 0.21 eV. (An MAE of 0.24 eV for the lowest CCSD(T) IEs of the GW100 data set was reported recently.[33]) The MAEs of P3+, Q3+ and L3+B (0.13, 0.12, 0.09 eV, respectively) decline in RP3, RQ3 and RL3 (0.12, 0.10, 0.08 eV, respectively). Improved accuracy (by approximately 0.1 eV) in Q3+ and RQ3 relative to HF-TDA G_0W_0 may be obtained at the cost of an OV^3 integral storage requirement, but the number of bottleneck steps is reduced (see Table 1). The most accurate diagonal methods, L3+B and RL3, have the same arithmetic bottleneck scaling as that of D3 and OVGF. The L3+ methods have fewer bottleneck steps than RL3. L3+B and RL3 have lower MAEs than all other methods for this data set. These two methods bear close resemblance to third-order formulae that

Table 3 Errors (eV) with respect to cc-pVTZ/IP-EOM-CCSDT standards[42] for 80 IEs.

Σ(E) Classification	Method[a]	Mean absolute error[b]	Standard deviation	Maximum absolute error
Uncorrelated	KT	1.421	0.834	2.839
Second order, diagonal	D2	0.628	0.349	1.389
	os-D2	0.237	0.151	0.643
	nD-D2	0.660	0.363	1.442
	os-nD-D2	**0.226**	0.148	0.634
Post-second order, diagonal	HF-TDA G_0W_0	0.205	0.141	0.592
	D3	0.647	0.424	1.503
	OVGF-A	0.193	0.137	0.515
	OVGF-B	0.313	0.208	0.734
	OVGF-C	0.320	0.212	0.716
	P3	0.213	0.147	0.762
	P3+	0.134	0.087	0.425
	Q3	0.201	0.144	0.710
	Q3+	**0.117**	0.077	0.370
	L3	0.174	0.129	0.630
	L3+A	0.110	0.086	0.404
	L3+B	**0.090**	0.063	0.310
	L3+C	0.090	0.064	0.305
	RP3	0.119	0.090	0.368
	RQ3	**0.104**	0.078	0.306
	RL3	**0.077**	0.062	0.324
Non-diagonal	ND2	0.670	0.427	3.021
	2ph-TDA	0.626	0.203	1.384
	3+	0.360	0.236	0.792
	ADC(3)	0.323	0.205	0.761
	NR2	0.101	0.072	0.335
	BD-T1	0.092	0.063	0.243
	IP-EOM-CCSD	0.119	0.071	0.352

[a]New methods in **bold type**.
[b]Optimal methods in **bold type**.

can be derived from cluster perturbation theory of excitation energies.[72–76] Among the non-diagonal methods, only the NR2 and BD-T1 self-energies are competitive for valence VIEs. Their sixth-power arithmetic dependence is compensated when improved DOs are needed.

Inspection of unsigned errors reveals that most of the methods in Table 3 are more likely to overestimate than to underestimate IEs. Only D2, nD-D2, L3+A, RP3, RQ3, RL3, ND2, 2ph-TDA and NR2 produce negative mean signed errors. Mean unsigned errors are especially close to zero for RQ3, RL3 and NR2 (−0.04, −0.02 and −0.03 eV, respectively).

Fig. 2 summarizes error data generated by five, new diagonal self-energy methods (os-nD-D2, Q3+, L3+B, RQ3 and RL3), HF-TDA G_0W_0, two,

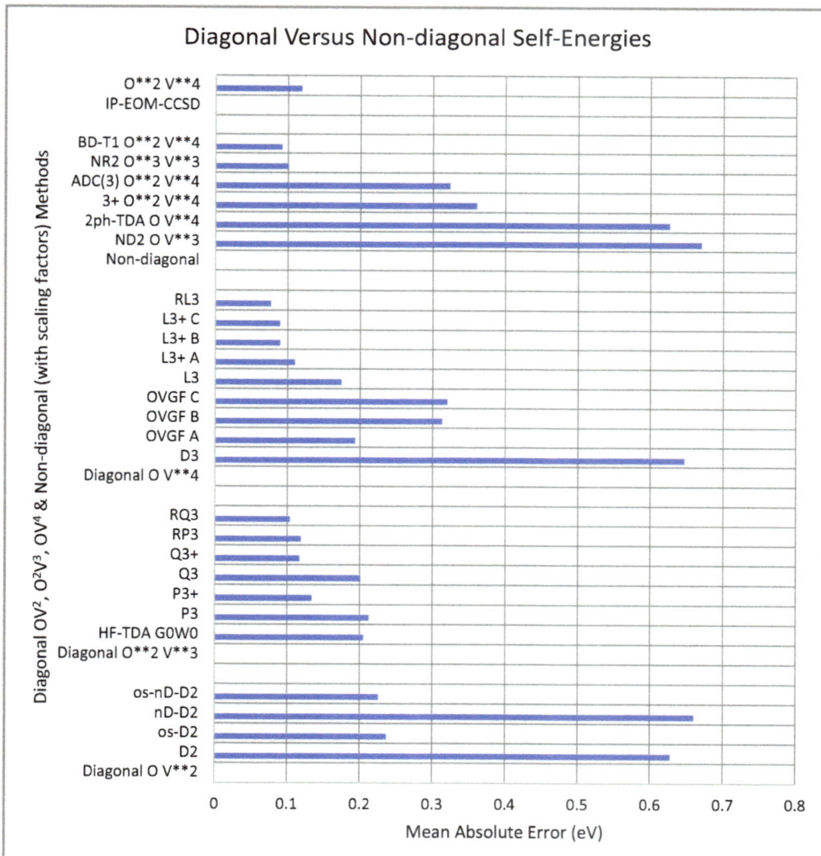

Fig. 2 Mean absolute errors[a] (eV) and arithmetic scaling factors[b] of diagonal vs non-diagonal self-energies for 80 VIEs with $\Pi > 0.85$ and convergent renormalizations. [a]Errors are calculated with respect to cc-pVTZ/IP-EOM-CCSDT results.[42] [b]Powers of occupied (O) and virtual (V) orbitals are preceded by **.

previously developed non-diagonal methods (NR2 and BD-T1) and IP-EOM-CCSD for the 80 VIEs where probability factors are greater than or equal to 0.85 for all methods and where ring-ladder renormalizations are converged via an order-by-order expansion of an inverse matrix. *For this data set, os-nD-D2, RQ3 and RL3 provide the highest accuracy for a given bottleneck scaling factor (OV^2, O^2V^3 and OV^4, respectively), but Q3+ and L3+B obtain similarly accuracy in the last two classifications of arithmetic efficiency*. L3+B has fewer OV^4 contractions than RL3. The P3+, Q3+ and L3+ methods provide fast, accurate estimates of ring-ladder renormalizations.

2. Approaching exact results with the extended Koopmans theorem

The extended Koopmans theorem[77–80] (EKT) is the foundation of attempts to calculate electron removal energies that are based on correlated wavefunctions and the solution of generalized secular equations with Fock and overlap matrices that generalize those assumed in the original Koopmans theorem.[9] (There has been extensive debate on the EKT's exactitude.[81–92]) In this section, the advantages of a new approach, the complete-active-space EKT (CAS-EKT) method,[93] are discussed.

The characteristic assumption of the EKT is expansion of the *k-th*, final state of a system with N − 1 electrons as a linear combination of annihilators acting on the initial, N-electron state, Ψ^N. The annihilation operator thus expressed,

$$r_k = \sum_j c_{jk} b_j, \tag{49}$$

defines the EKT orbital, $r_k(x)$, which connects the *k-th* final state to the initial state such that

$$\Psi_k^{N-1} = r_k \Psi^N. \tag{50}$$

Electron binding energies therefore equal differences between the initial state energy,

$$E^N = \langle \Psi^N | H \Psi^N \rangle, \tag{51}$$

and final state energies labeled by k,

$$E_k^{N-1} = \langle \Psi_k^{N-1} | H \Psi_k^{N-1} \rangle. \tag{52}$$

The hole-configuration-interaction (HCI) Ansatz is employed in the minimization with respect to the c coefficients of the Ritz quotient,

$$E_k^{N-1}(\mathbf{c}) = \frac{\left\langle \Psi^N | r_k^\dagger H r_k | \Psi^N \right\rangle}{\left\langle \Psi^N | r_k^\dagger r_k | \Psi^N \right\rangle}. \tag{53}$$

In terms of the elements of the Hamiltonian ($\boldsymbol{H}$) and one-electron reduced-density ($\boldsymbol{P}$) matrices, final-state energies read

$$E_k^{N-1} = \frac{\mathbf{c_k^\dagger H c_k}}{\mathbf{c_k^\dagger P c_k}}, \tag{54}$$

where

$$H_{pq} = \left\langle \Psi^N | a_p^\dagger H a_q | \Psi^N \right\rangle \tag{55}$$

$$P_{pq} = \left\langle \Psi^N | a_p^\dagger a_q | \Psi^N \right\rangle. \tag{56}$$

The three-electron, reduced-density matrix that appears in the Hamiltonian matrix elements may be avoided by making an additional assumption: Ψ^{N} is an eigenfunction of H. Subtraction of

$$\left\langle \Psi^N | r_k^\dagger r_k H | \Psi^N \right\rangle = E^N \left\langle \Psi^N | r_k^\dagger r_k | \Psi^N \right\rangle \tag{57}$$

from the Ritz quotient for final states produces an electron removal energy[77,78]

$$I_k^{EKT} = E_k^{N-1} - E^N = \frac{\left\langle \Psi^N | r_k^\dagger [H, r_k] | \Psi^N \right\rangle}{\left\langle \Psi^N | r_k^\dagger r_k | \Psi^N \right\rangle}. \tag{58}$$

Optimization of the c coefficients used to define r_k leads to the EKT secular equation,

$$\mathbf{K c}_k = I_k^{EKT} \mathbf{P c}_k, \tag{59}$$

where the elements of the so-called Koopmans matrix read

$$K_{pq} = \left\langle \Psi^N | a_p^\dagger [H, a_q] | \Psi^N \right\rangle. \tag{60}$$

The two-electron reduced density matrix of Ψ^N suffices to evaluate $\boldsymbol{K}$, which is Hermitian provided that the N-electron wavefunction is stationary with respect to variations within the space of natural orbitals that have non-vanishing occupation numbers. The negative of $\boldsymbol{K}$ is a generalized Fock matrix that also appears in non-Dyson electron propagators whose poles are electron removal energies.[6]

The difference between the HCI variational energy for final states and the initial-state's energy expectation value, I_k^{HCI}, is not in general equal to the EKT result. In fact, the discrepancy between the two estimates may be expressed as

$$I_k^{HCI} = I_k^{EKT} + \left\langle \Psi^N | r_k^\dagger r_k \left(H - E^N \right) | \Psi^N \right\rangle. \tag{61}$$

The discrepancy is consequence of assuming that Ψ^N is an eigenfunction of H, but this term also vanishes for Roothaan-Hartree-Fock wavefunctions[94] in an incomplete, finite basis. The equivalence of HCI and EKT results holds for the closed-shell case and may be extended to various classes of open-shell HF wavefunctions.[95–97]

Complete-active-space (CAS) configuration-interaction (CI) wavefunctions also force the discrepancy to vanish. The Slater determinants in CAS wavefunctions,

$$\Psi_{CAS}^N = \sum_r C_r \Phi_r^N, \tag{62}$$

have the same doubly occupied core, or inactive, orbitals and comprise all possible assignments of the remaining, N_{active}, electrons to M_{active} orbitals in the active orbital space. Variational optimization of CI coefficients, $\boldsymbol{C}$, requires that

$$\langle \Phi_r^N | \left(H - E_{CAS}^N \right) | \Psi^N \rangle = 0 \tag{63}$$

for a CAS (N_{active},M_{active}) wavefunction. For p and q within the active space,

$$\left\langle a_q^\dagger a_p \Psi^N | \left(H - E_{CAS}^N \right) | \Psi^N \right\rangle = \left\langle \Psi^N | a_p^\dagger a_q \left(H - E_{CAS}^N \right) | \Psi^N \right\rangle = 0, \tag{64}$$

for each $a_q^\dagger a_p \Psi^N$ is a linear combination of the active determinants. The discrepancy between HCI and EKT results therefore vanishes when the EKT orbital is expressed in terms of active orbitals only. This conclusion also applies to FCI wavefunctions wherein the active space comprises all orbitals that occur in the second-quantized expression for H.

The HCI wavefunction is not exact, even in the case of an exact Ψ^N for $N > 2$. Every HCI wavefunction has a (i.e., EKT) spin-orbital with a vanishing occupation number (i.e., zero expectation value of its number operator) and therefore the dimension of the spin-orbital space needed for the final-state is lower than the dimension needed for the initial state.[93] The HCI wavefunction cannot equal FCI for the final state when $N > 2$, otherwise any orthonormal basis chosen for the initial state would have to be larger than needed for the final state. The same objection applies to removal of additional electrons that results in states with N-2 or fewer electrons. Only when $N = 2$ can one ascribe exactness to the EKT result, for the final state has only one non-zero occupation number.

An exact electron removal energy for a CAS(N,M) wavefunction may be defined as

$$\Delta E^{CAS-CI} = E^{N-1}_{CAS-CI} - E^{N}_{CAS}. \tag{65}$$

The final-state energy corresponds to a CAS($N-1$,M) calculation performed with the CAS(N,M) spin-orbitals of the initial-state. Whereas this final-state wavefunction is variationally superior to the corresponding HCI wavefunction,

$$I^{CAS-EKT} \geq \Delta E^{CAS-CI}. \tag{66}$$

The equality holds for CAS(N,N/2) (i.e., the HF case) and for $N = 2$.

Despite the formal discrepancy between EKT results based on a CAS CI reference state and FCI results obtained within the active orbital space, it is possible for this approximation to approach the latter limit with arbitrarily small errors.[81,84,85,98–103] This goal is achieved by augmenting the basis with diffuse (i.e., remote) functions allowed by symmetry considerations to contribute to the EKT orbital.[93] Because the FCI wavefunction of the final state decays faster than its initial-state counterpart,[92] the additional functions can be chosen such that their contributions to the final state are arbitrarily small relative to their contributions to the initial state. Should the removal of a function from Ψ^N have an arbitrarily small effect on the final-state, then the EKT wavefunction for the final state can be arbitrarily close to its FCI limit. With increasing diffuseness of the additional functions, their weights in the EKT orbital increase. In numerical tests performed on the Be atom with FCI reference wavefunctions and diffuse s or Cartesian 6d functions, discrepancies between EKT and exact results were easily reduced to millihartrees.[93] Whereas the diffuse functions contribute little to the reference wavefunction, they can make major contributions to the EKT orbital.

The EKT orbital functions as a probe of electronic structure, but it is only indirectly related to orbitals with physical significance in electron removal experiments, such as the DO,[4] which may be re-expressed as

$$\phi_k^{Dyson}(x_1) = N^{\frac{1}{2}} \int \Psi^N(x_1, x_2, x_3, \cdots x_N) \left[\Psi_k^{N-1}(x_2, x_3, x_4, \cdots x_N)\right]^* dx_2 dx_3 dx_4, \cdots, dx_N. \tag{67}$$

The DO may be inferred from the EKT orbital and the 1-RDM:

$$\phi_k^{Dyson}(x) = \int \rho(x, x') r_k(x') dx'. \tag{68}$$

From the natural-orbital (χ) expansion of the 1-RDM,

$$\rho(x, x') = \sum_r n_r \chi_r(x) \chi_r^*(x'), \tag{69}$$

it follows that

$$n_s = \frac{\langle \chi_s | \phi_k^{Dyson} \rangle}{\langle \chi_s | r_k \rangle}. \tag{70}$$

Therefore, a natural orbital, χ_s, with a small occupation number, n_s, can make a small contribution to the DO or a large contribution to the EKT orbital.

Solution of the CAS-EKT secular equations produces a set of eigenvalues that equal those of HCI shifted by the reference-state's variational energy.[93] The appearance of small eigenvalues in the $\boldsymbol{P}$ matrix need not pose numerical difficulties unless a symmetric orthogonalization involving $\boldsymbol{P}^{-1/2}$ is invoked.[98]

The CAS-EKT approximation is equivalent to HCI provided that the EKT orbital is confined to the active orbital space. For $N > 2$, electron removal energies calculated with CAS-EKT are not equal to the difference between the reference state's FCI variational energy and its final-state counterpart. In previous works, the approach of the EKT result to exact results was held to obtain in the limit of an exact solution of the time-independent Schrödinger equation.[91,92] A more general conclusion follows from the CAS-EKT method, whose electron removal energy approaches the ΔFCI result in an incomplete basis provided that the latter set contains a function that contributes to the final-state wavefunction much less than to the reference-state wavefunction, Ψ^N. This conclusion does not arise from the asymptotic behavior of Ψ^N and therefore is applicable when basis functions with incorrect limits (e.g., Gaussian functions) are employed.[80,82,104,105]

3. Interpretation of Δ-self-consistent-field calculations

Since the 1920s, Slater determinants have occupied a prominent position in the theory and practice of quantum chemistry.[106] A determinant with elements defined in terms of spin-orbitals fulfills Pauli's exclusion principle and provides a variational Ansatz for one of quantum chemistry's most powerful ideas: the self-consistent-field (SCF). Slater determinants built with orthonormal spin-orbitals provide a convenient basis for many varieties of configuration-interaction wavefunctions.[107] Even theories that avoid the evaluation of many-electron wavefunctions[1,108] typically impose constraints inspired by the properties of Slater determinants.

Hartree-Fock or Kohn-Sham calculations of electron binding energies, excitation energies and corresponding transition intensities may be based on the optimization of spin-orbitals for each state of interest. ΔSCF methods[109] have enjoyed enduring success in assigning and interpreting core photoelectron spectra of isolated molecules, solids and surfaces.[110,111] Generalized variational principles[112] and improved algorithms that prevent convergence to ground states have been invoked in recent efforts to apply SCF methods to excited states.[113,114] Transitions that involve strong relaxation of spin-orbitals between initial and final states are natural targets for ΔSCF techniques. Core IEs, their spectral satellites that involve several qualitative changes in spin-orbital occupation numbers, charge-transfer transitions and double electron binding energies measured in Auger spectroscopy[115] are cases where calculations must give ample scope to the effects of spin-orbital relaxation.

Constraints applied to the SCF that pertain to point-group symmetry, overlap matching between consecutive iterations or orbital localization can succeed in producing reasonable descriptions of excited states.[116,117] Whereas the usual criterion of self-consistency is fulfillment of the Brillouin condition,[118] there are some cases where this standard cannot be applied.[119,120] Strong relaxation may result in initial-state and final-state canonical spin-orbitals that bear little resemblance to each other despite satisfying the Brillouin and Koopmans[9] conditions. Formally canonical unrestricted HF spin-orbitals[121,122] also may require circumspect interpretation due to spin contamination. In many cases, comparisons of initial-state and final-state canonical spin-orbitals reveal a need for analytical tools that reduce the copious information generated by orbital relaxation to a manageable form.

Alternative sets of one-electron functions, founded on the invariance of Slater determinants with respect to linear transformations of their occupied spin-orbitals,[123] often provide a clear analysis of electronic structure. The corresponding orbitals of Amos, Hall and Löwdin[124,125] are produced by unitary transformations of α and β spin-orbitals that involve a singular-value decomposition of their spatial overlap matrix. Orbital occupation numbers obtained from a spin-integrated, unrestricted HF wavefunction have successfully served as inclusion criteria in unrestricted-natural-orbital, complete-active-space wavefunctions.[126] These data have a paired structure and the density matrix in natural form therefore resembles that of a perfect-pairing, generalized-valence-bond wavefunction.[127] Succinct generalizations[128] of the Slater-Condon rules[106,129] have been generated with corresponding orbitals. The latter functions also have facilitated energetic analysis of rotational barriers[128] and clarified substituent[130] and relaxation[131] effects that appear in photoelectron spectra.

There is a close relationship between corresponding orbitals and the natural orbitals of an unrestricted, determinantal wavefunction. The latter functions have occupation numbers that may equal unity or may occur in $1 \pm t$ pairs, where t is a nonvanishing spatial overlap integral between α and β corresponding orbitals. Corresponding orbitals that have no nonvanishing overlaps are equal to natural orbitals with unit occupation numbers. Normalized sums or differences of pairs of corresponding orbitals with nonvanishing overlaps equal natural orbitals with paired occupation numbers. Whereas the Slater determinant Ansatz requires that the α and β density matrices have eigenvalues that equal zero or unity, the procurement of natural orbitals is equivalent to diagonalizing the sum of two idempotent matrices.

The overlaps of paired orbitals are the chief criteria of similarity when corresponding orbitals are calculated to enable comparisons between Slater determinants or between the α and β orbitals of a different-orbitals-for-different-spins wavefunction. An unpaired corresponding orbital whose occupation number changes from zero to unity or vice versa may be understood as having been added or removed in ΔSCF calculations of electron-binding energies. By examining differences of the corresponding idempotent matrices, a more thorough analysis of differences between states, molecules or spin-component densities can be facilitated.

Two theorems proven recently by Ortiz and Zalik (OZ) that determine the eigenvalues of the sum of two idempotent matrices (and that therefore confirm the conclusions of Amos and Hall[124]) pave the way for two additional theorems that perform the same function for the difference of two idempotent matrices. Corresponding orbitals are essential intermediates in deriving the eigenvalues in both cases.

The natural occupation numbers of an unrestricted Slater determinant equal the eigenvalues of the *sum* of the idempotent α and β density matrices. OZ Theorem 1 is concerned with the eigenvalue spectrum of the overlap matrix between α and β corresponding orbitals. The eigenvalues of the sum of the α and β density matrices are shown to have the same values in OZ Theorem 2. Relationships between α–β corresponding orbital overlaps and natural orbitals, their occupation numbers, density matrices and the S^2 criterion of spin contamination follow from these theorems.

Differences of idempotent matrices are considered in OZ Theorems 3 and 4, which employ arguments that resemble those of OZ Theorems 1 and 2. The eigenvalue spectrum of a matrix with off-diagonal blocks that contain overlaps between two sets (e.g., α and β, initial-state and final-state) of corresponding orbitals is disclosed by OZ Theorem 3. Eigenvalues of the difference of idempotent density matrices are shown to be the same in OZ Theorem 4. Overlaps between sets of corresponding orbitals are related to natural orbitals, occupation numbers and density matrices that pertain to *differences* between initial and final states of ionization or excitation. Because spin densities also pertain to differences, these results also are relevant to the analysis of spin-dependent properties.

ΔSCF calculations also may be used to calculate DOs, their probability factors, Fukui functions (ground-state, electron-density differences) and changes in orbital occupation numbers that accompany single-determinant descriptions of ionizing transitions. Overlaps between sets of corresponding orbitals also appear in the formulae of Table 6 for probability factors and Fukui functions. The present results also are applicable to electron-conserving excitations and to detachments and attachments of more than one electron.

3.1 Eigenvalues of the sum of two idempotent matrices

OZ Theorems 1 and 2 confirm the principal conclusions of Amos and Hall[124] and also apply to the eigenvalues of any sum of two idempotent matrices. They also suggest a solution to the subsequent problem that pertains to differences of idempotent matrices.

Amos and Hall[124] analyzed unrestricted HF wavefunctions, which had been introduced only a few years before,[121,122] wherein N_α α orbitals and

N_β β orbitals (with $N_\alpha \geq N_\beta$) are expanded in an orthonormal basis of m functions (ω_s, $s=1, \ldots, m$) according to

$$\psi_r^\alpha = \sum_{s=1}^{m} \omega_s A_{sr}, r = 1, \ldots, N_\alpha \tag{71}$$

$$\psi_r^\beta = \sum_{s=1}^{m} \omega_s B_{sr}, r = 1, \ldots, N_\beta. \tag{72}$$

The density matrices corresponding to the α and β spin-orbitals, **P** and **Q**,

$$\mathbf{P} = \mathbf{A}\mathbf{A}^\dagger \tag{73}$$

$$\mathbf{Q} = \mathbf{B}\mathbf{B}^\dagger \tag{74}$$

are idempotent, i.e.,

$$\mathbf{P}^2 = \mathbf{P} \tag{75}$$

$$\mathbf{Q}^2 = \mathbf{Q} \tag{76}$$

and their traces yield the number of electrons of each spin type, where

$$Tr\mathbf{P} = p = N_\alpha \tag{77}$$

$$Tr\mathbf{Q} = q = N_\beta. \tag{78}$$

The orthonormality that obtains within each of the two sets of spin-orbitals implies that

$$\mathbf{A}^\dagger\mathbf{A} = \mathbf{I}_p \tag{79}$$

$$\mathbf{B}^\dagger\mathbf{B} = \mathbf{I}_q, \tag{80}$$

but the non-orthogonality between the two sets of orbitals produces a rectangular overlap matrix, **S**, so that

$$\mathbf{A}^\dagger\mathbf{B} = \mathbf{S}. \tag{81}$$

Now let $\mathbf{C}$ be a matrix with $p+q$ columns:

$$\mathbf{C} = [\mathbf{A}\ \mathbf{B}]. \tag{82}$$

Therefore, the product $\mathbf{C}^\dagger\mathbf{C}$ has a block structure given by

$$\mathbf{C}^\dagger\mathbf{C} = \begin{bmatrix} \mathbf{I}_p & \mathbf{S} \\ \mathbf{S}^\dagger & \mathbf{I}_q \end{bmatrix}. \tag{83}$$

The corresponding orbitals are expressed as unitary transformations of the α and β orbitals,

$$\phi_r^\alpha = \sum_{s=1}^{N_\alpha} \psi_s^\alpha U_{sr} \tag{84}$$

$$\phi_r^\beta = \sum_{s=1}^{N_\beta} \psi_s^\beta V_{sr}, \tag{85}$$

which are produced in the singular-value decomposition (SVD) of the α-β overlap matrix,

$$\mathbf{U}^\dagger \mathbf{S} \mathbf{V} = \mathbf{T}. \tag{86}$$

T is the rectangular, diagonal matrix of overlaps between α and β corresponding orbitals. As a result of these transformations, there are *p-q* corresponding orbitals in the α set that are orthogonal to the β orbitals and *q* pairs of α and β corresponding orbitals with maximized overlaps. Elements of **T** read

$$T_{rs} = \delta_{rs} t_{rr}. \tag{87}$$

In the SVD of **T**, nonzero t_{ii}'s are the positive square roots of the nonzero eigenvalues of $\mathbf{T}\mathbf{T}^\dagger$, which are identical to the nonzero eigenvalues of $\mathbf{T}^\dagger\mathbf{T}$. Now let the number of diagonal elements with absolute values of zero and one be denoted respectively by ℓ and δ. In the corresponding orbital basis, $\mathbf{C}^\dagger\mathbf{C}$ has fewer non-vanishing elements, for

$$\begin{bmatrix} \mathbf{U}^\dagger & \mathbf{0} \\ \mathbf{0} & \mathbf{V}^\dagger \end{bmatrix} \mathbf{C}^\dagger \mathbf{C} \begin{bmatrix} \mathbf{U} & \mathbf{0} \\ \mathbf{0} & \mathbf{V} \end{bmatrix} = \begin{bmatrix} \mathbf{I}_p & \mathbf{T} \\ \mathbf{T}^\dagger & \mathbf{I}_q \end{bmatrix}. \tag{88}$$

The eigenvalues of $\mathbf{C}^\dagger\mathbf{C}$ and their multiplicities (i.e., degeneracies) are given by OZ Theorem 1. $\mathbf{C}\mathbf{C}^\dagger$ has the same eigenvalues and multiplicities and $m-(p+q)$ eigenvalues that equal zero as well.[124] Since

$$\mathbf{C}\mathbf{C}^\dagger = \mathbf{A}\mathbf{A}^\dagger + \mathbf{B}\mathbf{B}^\dagger = \mathbf{P} + \mathbf{Q}, \tag{89}$$

the eigenvalues of the sum of the idempotent α and β orbital-density matrices can be classified according to Table 4, where the last two columns pertain to the paired eigenvalues that are not equal to those of the second, third and fourth columns.

Table 4 Eigenvalues of **P**+**Q** and their multiplicities.

Eigenvalue	**1**	**0**	**2**	**1+t_{rr}**	**1 − t_{rr}**
Multiplicity	$p-q+2\ell$	$m-p-q+\delta$	δ	$q-\ell-\delta$	$q-\ell-\delta$

Deviations of the corresponding-orbital overlaps from unity are associated with spin contamination. In the corresponding basis, the difference between the exact expectation value of S^2 and the unrestricted, HF expectation value reads

$$\left\langle S^2 \right\rangle_{exact} - \left\langle S^2 \right\rangle_{UHF} = N_\beta - \sum_{r=1}^{N_\beta} t_{rr}^2. \tag{90}$$

Deviations from the exact value vanish only when all the overlaps equal unity.

The normalized sums and differences of pairs of overlapping corresponding orbitals,

$$\mu_r(1) = (2 + 2t_{rr})^{-\frac{1}{2}}\left[\phi_r^\alpha(1) + \phi_r^\beta(1)\right] \tag{91}$$

$$\nu_r(1) = (2 - 2t_{rr})^{-\frac{1}{2}}\left[\phi_r^\alpha(1) - \phi_r^\beta(1)\right], \tag{92}$$

are present in the spin-integrated density matrix,

$$\rho(1|1') = \sum_{r=1}^{p-q} \phi_r^\alpha(1)\phi_r^{\alpha*}(1') + \sum_{r=1}^{q}\left[(1 + t_{rr})\mu_r(1)\mu_r^*(1')\right.$$
$$\left. + (1 - t_{rr})\nu_r(1)\nu_r^*(1')\right], \tag{93}$$

where the φ, μ and ν functions constitute an orthogonal set of natural orbitals. There are $N_\alpha - N_\beta$ unpaired, high-spin contributions to the first summation from corresponding orbitals. N_β pairs of natural orbitals reside in the second term.

3.2 Eigenvalues of the difference of two idempotent matrices

OZ Theorems 3 and 4 employ similar strategies as those of the previous section and apply to the eigenvalues of any difference of two idempotent matrices.

After defining $\mathbf{C}_1$ and $\mathbf{C}_2$ according to

$$\mathbf{C}_1 = [\mathbf{A} - \mathbf{B}] \tag{94}$$

and

$$\mathbf{C}_2 = \begin{bmatrix} \mathbf{A}^\dagger \\ \mathbf{B}^\dagger \end{bmatrix}, \tag{95}$$

$\mathbf{C_2C_1}$ has the following block structure:

$$\mathbf{C}_2\mathbf{C}_1 = \begin{bmatrix} \mathbf{I}_p & -\mathbf{S} \\ \mathbf{S}^\dagger & -\mathbf{I}_q \end{bmatrix}. \tag{96}$$

Transformation to corresponding orbitals produces

$$\begin{bmatrix} \mathbf{U}^\dagger & \mathbf{0} \\ \mathbf{0} & \mathbf{V}^\dagger \end{bmatrix} \mathbf{C}_2\mathbf{C}_1 \begin{bmatrix} \mathbf{U} & \mathbf{0} \\ \mathbf{0} & \mathbf{V} \end{bmatrix} = \begin{bmatrix} \mathbf{I}_p & -\mathbf{T} \\ \mathbf{T}^\dagger & -\mathbf{I}_q \end{bmatrix}. \tag{97}$$

Eigenvalues of $\mathbf{C_2C_1}$ and their multiplicities are given by OZ Theorem 3. $\mathbf{C_1C_2}$ has the same eigenvalues and multiplicities and $m - (p+q)$ eigenvalues that equal zero as well.[124] Because

$$\mathbf{C}_1\mathbf{C}_2 = \mathbf{AA}^\dagger - \mathbf{BB}^\dagger = \mathbf{P} - \mathbf{Q}, \tag{98}$$

the eigenvalues of the difference of the idempotent α and β orbital-density matrices can be classified according to Table 5, where the last two columns refer to the paired eigenvalues that are not equal to integers. Whereas the results of the previous section pertain to the charge density (i.e., to the sum of the two density matrices), those of this section pertain to the spin density (i.e., to the difference of the two density matrices).

The normalized, symmetrically orthogonalized[132] combinations of pairs of overlapping corresponding orbitals,

$$\kappa_r(1) = a_r\phi_r^\alpha(1) + b_r\phi_r^\beta(1) \tag{99}$$

$$\lambda_r(1) = b_r\phi_r^\alpha(1) + a_r\phi_r^\beta(1), \tag{100}$$

where

$$a_r = \frac{1}{2}\left(\sqrt{1+|t_{rr}|} + \sqrt{1-|t_{rr}|}\right) \tag{101}$$

Table 5 Eigenvalues of $\mathbf{P}-\mathbf{Q}$ and their multiplicities.

Eigenvalue	1	-1	0	$+\sqrt{1-t_{rr}^2}$	$-\sqrt{1-t_{rr}^2}$
Multiplicity	$p-q+\ell$	ℓ	$m-p-q+2\delta$	$q-\ell-\delta$	$q-\ell-\delta$

$$b_r = \frac{1}{2}\left(\sqrt{1+|t_{rr}|} - \sqrt{1-|t_{rr}|}\right), \tag{102}$$

appear in the spin-integrated, density-difference matrix,

$$\Delta\rho(1|1') = \sum_{r=1}^{p-q} \phi_r^{\alpha}(1)\phi_r^{\alpha*}(1') + \sum_{r=1}^{q}\left[\sqrt{1-t_{rr}^2}\kappa_r(1)\kappa_r^*(1') - \sqrt{1-t_{rr}^2}\lambda_r(1)\lambda_r^*(1')\right], \tag{103}$$

where the φ, κ and λ functions constitute an orthogonal set of natural orbitals. The first term has $N_\alpha - N_\beta$ members that involve unpaired, corresponding orbitals. N_β pairs of natural orbitals are present in the second term.

3.3 Dyson orbitals, probability factors and Fukui functions

Differences of initial-state and final-state energies evaluated variationally with Slater determinants may be based on a fixed set of spin-orbitals, as in Koopmans's identity,[9] or may include the effects of orbital relaxation, as in the ΔSCF method.[109] Differences between Koopmans and ΔSCF results arise because of relaxation of the occupied spin-orbitals in the final state.

The DSOs of Eqs. (5) and (6) equal canonical, HF spin-orbitals and the probability factors equal unity when Koopmans's identity is employed. This conclusion is formally valid in cases where an unrestricted HF wavefunction is chosen for the initial state and frozen-orbital approximations are made for final states. Provided that the initial and final states are well approximated by single Slater determinants and spin contamination is low, this formal extension can be physically meaningful. Spin contamination may be eliminated with restricted, open-shell Hartree-Fock (ROHF) wavefunctions and their canonical orbitals that pertain to high-spin final states.[95,96] In these generalizations of Koopmans's identity, overlaps between initial and final spin-orbitals are equal to zero or unity. It follows that there is only one natural ionization orbital[133] (i.e., an eigenfunction of the density-difference matrix) with a non-zero eigenvalue. This approach is inapplicable to most final states of low spin (i.e., those that pertain to lower, occupied orbitals) where more than one Slater determinant is usually needed to construct a spin eigenstate. In such cases, the single-determinant approximation, having components of more than one multiplicity, will correspond to a spin-contaminated final state. For example, detachment of an α electron from the core (K shell) orbital of the doublet (S=½) ROHF wavefunction of ground-state CH_3 with M_S =½ yields a final-state determinant that is half singlet and half triplet. Some definitions of orbital energies correspond to electron binding energies that are averaged over more than one final state.[134]

One may procure DSOs that pertain to ΔSCF calculations by expressing Slater determinants in terms of corresponding spin-orbitals that arise from the SVD of the overlap matrix between initial-state and final-state spin-orbitals. (Note the similarity with procedures discussed above for determining charge or spin densities and the role of overlaps between α and β orbitals of the Slater determinant of a single state.) The DSO is proportional to the unpaired, corresponding spin-orbital, i.e., the natural ionization spin-orbital with an eigenvalue of unity. Information about orbital relaxation is provided by non-integer eigenvalues of differences of idempotent, orbital-density matrices. The N − 1 or N integrations that appear in the two definitions (see Eqs. 5 and 6) of the DSOs yield a non-vanishing term only when there are no permutationally induced mismatches between the paired, corresponding spin-orbitals. The DSO that emerges from these integrations therefore equals the unpaired, corresponding spin-orbital times the product of the overlap matrices of the paired spin-orbitals. For final states with N − 1 electrons, the probability factor, Π (see Eq. 24), equals the product of the squares of the N-1 overlaps between initial-state and final-state orbitals. OZ Theorem 4 implies the following result that involves N − 1 ($\pm w$) pairs of density-difference eigenvalues:

$$\Pi = \prod_{r=1}^{N-1} t_{rr}^2 = \prod_{r=1}^{N-1} \left(1 - w_r^2\right). \tag{104}$$

Paired eigenvalues has been confirmed in numerical calculations of natural ionization orbitals.[133] When the Koopmans description is qualitatively valid, values of w will be close to zero. On the other hand, when one or more of the overlaps between sets of corresponding orbitals approaches zero, pairs of density-difference eigenvalues that approach ±1 will appear. Such results characterize shake-up (correlation) final state for which the probability factor will be close to zero.

In Koopmans and ΔSCF calculations with non-zero probability factors, the expectation value of a single spin-orbital's number operator (i.e., its occupation number) changes from unity to zero for electron detachments and from zero to unity for electron attachments. Whereas no other changes in occupation numbers occur at the Koopmans level, ΔSCF calculations

Table 6 Approximate and exact Dyson orbitals, probability factors and changes in occupation numbers.

Theory	Dyson orbital	Probability factor	Δn^{a}_{Dyson}	$\Delta n^{a}_{non-Dyson}$
Koopmans	ψ	1	± 1	0
ΔSCF	$\sqrt{P}\phi^{b}$	$0 \leq P \leq 1$	± 1	$0 \leq \lvert \Delta n_{non-Dyson} \rvert \leq 1$
Diagonal Σ	$\sqrt{P}\psi$	$0 \leq P \leq 1$	$0 \leq \lvert \Delta n_{Dyson} \rvert \leq 1$	$0 \leq \lvert \Delta n_{non-Dyson} \rvert \leq 1$
Exact	$\sum_{r}^{all} c_r \psi_r$	$0 \leq P \leq 1$	$0 \leq \lvert \Delta n_{Dyson} \rvert \leq 1$	$0 \leq \lvert \Delta n_{non-Dyson} \rvert \leq 1$

[a]Changes in occupation number for spin-orbital $p = \Delta\langle p^{\dagger} p\rangle$.
[b]ϕ may be expressed as a linear combination of canonical, Hartree-Fock orbitals (ψ) using Eqs. (84) and (85).

allow occupation numbers of the other initial-state spin-orbitals to change from 1 or 0 to other values in the final state.

These simple rules are obliterated when initial and final states are correlated. For example, in correlated electron propagator calculations,[1,7,8] changes in occupation numbers inferred from density-difference matrices[135,136] usually do not equal integers. Self-energy approximations that improve ΔSCF results with selected correlation terms have been applied to EAs.[137,138] Table 6 provides a summary of DOs expressed in terms of canonical HF (ψ) or corresponding (φ) orbitals, probability factors and changes in occupation numbers that emerge in various levels of theory.

ΔSCF DOs may be expressed as a linear combination of occupied orbitals, but exact DOs involve summations over all (occupied or virtual) orbitals. Probability factors of unity and changes of ± 1 in occupation numbers at the Koopmans and ΔSCF levels of theory have a simple interpretation: an electron assigned to a DO is completely annihilated (or created) in states with $N-1$ (or $N+1$) electrons. Such certainty disappears at higher levels of theory. Changes in occupation numbers of the other orbitals equal zero only at the frozen-orbital level of theory.

Because the Koopmans approximation corresponds to a single, non-vanishing eigenvalue of unity for the density-difference matrix, the accompanying electron-density differences pertaining to the lowest electron detachment energy or the largest electron attachment energy, also known as Fukui functions,[139] have non-negative values at all points in three-dimensional space. Negative values of a detachment or attachment Fukui function,[140] f^{Fukui}, can appear in ΔSCF calculations, for negative eigenvalues of the density-difference matrix are usually present. Such conclusions become more likely when non-zero overlaps between paired corresponding orbitals are low (e.g., due to charge-transfer relaxation), leading to one or more large λ terms in

$$f^{Fukui}(1) = |\phi(1)|^2 + \sum_{r=1}^{q} \left[\sqrt{1 - t_{rr}^2} |\kappa_r(1)|^2 - \sqrt{1 - t_{rr}^2} |\lambda_r(1)|^2 \right]. \quad (105)$$

For similar reasons, negative spin-densities may appear in SCF calculations on a single state or in ΔSCF calculations in which differences of α or β density matrices are evaluated.

3.4 Other classes of transitions

Double or multiple electron-binding energies that correspond to initial and final states that are well-approximated by single Slater determinants (e.g., transitions to triplets with two open shells that are measured in Auger spectroscopy[115] on closed-shell molecules) can be expected to produce two or more density-difference eigenvalues of unity. The accompanying unpaired orbitals may be used obtain Dyson-geminal (which occur in expressions for Auger intensities) or multiple-electron-overlap amplitudes. Values of w that occur in the remaining pairs of $\pm w$ eigenvalues are likely to become larger because of the enhanced importance of orbital relaxation when two or more electrons are added or removed.

The results of Tables 5 and 6 are equally applicable to ΔSCF calculations of excitation energies, i.e., transitions that conserve the number of electrons. Final states that describe core-valence excitation, for example, can be expected to yield paired density-difference eigenvalues that are close to unity and other pairs that describe strong relaxation effects.

4. Conclusions

A new generation of electron-propagator self-energies that are free of adjustable parameters and that employ HF orbitals has been presented. The advantages of the new, diagonal self-energies may be summarized as follows:

- The cubically scaling os-D2 and os-nD-D2 methods reduce the Koopmans MAE from over 1 eV to approximately 0.2–0.3 eV at essentially no cost after a partial integral transformation.
- RQ3 and Q3+ have MAEs of approximately 0.1 eV and introduce no steps that are more demanding than the preceding integral transformation.

- RL3, with its MAE below 0.1 eV, is a competitive alternative to non-diagonal self-energy methods such as NR2 and BD-T1 and to IP-EOM-CCSD.
- P3, Q3 and L3 calculations lead to computationally trivial P3+, Q3+ and L3+ improvements that provide fast, accurate approximations to RP3, RQ3 and RL3.
- *The following hierarchy of increasing accuracy (indicated by inequalities) accompanied by decreasing efficiency is indicated by numerical tests: os-nD-D2 < Q3+ < RQ3 < L3+ < RL3. These methods are the best compromises of accuracy, efficiency, and feasibility.*

The CAS-EKT method is founded on a reference state that complies with the assumptions made in the derivation of the EKT. With annihilation operators confined to the active space and the inclusion of basis functions that make much greater contributions to initial than to final states, this method can generate results that systematically approach those of full configuration interaction in the active space.

Two theorems on the eigenvalue spectrum of the difference of two idempotent matrices have implications for the natural orbitals and occupation numbers that pertain to ΔSCF calculations. Corresponding orbital overlaps determine the DOs, probability factors, paired natural transition orbitals that describe orbital relaxation and Fukui functions that may be used to interpret the results of SCF calculations.

Acknowledgments

As the instructor in a graduate class entitled Quantum Theory of Matter in the late 1970s, Professor John R. Sabin introduced many of the concepts that underlie the advances discussed above to the author. Professor Sabin also was a member of the oral examination committee that approved the author's doctoral candidacy. The author is grateful for his friendship and dedication to the aims of the scientific community.

The author acknowledges indispensable collaborations with Ernest Opoku, Filip Pawłowski, Viktor N. Staroverov, Ernest R. Davidson and Richard Zalik.

This work was supported by grant CHE-1565760 to Auburn University.

References

1. Linderberg, J.; Öhrn, Y. *Propagators in Quantum Chemistry*, 2nd ed.; Wiley-Interscience: Hoboken, NJ, 2004.
2. Jørgensen, P.; Simons, J. *Second Quantization-Based Methods in Quantum Chemistry*; Academic Press: New York, 1981; p. 172.
3. Schirmer, J. *Many-Body Methods for Atoms, Molecules and Clusters*; Springer International Publishing AG, 2018; p. 332.
4. Ortiz, J. V. Dyson-Orbital Concepts for Description of Electrons in Molecules. *J. Chem. Phys.* **2020**, *153* (7), 070902.

5. Ortiz, J. V. Concepts of Chemical Bonding From Electron Propagator Theory. *AIP Conf. Proc.* **2012**, *1456*, 73–80 (Theory and Applications of Computational Chemistry: The First Decade of the Second Millennium).
6. Ortiz, J. V. Electron Propagator Theory: An Approach to Prediction and Interpretation in Quantum Chemistry. *Wiley Interdiscip. Rev.: Comput. Mol. Sci.* **2013**, *3* (2), 123–142.
7. Ortiz, J. V.; Dixon, D. A., Eds. Chapter Four - Interpreting Bonding and Spectra With Correlated, One-Electron Concepts From Electron Propagator Theory. In *Annual Reports in Computational Chemistry*; Vol. 13; Elsevier, 2017; pp. 139–182.
8. Corzo, H. H.; Ortiz, J. V. Chapter Thirteen—Electron Propagator Theory: Foundations and Predictions. In *Advances in Quantum Chemistry*; Sabin, J. R., Brändas, E. J., Eds.; Vol. 74; Academic Press, 2017; pp. 267–298.
9. Koopmans, T. Über die Zuordnung von Wellenfunktionen und Eigenwerten zu den Einzelnen Elektronen Eines Atoms. *Physica* **1934**, *1* (1–6), 104–113.
10. Díaz-Tinoco, M.; Corzo, H. H.; Pawłowski, F.; Ortiz, J. V. Do Dyson Orbitals Resemble Canonical Hartree–Fock Orbitals? *Mol. Phys.* **2019**, *117* (17), 2275–2283.
11. Cederbaum, L. S.; Domcke, W. Theoretical Aspects of Ionization Potentials and Photoelectron Spectroscopy: A Green's Function Approach. *Adv. Chem. Phys.* **1977**, *36*, 205–344.
12. von Niessen, W.; Schirmer, J.; Cederbaum, L. S. Computational Methods for the One-Particle Green's Function. *Comput. Phys. Rep.* **1984**, *1*, 57–125.
13. Ortiz, J. V. The Electron Propagator Picture of Molecular Electronic Structure. In *Computational Chemistry: Reviews of Current Trends*; 1997; pp. 1–61.
14. Körzdörfer, T.; Marom, N. Strategy for Finding a Reliable Starting Point for G_0W_0 Demonstrated for Molecules. *Phys. Rev. B* **2012**, *86* (4), 041110.
15. Gallandi, L.; Marom, N.; Rinke, P.; Körzdörfer, T. Accurate Ionization Potentials and Electron Affinities of Acceptor Molecules II: Non-Empirically Tuned Long-Range Corrected Hybrid Functionals. *J. Chem. Theory Comput.* **2016**, *12* (2), 605–614.
16. Knight, J. W.; Wang, X.; Gallandi, L.; Dolgounitcheva, O.; Ren, X.; Ortiz, J. V.; Rinke, P.; Körzdörfer, T.; Marom, N. Accurate Ionization Potentials and Electron Affinities of Acceptor Molecules III: A Benchmark of GW Methods. *J. Chem. Theory Comput.* **2016**, *12* (2), 615–626.
17. Cederbaum, L. S. One-Body Green's Function for Atoms and Molecules. Theory and Application. *J. Phys. B* **1975**, *8*, 290–303.
18. Ortiz, J. V. Partial Third-Order Quasiparticle Theory: Comparisons for Closed-Shell Ionization Energies and an Application to the Borazine Photoelectron Spectrum. *J. Chem. Phys.* **1996**, *104* (19), 7599–7605.
19. Ferreira, A. M.; Seabra, G.; Dolgounitcheva, O.; Zakrzewski, V. G.; Ortiz, J. V. Application and Testing of Diagonal, Partial Third-Order Electron Propagator Approximations. In *Quantum-Mechanical Prediction of Thermochemical Data*; Cioslowski, J., Ed.; Springer Netherlands: Dordrecht, 2001; pp. 131–160.
20. Ortiz, J. V. An Efficient, Renormalized Self-Energy for Calculating the Electron Binding Energies of Closed-Shell Molecules and Anions. *Int. J. Quantum Chem.* **2005**, *105* (6), 803–808.
21. Pérez-González, A.; Galano, A.; Ortiz, J. V. Vertical Ionization Energies of Free Radicals and Electron Detachment Energies of their Anions: A Comparison of Direct and Indirect Methods Versus Experiment. *J. Phys. Chem. A* **2014**, *118* (31), 6125–6131.
22. Corzo, H. H.; Galano, A.; Dolgounitcheva, O.; Zakrzewski, V. G.; Ortiz, J. V. NR2 and P3+: Accurate, Efficient Electron-Propagator Methods for Calculating Valence, Vertical Ionization Energies of Closed-Shell Molecules. *J. Phys. Chem. A* **2015**, *119* (33), 8813–8821.

23. Díaz-Tinoco, M.; Corzo, H. H.; Ortiz, J. V. Electron Propagator Methods for Vertical Electron Detachment Energies of Anions: Benchmarks and Case Studies. *J. Chem. Theory Comput.* **2018**, *14* (11), 5881–5895.
24. Dolgounitcheva, O.; Díaz-Tinoco, M.; Zakrzewski, V. G.; Richard, R. M.; Marom, N.; Sherrill, C. D.; Ortiz, J. V. Accurate Ionization Potentials and Electron Affinities of Acceptor Molecules IV: Electron-Propagator Methods. *J. Chem. Theory Comp.* **2016**, *12* (2), 627–637.
25. Opoku, E.; Pawłowski, F.; Ortiz, J. V. A New Generation of Diagonal Self-Energies for the Calculation of electron Removal Energies. *J. Chem. Phys.* **2021**, *155* (20), 204107.
26. Zakrzewski, V. G.; Ortiz, J. V.; Nichols, J. A.; Heryadi, D.; Yeager, D. L.; Golab, J. T. Comparison of Perturbative and Multiconfigurational Electron Propagator Methods. *Int. J. Quantum Chem.* **1996**, *60* (1), 29–36.
27. Ortiz, J. V. The electron Propagator Picture of Molecular Electronic Structure. *Comput. Chem. (Singapore)* **1997**, *2*, 1–61.
28. Reining, L. The GW Approximation: Content, Successes and Limitations. *Wiley Interdiscip. Rev.: Comput. Mol. Sci.* **2018**, *8* (3), e1344.
29. Schirmer, J.; Cederbaum, L. S.; Walter, O. New Approach to the One-Particle Green's Function for Finite Fermi Systems. *Phys. Rev. A* **1983**, *28* (3), 1237–1259.
30. Tamm, I. Relativistic Interaction of Elementary Particles. *J. Phys. USSR* **1945**, *9*, 449.
31. Dancoff, S. M. Non-Adiabatic Meson Theory of Nuclear Forces. *Phys. Rev.* **1950**, *78* (4), 382–385.
32. Schirmer, J.; Cederbaum, L. S. The Two-Particle-Hole Tamm-Dancoff Approximation (2ph-TDA) Equations for Closed-Shell Atoms and Molecules. *J. Phys. B* **1978**, *11*, 1889–1900.
33. Bintrim, S. J.; Berkelbach, T. C. Full-Frequency GW Without Frequency. *J. Chem. Phys.* **2021**, *154* (4), 041101.
34. Schirmer, J.; Trofimov, A. B.; Stelter, G. A Non-Dyson Third-Order Approximation Scheme for the Electron Propagator. *J. Chem. Phys.* **1998**, *109* (12), 4734–4744.
35. Hu, C.-H.; Chong, D. P.; Casida, M. E. The Parametrized Second-Order Green Function Times Screened Interaction (pGW2) Approximation for Calculation of Outer Valence Ionization Potentials. *J. Electron Spectrosc. Relat. Phenom.* **1997**, *85* (1–2), 39–46.
36. Lim, F.-H.; Nishida, M.; Hori, Y.; Ida, T.; Mizuno, M. Optimization of the Spin-Component-Scaled Factor for Electron Propagator Method. *Chem. Phys. Lett.* **2017**, *678*, 159–166.
37. Romero, J.; Charry, J. A.; Nakai, H.; Reyes, A. Improving Quasiparticle Second Order Electron Propagator Calculations With the Spin-Component-Scaled Technique. *Chem. Phys. Lett.* **2014**, *591*, 82–87.
38. Śmiga, S.; Grabowski, I. Spin-Component-Scaled ΔMP2 Parametrization: Toward a Simple and Reliable Method for Ionization Energies. *J. Chem. Theory Comput.* **2018**, *14* (9), 4780–4790.
39. Raghavachari, K.; Trucks, G. W.; Pople, J. A.; Head-Gordon, M. A Fifth-Order Perturbation Comparison of Electron Correlation Theories. *Chem. Phys. Lett.* **1989**, *157* (6), 479–483.
40. Richard, R. M.; Marshall, M. S.; Dolgounitcheva, O.; Ortiz, J. V.; Brédas, J.-L.; Marom, N.; Sherrill, C. D. Accurate Ionization Potentials and Electron Affinities of Acceptor Molecules I. Reference Data at the CCSD(T) Complete Basis Set Limit. *J. Chem. Theory Comput.* **2016**, *12* (2), 595–604.
41. Bartlett, R. J. Coupled-Cluster Theory and Its Equation-of-Motion Extensions. *WIREs Comput. Mol. Sci.* **2012**, *2* (1), 126–138.

42. Ranasinghe, D. S.; Margraf, J. T.; Perera, A.; Bartlett, R. J. Vertical Valence Ionization Potential Benchmarks From Equation-of-Motion Coupled Cluster Theory and QTP Functionals. *J. Chem. Phys.* **2019**, *150* (7), 074108.
43. Dunning, T. H. Gaussian Basis Sets for Use in Correlated Molecular Calculations. I. the Atoms Boron Through Neon and Hydrogen. *J. Chem. Phys.* **1989**, *90* (2), 1007–1023.
44. Manne, R. A Completeness Theorem for Operator Spaces. *Chem. Phys. Lett.* **1977**, *45* (3), 470–472.
45. Goscinski, O.; Lukman, B. Moment-Conserving Decoupling of Green Functions Via Pade Approximants. *Chem. Phys. Lett.* **1970**, 7, 573–576.
46. Löwdin, P. O. Studies in Perturbation Theory. X. Lower Bounds to Energy Eigenvalues in Perturbation-Theory Ground State. *Phys. Rev.* **1965**, *139* (2A), 357–372.
47. Møller, C.; Plesset, M. S. Note on an Approximation Treatment for Many-Electron Systems. *Phys. Rev.* **1934**, *46* (7), 618–622.
48. Hirata, S.; Hermes, M. R.; Simons, J.; Ortiz, J. V. General-Order Many-Body Green's Function Method. *J. Chem. Theory Comput.* **2015**, *11* (4), 1595–1606.
49. Hirata, S.; Doran, A. E.; Knowles, P. J.; Ortiz, J. V. One-Particle Many-Body Green's Function Theory: Algebraic Recursive Definitions, Linked-Diagram Theorem, Irreducible-Diagram Theorem, and General-Order Algorithms. *J. Chem. Phys.* **2017**, *147* (4), 044108.
50. Chong, D. P.; Herring, F. G.; McWilliams, D. Perturbation Corrections to Koopmans' Theorem. I. Double-Zeta Slater-Type-Orbital Basis. *J. Chem. Phys.* **1974**, *61* (1), 78–84.
51. Schirmer, J.; Angonoa, G. On Green-Function Calculations of the Static Self-Energy Part, the Ground State Energy, and Expectation Values. *J. Chem. Phys.* **1989**, *91*, 1754–1761.
52. Trofimov, A. B.; Schirmer, J. Molecular Ionization Energies and Ground- and Ionic-State Properties Using a Non-Dyson Electron Propagator Approach. *J. Chem. Phys.* **2005**, *123* (14), 144115.
53. Dempwolff, A. L.; Schneider, M.; Hodecker, M.; Dreuw, A. Efficient Implementation of the Non-Dyson Third-Order Algebraic Diagrammatic Construction Approximation for the Electron Propagator for Closed- and Open-Shell Molecules. *J. Chem. Phys.* **2019**, *150* (6), 064108.
54. Banerjee, S.; Sokolov, A. Y. Third-Order Algebraic Diagrammatic Construction Theory for Electron Attachment and Ionization Energies: Conventional and Green's Function Implementation. *J. Chem. Phys.* **2019**, *151* (22), 224112.
55. Banerjee, S.; Sokolov, A. Y. Efficient Implementation of the Single-Reference Algebraic Diagrammatic Construction Theory for Charged Excitations: Applications to the TEMPO Radical and DNA Base Pairs. *J. Chem. Phys.* **2021**, *154* (7), 074105.
56. Bartlett, R. J.; Musiał, M. Coupled-Cluster Theory in Quantum Chemistry. *Rev. Mod. Phys.* **2007**, *79* (1), 291–352.
57. Ortiz, J. V. A Nondiagonal, Renormalized Extension of Partial Third-Order Quasiparticle Theory: Comparisons for Closed-Shell Ionization Energies. *J. Chem. Phys.* **1998**, *108*, 1008–1014.
58. Storchi, L.; Vitillaro, G.; Tarantelli, F. Implementation and Use of a Direct, Partially Integral-Driven Non-Dyson Propagator Method for Molecular Ionization. *J. Comput. Chem.* **2009**, *30* (5), 818–825.
59. Nooijen, M.; Snijders, J. G. Second Order Many-Body Perturbation Approximations to the Coupled Cluster Green's Function. *J. Chem. Phys.* **1995**, *102* (4), 1681–1688.
60. Stanton, J. F.; Gauss, J. Perturbative Treatment of the Similarity Transformed Hamiltonian in Equation-of-Motion Coupled-Cluster Approximations. *J. Chem. Phys.* **1995**, *103* (3), 1064–1076.

61. Prasad, M. D.; Pal, S.; Mukherjee, D. Some Aspects of Self-Consistent Propagator Theories. *Phys. Rev. A* **1985**, *31* (3), 1287–1298.
62. Ortiz, J. V. Electron Detachment Energies of Closed-Shell Anions Calculated with a Renormalized Electron Propagator. *Chem. Phys. Lett.* **1998**, *296* (5), 494–498.
63. Ortiz, J. V. Single-Reference Electron Propagator Calculations on Vertical Ionization Energies of Ozone. *Chem. Phys. Lett.* **1998**, *297* (3), 193–199.
64. Ortiz, J. V. Approximate Brueckner Orbitals and Shakeup Operators in Electron Propagator Calculations: Applications to F- and OH. *Int. J. Quantum Chem.* **1998**, *70* (4/5), 651–658.
65. Ortiz, J. V. Orbital and Shakeup Operator Renormalizations in Electron Propagator Theory. *J. Chem. Phys.* **1998**, *109*, 5741–5746.
66. Ortiz, J. V. Approximate Brueckner Orbitals in Electron Propagator Calculations. *Int. J. Quantum Chem.* **1999**, *75* (4/5), 615–621.
67. Stanton, J. F.; Gauss, J. Analytic Energy Derivatives for Ionized States Described by the Equation-of-Motion Coupled Cluster Method. *J. Chem. Phys.* **1994**, *101* (10), 8938–8944.
68. Baker, J.; Pickup, B. T. A Method for Molecular Ionization Potentials. *Chem. Phys. Lett.* **1980**, *76*, 537–541.
69. Baker, J. EOM/Propagator Ionization Potentials: Prediction of Complete Valence-Shell Ionization Spectra Using Direct Diagonalization. *Chem. Phys. Lett.* **1983**, *101* (2), 136–140.
70. Zakrzewski, V. G.; Ortiz, J. V. Semidirect algorithms in electron propagator calculations. *Int. J. Quantum Chem.* **1994**, *28*, 23–27 (Proceedings of the International Symposium on Atomic, Molecular, and Condensed Matter Theory and Computational Methods, 1994).
71. Zakrzewski, V. G.; Ortiz, J. V. Semidirect Algorithms for Third-Order electron Propagator Calculations. *Int. J. Quantum Chem.* **1995**, *53* (6), 583–590.
72. Pawłowski, F.; Olsen, J.; Jørgensen, P. Cluster Perturbation Theory. I. Theoretical Foundation for a Coupled Cluster Target State and Ground-State Energies. *J. Chem. Phys.* **2019**, *150* (13), 134108.
73. Pawłowski, F.; Olsen, J.; Jørgensen, P. Cluster Perturbation Theory. II. Excitation Energies for a Coupled Cluster Target State. *J. Chem. Phys.* **2019**, *150* (13), 134109.
74. Baudin, P.; Pawłowski, F.; Bykov, D.; Liakh, D.; Kristensen, K.; Olsen, J.; Jørgensen, P. Cluster Perturbation Theory. III. Perturbation Series for Coupled Cluster Singles and Doubles Excitation Energies. *J. Chem. Phys.* **2019**, *150* (13), 134110.
75. Pawłowski, F.; Olsen, J.; Jørgensen, P. Cluster Perturbation Theory. IV. Convergence of Cluster Perturbation Series for Energies and Molecular Properties. *J. Chem. Phys.* **2019**, *150* (13), 134111.
76. Pawłowski, F.; Olsen, J.; Jørgensen, P. Cluster Perturbation Theory. V. Theoretical Foundation for Cluster Linear Target States. *J. Chem. Phys.* **2019**, *150* (13), 134112.
77. Day, O. W.; Smith, D. W.; Garrod, C. Generalization of the Hartree-Fock one-particle potential. *Int. J. Quantum Chem.* **1974**, *8*, 501–509 (Proc. Int. Symp. At., Mol. Solid-State Theory Quantum Stat.).
78. Smith, D. W.; Day, O. W. Extension of Koopmans' Theorem. I. Derivation. *J. Chem. Phys.* **1975**, *62* (1), 113–114.
79. Morrison, R. C.; Day, O. W.; Smith, D. W. Extension of Koopmans' Theorem. III. Ionization Energies of the Open-Shell Atoms Lithium and Boron. *Int. J. Quantum Chem.* **1975**, *9*, 229–235.
80. Morrell, M. M.; Parr, R. G.; Levy, M. Calculation of Ionization Potentials From Density Matrixes and Natural Functions, and the Long-Range Behavior of Natural Orbitals and Electron Density. *J. Chem. Phys.* **1975**, *62* (2), 549–554.

81. Day, O. W.; Smith, D. W.; Morrison, R. C. Extension of Koopmans' Theorem. II. Accurate Ionization Energies From Correlated Wavefunctions for Closed-Shell Atoms. *J. Chem. Phys.* **1975**, *62* (1), 115–119.
82. Katriel, J.; Davidson, E. R. Asymptotic Behavior of Atomic and Molecular Wavefunctions. *Proc. Natl. Acad. Sci. U. S. A.* **1980**, 77 (8), 4403–4406.
83. Pickup, B. T.; Snijders, J. G. On the Exactness of Extended Koopmans Eigenvalues. *Chem. Phys. Lett.* **1988**, *153*, 69–75.
84. Morrison, R. C. The Extended Koopmans Theorem and its Exactness. *J. Chem. Phys.* **1992**, *96* (5), 3718–3722.
85. Sundholm, D.; Olsen, J. The Exactness of the Extended Koopmans' Theorem: A Numerical Study. *J. Chem. Phys.* **1993**, *98* (5), 3999–4002.
86. Morrison, R. C. Comment on "The exactness of the extended Koopmans' theorem: a numerical study" [J. Chem. Phys. 98, 3999 (1993)]. *J. Chem. Phys.* **1993**, *99* (8), 6221.
87. Sundholm, D.; Olsen, J. Response to "Comment on 'The Exactness of the Extended Koopmans' Theorem: A Numerical study'" [J. Ch. Phys. 99.6221 (1993)]. *J. Chem. Phys.* **1993**, *99* (8), 6222–6223.
88. Morrison, R. C.; Zhou, Z.; Parr, R. G. The Question of the Completeness of the Natural Orbitals With Nonzero Occupation Numbers for Atoms and Molecules. *Theor. Chim. Acta* **1993**, *86* (1–2), 3–11.
89. Olsen, J.; Sundholm, D. On Perturbation Expansions of the Extended Koopmans Theorem. *Chem. Phys. Lett.* **1998**, *288* (2–4), 282–288.
90. Pernal, K.; Cioslowski, J. On the Validity of the Extended Koopmans' Theorem. *J. Chem. Phys.* **2001**, *114* (10), 4359–4361.
91. Vanfleteren, D.; Van, N. D.; Ayers, P. W.; Morrison, R. C.; Bultinck, P. Exact Ionization Potentials From Wavefunction Asymptotics: The Extended Koopmans' Theorem, Revisited. *J. Chem. Phys.* **2009**, *130*, 194104.
92. Ernzerhof, M. Validity of the Extended Koopmans' Theorem. *J. Chem. Theory Comput.* **2009**, *5*, 793–797.
93. Davidson, E. R.; Ortiz, J. V.; Staroverov, V. N. Complete-Active-Space Extended Koopmans Theorem Method. *J. Chem. Phys.* **2021**, *155* (5), 051102.
94. Roothaan, C. C. J. New Developments in Molecular Orbital Theory. *Rev. Mod. Phys.* **1951**, *23* (2), 69–89.
95. Plakhutin, B. N.; Davidson, E. R. Koopmans' Theorem in the Restricted Open-Shell Hartree-Fock Method. 1. A Variational Approach. *J. Phys. Chem. A* **2009**, *113*, 12386–12395.
96. Davidson, E. R.; Plakhutin, B. N. Koopmans's Theorem in the Restricted Open-Shell Hartree-Fock Method. II. The Second Canonical Set for Orbitals and Orbital Energies. *J. Chem. Phys.* **2010**, *132*, 184110.
97. Plakhutin, B. N.; Davidson, E. R. Canonical Form of the Hartree-Fock Orbitals in Open-Shell Systems. *J. Chem. Phys.* **2014**, *140* (1), 014102.
98. Morrison, R. C.; Liu, G. Extended Koopman's Theorem: Approximate Ionization Energies From MCSCF Wave Functions. *J. Comput. Chem.* **1992**, *13* (8), 1004–1010.
99. Morrison, R. C. Extended Koopmans' Theorem Ionization Potentials for Beryllium Atom Shake-up Transitions. *Int. J. Quantum Chem.* **1994**, *49* (5), 649–656.
100. Morrison, R. C.; Dixon, C. M.; Mizell, J. R., Jr. Examination of the Limits of Accuracy of the Extended Koopman's Theorem Ionization Potentials in Excited States of Ions of LiH, He2, and Li2. *Int. J. Quantum Chem.* **1994**, *28*, 309–314 (Proceedings of the International Symposium on Atomic, Molecular, and Condensed Matter Theory and Computational Methods, 1994).
101. Cioslowski, J.; Piskorz, P.; Liu, G. Ionization Potentials and Electron Affinities From the Extended Koopmans' Theorem Applied to Energy-Derivative Density Matrixes: The EKTMPn and EKTQCISD Methods. *J. Chem. Phys.* **1997**, *107*, 6804–6811.

102. Bozkaya, U.; Ünal, A. State-of-the-Art Computations of Vertical Ionization Potentials With the Extended Koopmans' Theorem Integrated With the CCSD(T) Method. *J. Phys. Chem. A* **2018**, *122* (17), 4375–4380.
103. Gu, Y.; Xu, X. Extended Koopmans' Theorem at the Second-Order Perturbation Theory. *J. Comput. Chem.* **2020**, *41* (12), 1165–1174.
104. El-Samman, A. M.; Staroverov, V. N. Asymptotic Behavior of the Average Local Ionization Energy in Finite Basis Sets. *J. Chem. Phys.* **2020**, *153* (13), 134109.
105. El-Samman, A. M.; Ospadov, E.; Staroverov, V. N. First Ionization Energy as the Asymptotic Limit of the Average Local Electron Energy. *J. Chem. Theory Comput.* **2020**, *16* (11), 6886–6893.
106. Slater, J. C. The Theory of Complex Spectra. *Phys. Rev.* **1929**, *34* (10), 1293–1322.
107. Helgaker, T.; Jørgensen, P.; Olsen, J. *Molecular Electronic-Structure Theory*; John Wiley & Sons, Inc., 2000; p. 908.
108. Kohn, W.; Sham, L. J. Self-Consistent Equations Including Exchange and Correlation Effects. *Phys. Rev.* **1965**, *140* (4A), A1133–A1138.
109. Bagus, P. S. Self-Consistent-Field Wave Functions for Hole States of Some Ne-Like and Ar-Like Ions. *Phys. Rev.* **1965**, *139* (3A), A619–A634.
110. Bagus, P. S.; Ilton, E.; Nelin, C. J. Extracting Chemical Information From XPS Spectra: A Perspective. *Catal. Lett.* **2018**, *148* (7), 1785–1802.
111. Bagus, P. S.; Illas, F. Orbitals Permit the Interpretation of Core-Level Spectroscopies in Terms of Chemistry. *Catal. Lett.* **2020**, *150*, 2457–2463.
112. Shea, J. A. R.; Gwin, E.; Neuscamman, E. A Generalized Variational Principle With Applications to Excited State Mean Field Theory. *J. Chem. Theory Comput.* **2020**, *16* (3), 1526–1540.
113. Barca, G. M. J.; Gilbert, A. T. B.; Gill, P. M. W. Simple Models for Difficult Electronic Excitations. *J. Chem. Theory Comput.* **2018**, *14* (3), 1501–1509.
114. Kaduk, B.; Kowalczyk, T.; Van Voorhis, T. Constrained Density Functional Theory. *Chem. Rev. (Washington, DC, U. S.)* **2012**, *112* (1), 321–370.
115. Ågren, H.; Cesar, A.; Liegener, C. M. Theory of Molecular Auger Spectra. *Adv. Quantum Chem.* **1992**, *23*, 1–82.
116. Bagus, P. S.; Liu, B.; McLean, A. D.; Yoshimine, M. *Alchemy Manual*; 1970.
117. Bagus, P. S.; Ilton, E. S.; Nelin, C. J. The Interpretation of XPS Spectra: Insights into Materials Properties. *Surf. Sci. Rep.* **2013**, *68* (2), 273–304.
118. Brillouin, L. *Actual. Sci. Ind.* **1934**, *159*, 28.
119. Coulson, C. A. Brillouin's Theorem and the Hellmann-Feynman Theorem for Hartree-Fock Wave Functions. *Mol. Phys.* **1971**, *20* (4), 687–694.
120. Froese-Fischer, C. *The Hartree-Fock Method for Atoms. A Numerical Approach*; John Wiley: New York, 1977.
121. Pople, J. A.; Nesbet, R. K. Self-Consistent Orbitals for Radicals. *J. Chem. Phys.* **1954**, *22* (3), 571–572.
122. Berthier, G. Extension de la methode du champ moleculaire self-consistent a l'etude des couches incompletes. *C. R. Acad. Sci.* **1954**, *238*, 91–93.
123. Fock, V. Näherungsmethode zur Lösung des quantenmechanischen Mehrkörperproblems. *Z. Phys.* **1930**, *61* (1), 126–148.
124. Amos, A. T.; Hall, G. G. Single Determinant Wave Functions. *Proc. R. Soc. Lond. A Math. Phys. Sci.* **1961**, *263* (1315), 483–493.
125. Lowdin, P. O. Band Theory, Valence Bond, and Tight-Bonding Calculations. *J. Appl. Phys.* **1962**, *33* (1), 251–280.
126. Bofill, J. M.; Pulay, P. The Unrestricted Natural Orbital-Complete Active Space (UNO-CAS) Method: An Inexpensive Alternative to the Complete Active Space-Self-Consistent-Field (CAS-SCF) Method. *J. Chem. Phys.* **1989**, *90* (7), 3637–3646.

127. Bobrowicz, F. W.; Goddard, W. A., III. The Self-Consistent Field Equations For Generalized Valence Bond And Open-Shell Hartree-Fock Wave Functions. *Modern Theor. Chem.* **1977**, *3*, 79–127 (Methods Electron. Struct. Theory).
128. King, H. F.; Stanton, R. E.; Kim, H.; Wyatt, R. E.; Parr, R. G. Corresponding Orbitals and the Nonorthogonality Problem in Molecular Quantum Mechanics. *J. Chem. Phys.* **1967**, *47* (6), 1936–1941.
129. Condon, E. U. The Theory of Complex Spectra. *Phys. Rev.* **1930**, *36* (7), 1121–1133.
130. Ågren, H.; Bagus, P. S. Corresponding Orbital Analysis of the Bonding of Some Monosubstituted Benzenes. *J. Am. Chem. Soc.* **1985**, *107* (1), 134–137.
131. Martin, R. L.; Davidson, E. R. Halogen Atomic and Diatomic 1s Hole States. *Phys. Rev. A* **1977**, *16* (4), 1341–1346.
132. Löwdin, P. O. On the Non-Orthogonality Problem Connected With the Use of Atomic Wave Functions in the Theory of Molecules and Crystals. *J. Chem. Phys.* **1950**, *18* (3), 365–375.
133. Thompson, L. M.; Harb, H.; Hratchian, H. P. Natural Ionization Orbitals for Interpreting Electron Detachment Processes. *J. Chem. Phys.* **2016**, *144* (20), 204117.
134. Cox, P. A. Ionization Energies of Open-Shell Molecules in the Frozen Orbital Approximation. *Mol. Phys.* **1975**, *30* (2), 389–403.
135. Cioslowski, J.; Ortiz, J. V. One-electron Density Matrixes and Energy Gradients in Second-Order Electron Propagator Theory. *J. Chem. Phys.* **1992**, *96* (11), 8379–8389.
136. Ortiz, J. V. Energy Gradients and Effective Density Differences in Electron Propagator Theory. *J. Chem. Phys.* **2000**, *112* (1), 56–68.
137. Ortiz B., J. V.; Öhrn, Y. Electron Propagator Theory of Molecular Anions. *J. Chem. Phys.* **1980**, *72* (10), 5744–5751.
138. Ortiz B., J. V.; Öhrn, Y. Electron Propagator Calculations of Molecular Electron Affinities. *Chem. Phys. Lett.* **1981**, *77*, 548–554.
139. Parr, R. G.; Yang, W. Density Functional Approach to the Frontier-Electron Theory of Chemical Reactivity. *J. Am. Chem. Soc.* **1984**, *106* (14), 4049–4050.
140. Melin, J.; Ayers, P. W.; Ortiz, J. V. Removing Electrons Can Increase the Electron Density: A Computational Study of Negative Fukui Functions. *J. Phys. Chem. A* **2007**, *111* (40), 10017–10019.

CHAPTER EIGHT

The electronic stopping power of heavy targets

Alejandra M.P. Mendez, Darío M. Mitnik, and Claudia C. Montanari*
Institute of Astronomy and Space Physics, CONICET and University of Buenos Aires, Buenos Aires, Argentina
*Corresponding author: e-mail address: mclaudia@iafe.uba.ar

Contents

Abstract

The aim of this work is to face the challenge of describing the stopping power in heavy transition and posttransition metal solids in an extended energy range. The present study examines the stopping of hydrogen in Hf, W, Pb, and Bi. The electronic structure description of the targets requires solving the many-electron Dirac Hamiltonian. We inquire the influence of employing various relativistic approximations, and its effect on the energy loss. The theoretical model of the stopping power considers separately the valence electrons of the metal-accounted for as a free-electron gas and the bound electrons. The former are described using a nonperturbative binary screening potential or the dielectric formalism, depending on the energy regions and plasmon excitation threshold. For the later, the approach uses the dielectric formalism for each subshell by means of the shellwise local plasma approximation. We inspect the importance of accurately describing the 4*f* and 5*d* electrons, the spin–orbit split, the intra- and intershell screening among electrons, and the different quantum dielectric response functions. The present results correctly describe the stopping cross sections in comparison with the data available, the theoretical models DPASS and CasP, and the semiempirical SRIM. Based on these results, we analyze the state of art of the experimental values and suggest the need for new measurements in certain energy ranges and targets.

Advances in Quantum Chemistry, Volume 85
ISSN 0065-3276
https://doi.org/10.1016/bs.aiq.2022.04.005
Copyright © 2022 Elsevier Inc.
All rights reserved.

1. Introduction

The electronic stopping power is a fundamental input of any simulation or experimental research that describes the passage of high energy charged particles through matter. These applications include, among others, ion beam analysis, deposition range and radiation damage; all of them with different technological or medical applications. Providing reliable stopping values for such purposes challenges modern science and encourages the development of accelerator technologies, theoretical modeling, and high performance computing.

Theoretical electronic stopping power calculations go back to the early times of atomic physics.[1–3] Pioneering works by Sabin and coworkers[4] provided useful tabulations for atoms, with shell-by-shell calculations. For molecular targets, the authors considered chemical bonds[5] to improve the usual Bragg rule (addition of the atomic constituents). Many theoretical approaches have been proposed since then. For instance, we can point out the electron–nuclear dynamics to solve the time-dependent Schrödinger equation[6] and the time-dependent density functional theory.[7] Within the binary collision formalism, stand out theories such as the unitary convolution approximation,[8] the binary stopping theory[9] and the self-consistent nonperturbative approach,[10] among others.

A different approach to the inelastic response of the target electrons to the ion passage comes from the quantum dielectric formalism.[11–14] This many-body approximation considers an homogeneous density of non-interacting electrons within the Fermi energy band, including dynamic screening and collective excitations within the limits of the linear response approximation. The dielectric formalism was extended to describe all the target electrons by means of the local plasma approximation (LPA).[15,16] Even though Lindhard's LPA, and its subsequent developments, assume a local response of the bound electrons, it considers the electronic cloud as a whole by using the total electronic density of the atom. Recent proposals[17–19] prove that the models based on the dielectric formalism are still quite efficient to describe the stopping power.

However, despite all the recent theoretical efforts, discrepancies between models and experimental measurements remain present.[20] These differences are originated primarily because the energy loss through matter is a many-body and many-collision phenomena. One of the open problems in the field is the stopping power modeling of systems with nuclear charge $Z_T \geq 55$:

lanthanides, heavy transition metals, and beyond. Recent measurements[21–23] show the limitations of the energy loss models for these heavy targets. A first step toward solving this problem presumes considering a relativistic formalism to describe the electronic structure. In fact, recent theoretical works support this assertion.[24]

In this contribution, we present and discuss the results of the shellwise local plasma approximation (SLPA)[18,25] combined with various relativistic electronic structure calculations. This approximation improves the full dielectric formalism by Chu and Powerd[16] considering independent subshells, as proposed by Meltzer and Sabin,[17] and the ionization threshold of each subshell by using the Levine dielectric function.[26] In fact, the present SLPA surpasses previous versions by introducing the Levine-Mermin dielectric function.[27] The SLPA is a collective model: it takes into account screening among electrons of the same subshell, and also inter-shell screening, when required.[24] This feature constitutes a valuable asset when dealing with the $4f$ and $5p$ subshells of the heavy transition metals. For these targets, the $4f$ and $5p$ electrons are so close in energy that, for impact ions above certain threshold, they respond as a single density of electrons and mutual screening must be considered.

The aim of the present contribution is to study the stopping power in relativistic targets. We examine four illustrative cases: two transition metals, Hf and W, and two posttransition metals, Pb and Bi. These targets include the closed $4f$-subshell, and open (Hf and W) or closed (Pb and Bi) $5d$-subshell, while the valence electrons are considered as a free-electron gas (FEG). Detailed electronic structure calculations were performed by implementing nonrelativistic, semirelativistic and fully relativistic approaches. The importance of these calculations to the final stopping values is discussed. Atomic units are employed throughout all this work, unless explicitly stated otherwise.

2. Theoretical description

2.1 The shell-wise local plasma approximation

The many-body consistent treatment for an ion traveling through an homogeneous FEG was developed by Lindhard[11] and by Ritchie[12,14] within the linear response approximation. For the valence electrons of metals we can consider a FEG of density $\rho_{\mathrm{FEG}} = N_e \rho_{at}$, with N_e and ρ_{at} being the number of valence electrons and the atomic density of the solid target, respectively.

The stopping power of a bare ion of charge Z_P moving at velocity v within the FEG is expressed as

$$S_{\mathrm{FEG}} = \frac{2\, Z_P^2}{\pi v^2} \int_0^{\infty} \frac{dk}{k} \int_0^{kv} \omega \,\mathrm{Im}\left[\frac{-1}{\varepsilon(k,\, \omega\, \rho_{\mathrm{FEG}})}\right] d\omega, \tag{1}$$

where $\varepsilon(k, \omega, \rho_{\mathrm{FEG}})$ is the dielectric function. In this work, we employed the Mermin–Lindhard dielectric function,[28] which improves Lindhard's[11] by considering a finite time between electron–electron collisions τ or, equivalently, a damping $\gamma = 1/\tau$. It is worth to underline that this many-electron formalism includes plasmon excitation, with the plasmon frequency being $\omega_p = \sqrt{4\pi\, \rho_{\mathrm{FEG}}}$, as introduced by Lindhard and Scharff.[15] Instead, the value of γ is experimental and related to the energy loss function of the target in the optical limit.[29,30]

Within this formalism, the SLPA[18] describes the response of each nl subshell of bound electrons as a gas of electrons of *local* density $\rho_{nl}(r)$ and ionization threshold E_{nl}. The SLPA expression for the energy loss by a bare ion within the atomic cloud of electrons of the nl subshell is given by

$$S_{nl} = \frac{2\, Z_P^2}{\pi v^2} \int_0^{\infty} \frac{dk}{k} \int_0^{kv} \omega \,\mathrm{Im}\left[\frac{-1}{\varepsilon_{nl}(k,\, \omega)}\right] d\omega, \tag{2}$$

with

$$\mathrm{Im}\left[\frac{-1}{\varepsilon_{nl}(k,\, \omega)}\right] = \int \mathrm{Im}\left[\frac{-1}{\varepsilon(k,\, \omega,\, \rho_{nl}(r),\, E_{nl})}\right] \mathbf{dr}. \tag{3}$$

The dielectric function $\varepsilon(k, \omega, \rho_{nl}(r), E_{nl})$ is the Levine–Mermin one,[26] which includes explicitly the energy gap E_{nl} to the Mermin dielectric function.[28] The selection of the dielectric function is key since it represents all possible inelastic processes of the bound electrons. Within the SLPA, a local density $\rho_{nl}(r)$ implies a local plasmon frequency $\omega_p^{nl}(r) = \sqrt{4\pi\, \rho_{nl}(r)}$. Following this methodology, a local damping is proposed,[27] given by $\gamma_{nl}(r) = \omega_p^{nl}(r)/2 \equiv \sqrt{\pi \rho_{nl}(r)}$. It is worth mentioning that all the parameters included in the SLPA depend on $\rho_{nl}(r)$ and E_{nl}, which are computed beforehand. This feature maintains the full-theoretical characteristic of the model. As a collective approach, no postmodel shell correction is necessary. The evaluation of Eqs. (2) and (3) is straightforward, with only two inputs values for each nl subshell: the electron density (derived from the wave functions) and the binding energy.

2.2 Electronic structure

Since electron densities and binding energies are the only input parameters of the SLPA model, obtaining an accurate description of the target is key. For species with heavy nuclei, whether neutral or ionized, relativistic effects must be included in the computation of the electronic structure. Many methods and numerical codes have been developed to this end; for example, see the reviews by Grant[31] and Fischer et al.[32] In this work, we describe the electronic structure of heavy targets by implementing a nonrelativistic method with relativistic corrections (semirelativistic) and a fully relativistic approach.

The first technique considers the Breit–Pauli (BP) effective Hamiltonian. For a target of N electrons and nuclear charge Z, the BP Hamiltonian is given by,

$$H_{\mathrm{BP}} = H_{\mathrm{NR}} + H_{\mathrm{RC}}, \tag{4}$$

where the first term is the usual nonrelativistic (NR) Hamiltonian

$$H_{\mathrm{NR}} = \sum_{i=1}^{N} h_i + \sum_{i<j} \frac{1}{r_{ij}}, \tag{5}$$

and

$$h_i = -\frac{1}{2}\nabla_i^2 - \frac{Z}{r_i}. \tag{6}$$

The second term correspond to relativistic corrections, composed by one- and two-body terms; see Refs. 33, 34 for more details.

In the second method, the Dirac-Coulomb equation is solved. This Hamiltonian is given by

$$H = \sum_i \left[h_i^{\mathrm{D}} - \frac{Z}{r_i} \right] + \sum_{i<j} \frac{1}{r_{ij}}, \tag{7}$$

where h_i^{D} is the single-electron kinetic Dirac Hamiltonian

$$h_i^{\mathrm{D}} = c\,\boldsymbol{\alpha}_i \cdot \mathbf{p}_i + (\beta_i - 1)c^2, \tag{8}$$

while the remaining terms correspond to the nuclei–electron and electron–electron interaction. Many approaches have been developed to solve Eq. (7); for example, see Refs. 35–37.

3. Results and discussion

For solving the Dirac–Coulomb Hamiltonian, we considered the relativistic version of the parametric potential method developed by Klapisch,[38] which is implemented in the RELAC code and included in the HULLAC suite of codes.[39] We call these calculations R. The full Breit interaction in the Pauli approximation, developed by Eissner et al.,[33] was implemented here by considering the AUTOSTRUCTURE code by Badnell.[40] Within this approach, we used the Thomas–Fermi–Dirac–Amaldi (TFDA) potential,[33,41,42] which features scaling λ_{nl} parameters, and allows adjusting the solutions of Eq. (5). These calculations are denoted as sR. We also implemented a nonrelativistic approach. For this, we used the Hartree–Fock code by Froese Fischer.[43]

The relativistic binding energies (R), the semirelativistic (sR), and the nonrelativistic values (NR) are shown in Fig. 1 for Hf, W, Pb, and Bi. We compare our theoretical calculations with experimental binding energies (measured in solids), compiled by Williams.[44]

Since Fig. 1 displays the binding energies in a five order magnitude range, we add Fig. 2 to discern the discrepancies between the present values and the

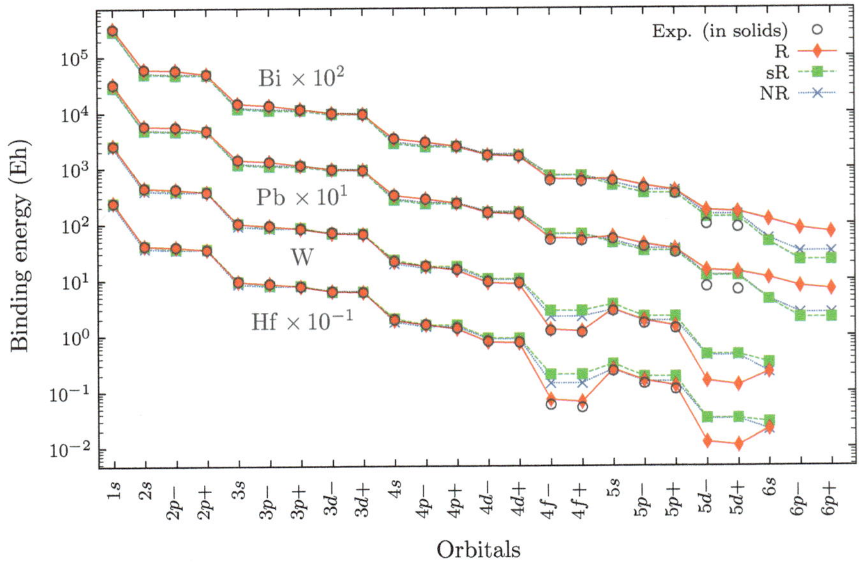

Fig. 1 Binding energy for Hf, W, Pb, and Bi. Symbols: ∘ Experimental data in solids,[44] ◆ relativistic (R), ■ semirelativistic (sR), and × nonrelativistic (NR) values.

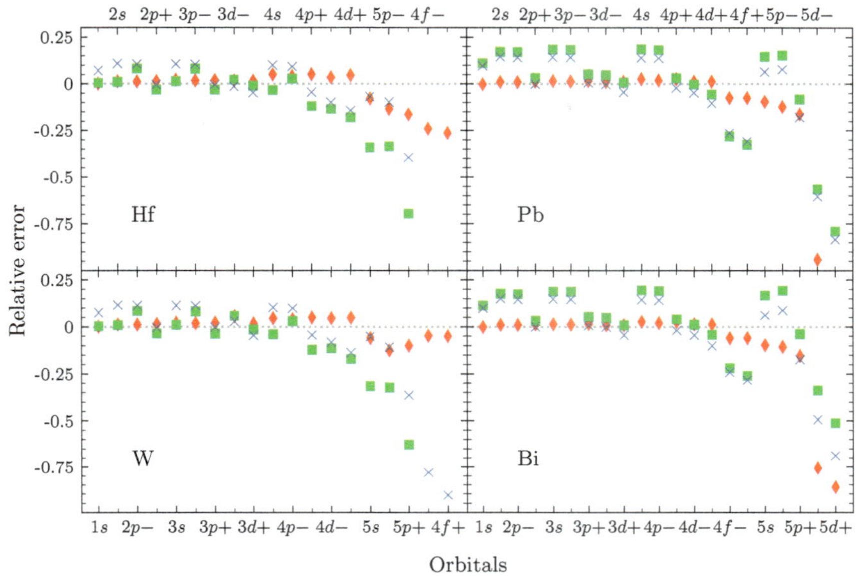

Fig. 2 Binding energy relative error with respect to experimental values for Hf, W, Pb, and Bi. Symbols: same as Fig. 1.

measurements. Binding energy relative errors are given in Fig. 2 for Hf, W, Pb, and Bi. The missing orbital data in the figure correspond to values smaller than −1. For the inner shells, i.e., 1*s*-5*s* for Hf and W and 1*s*-5*d* for Pb and Bi, the semirelativistic calculations and nonrelativistic ones are fairly similar, with a ~9% agreement, while the relativistic binding energy calculations coincide with the experimental values in about 3%. As expected, the inclusion of relativistic effects are responsible for these improvements. The overall good agreement of the relativistic calculations demonstrates the necessity of including properly the relativistic effects in the structure calculations. As it will be shown later, some of the outer electron wavefunctions have a non-negligible overlap with the internal ones. Therefore, the relativistics effects spread all over the full structure calculation, including the outer-shells. There are some discrepancies for the 5*p* and 4*f* electrons, in the case of Hf and W, and for the 5*p* and 5*d* subshells in Pb and Bi, which can be understood if one recalls that the theoretical calculations assume isolated atoms, whereas the experimental data were taken from the targets in solid state.

In Fig. 3, we display the most outer shells of Hf, W, Pb, and Bi. As can be noticed, the 5*d* and 6*s*, in Hf and W, and the 6*s* and 6*p* electrons, in Pb and Bi, are part of the FEG. For these subshells, the binding energies displayed in

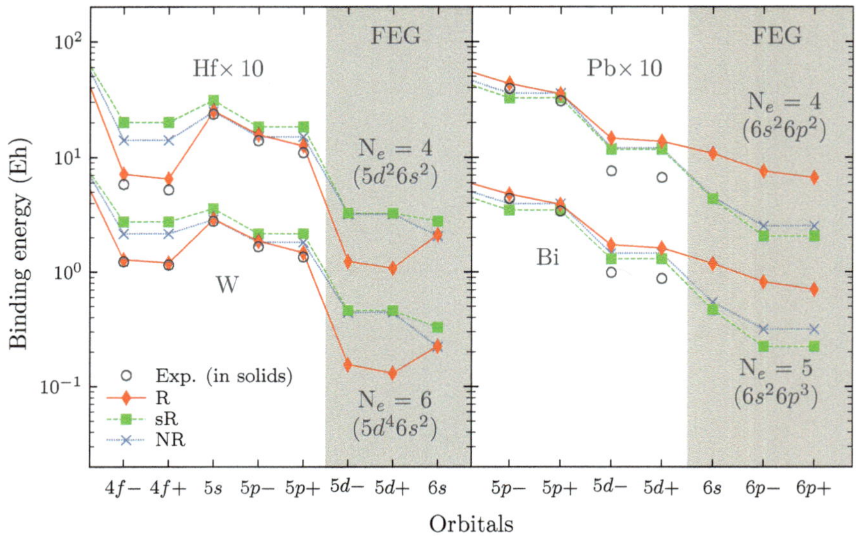

Fig. 3 Binding energy and number of electrons in FEG considered for Hf, W, Pb, and Bi. Symbols: same as Fig. 1.

Fig. 3 only have sense for atoms but not for solids. No experimental values in solids are available, as expected, because these electrons belong to the conduction band. Moreover, these results for the electronic structure are coherent with the expected number of electrons in the FEG, i.e., $N_e = 4$ for Hf and Pb, $N_e = 5$ for Bi, and $N_e = 6$ for W.

The semirelativistic (sR) calculation shown in Figs. 1 and 3 can be further improved by optimizing the λ_{nl} radial scaling parameters that define the TFDA potential employed. There are many approaches to this end; in particular, we manually adjusted the values of λ_{nl} so that the computed binding energies are the same as the experimental measurements. The optimized calculation, denoted sR*, is not shown in Figs. 1 and 3. To illustrate this result, we present in Fig. 4 the calculated densities and mean radius of the inner subshells of Pb. For the relativistic data, the reduced electron density is given by $d_{nl}(r) = P_{nl}(r)^2 + Q_{nl}(r)^2$, where P and Q are the strong and weak components of the reduced radial wavefunction ($\int_0^\infty d_{nl}(r)\, dr = 1$). To simplify this analysis, we have considered a split-orbit average for the nlj subshells. The right panel in Fig. 4 shows that the NR mean radius of the core electron's radial distributions are generally larger than the R, sR and sR* results, as expected. We can observe that the radial distribution of the sR* calculations are also influenced by the optimization. For most of the inner shells,

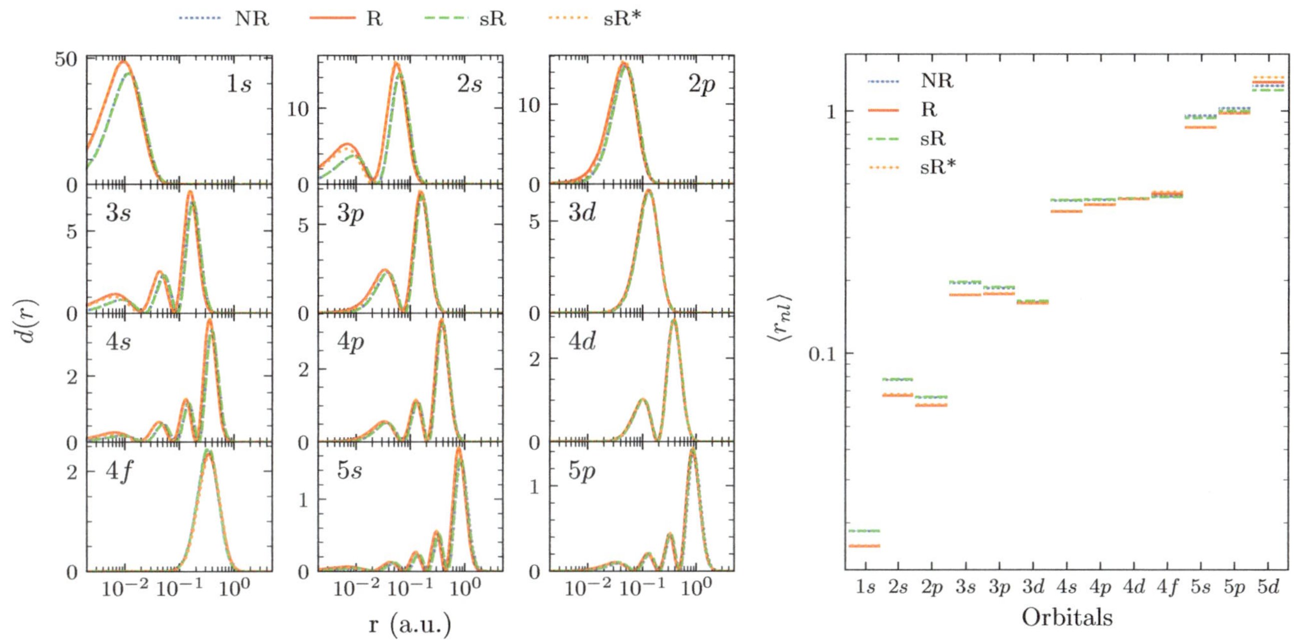

Fig. 4 Present relativistic (R), semirelativistic (sR), optimized semirelativistic (sR*), and nonrelativistic (NR) atomic structure results for Pb. *Left*: Reduced radial electron density for 1*s* to 5*p* subshells (5*d* not included for simplicity). *Right*: mean radius for each subshell.

these results follow the correct behavior: the optimized electron densities are pulled toward the nuclei. The analysis of the differences in the $4f$ and $5d$ subshells surpasses the scope of this work.

In what follows, we present stopping cross sections for Hf, W, Pb, and Bi. The total stopping is obtained by calculating separately the contribution of valence (FEG) and bound electrons, given by Eqs. (1) and (2), respectively. The density ρ_{nl} used in Eq. (3) is related to the reduced electron density d_{nl} by $\rho_{nl}(r) = N_{nl}d_{nl}(r)/r^2$, with N_{nl} being the number of electrons in the subshell nl. Three formalisms are combined to obtain the total stopping cross sections. The first one considers a nonperturbative description of the energy loss in the FEG, based on an screening potential SPCC.[25] This model relies on the binary collisional theory, so no collective excitations (plasmons) are included. The second one calculates the stopping power of the FEG by using Mermin–Lindhard[28] dielectric formalism (ML), which follows a perturbative approximation but includes collective excitations in the energy region, where the plasmon contribution is important. Finally, the third approach combines the SLPA and relativistic electronic structure computations described above.

The first step in the total stopping calculation consists on determining how many electrons are included in the FEG, while the rest are assumed bound to the atom. To make this decision, we considered the relativistic results displayed in Fig. 3. In previous works,[45,46] we based this choice on external experimental values from energy loss measurements in the optical limit. In this contribution, we decided to rely on our full-theoretical developments. Other important improvement constitutes the introduction of the Levine-Mermin dielectric function, which is critical for many target subshells. It is worth mentioning that, in all our calculations, the H projectiles are considered as protons. This characterization is based on two physical processes: (i) Inside the solid, at low impact energies, the dynamic screening of the ion by the FEG prevents any electron to occupy the $1s$ bound state of H.[47] (ii) At high energies, the probability of projectile ionization is more important than capture, so the hydrogen looses the only bound electron in few collisions and arrives to an equilibrium charge state of $Z_P = 1$. For hydrogen projectiles, this charge state is also considered in other calculations like SRIM[48,49] but these are based on the empirical stopping results.

In Fig. 5, we display the present results for H in Hf. The total stopping agrees quite well with the experimental data, suggesting a maximum around 70 keV/amu. Separate curves for the FEG and bound electrons contributions are displayed, together with the total value. As mentioned before,

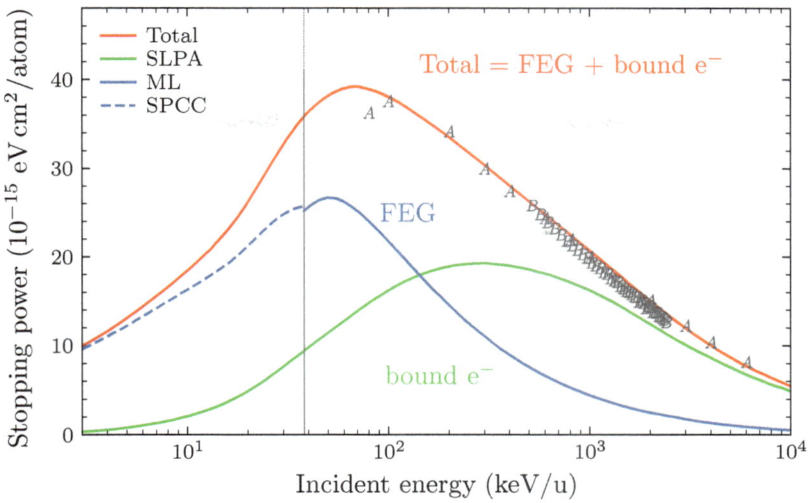

Fig. 5 Stopping power of H on Hf. The contribution of bound electrons and FEG is displayed separately (see acronyms from inset in text). Symbols, experimental data as in IAEA database[50]: A[51] and B.[24]

the FEG curve results from the combination of the SPCC at low energies (dashed line) and the ML at intermediate and high energies. The vertical line illustrates the threshold energy for plasmon excitations.[25,47] The recent version of the SLPA, which considers a local damping,[27] and the present relativistic calculations are employed, improving previous results obtained in Montanari et al.[24]

In Figs. 6–9, we show a set of four total stopping cross sections calculations for H on Hf, W, Pb, and Bi, respectively. In the following, we analyze the importance of the electronic structure calculations by comparing relativistic, semirelativistic and nonrelativistic results. We also include comparison with the semiempirical values by Ziegler's SRIM code,[48,49] the theoretical results by Schiwietz and Grande's code CasP version 6.0 for $Z_P = 1$,[52,53] and the Sigmund and Schinner's DPASS[54,55] curve.

For Hf, no experimental data are available for impact energies around the maximum and below. It can be noted in Fig. 6 that the differences between NR and R are not dramatic, being the former smaller than the latter and with a maximum value slightly shifted to higher energies. On the other hand, the sR and sR* curves are quite different, but neither of them agree with the experimental data. If one recalls Fig. 3, it becomes evident that the more binded are the electrons of the subshell below the FEG (following the experimental description), the larger the cross section.

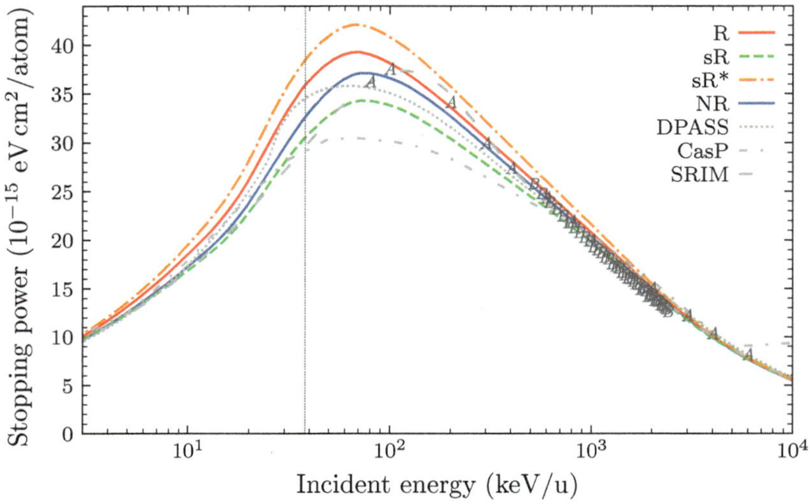

Fig. 6 Stopping power of H on Hf. Present results using relativistic and nonrelativistic electronic structure calculations for bound electrons. In gray theoretical curves CASP[52] and DPASS,[54] and semiempirical values by SRIM.[48] Details in the figure inset. Symbols, experimental data as in IAEA database[50]: A[51] and B.[24]

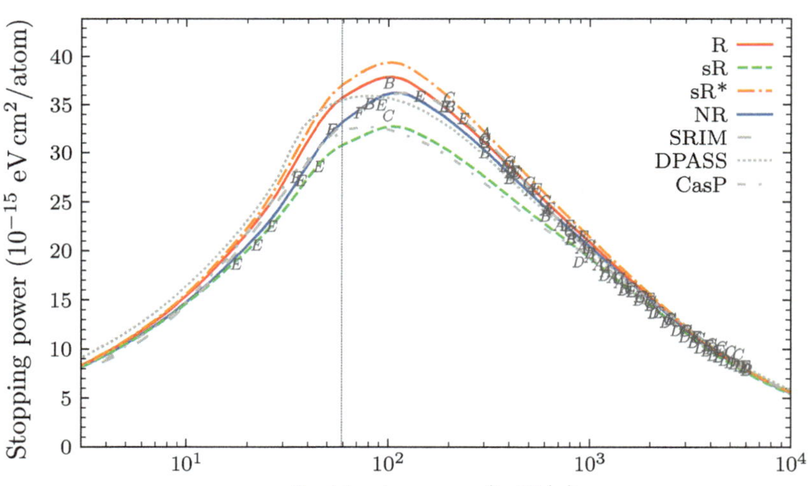

Fig. 7 Stopping power of H on W. Curves as in Fig. 6. Symbols, experimental data as in IAEA database[50]: A,[56] B,[51] C,[57] D,[58] E,[23] and F[23] (for D^+ in W).

Similar results are obtained for H on W, as displayed in Fig. 7. However, recent measurements by Moro et al.[23] (represented with letters E and F in Fig. 7) agree better with the NR results for energies below the maximum.

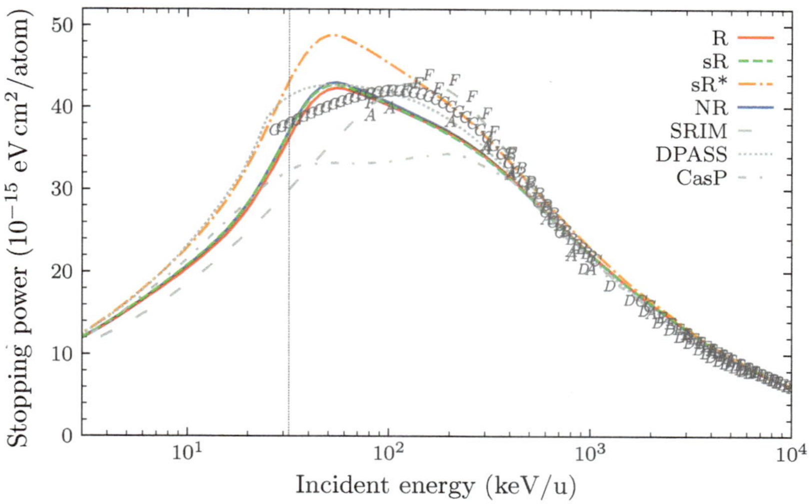

Fig. 8 Stopping power of H on Pb. Curves as in Fig. 6. Symbols, experimental data as in IAEA database[50]: A,[51] B,[59] C,[60] D,[58] E,[61] F,[62] G,[63] H,[64] I,[65] J,[66] and K.[67]

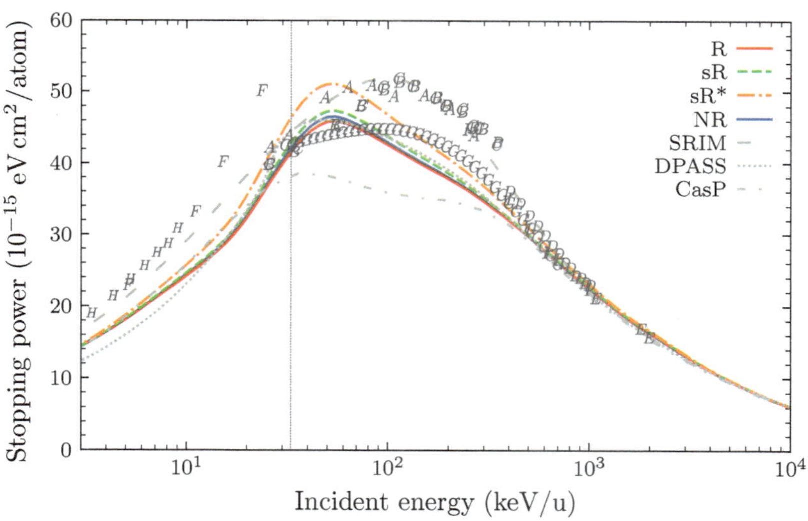

Fig. 9 Stopping power of H on Bi. Curves as in Fig. 6. Symbols, experimental data as in IAEA database[50]: A,[68] B,[69] C,[70] D,[59] E,[60] F,[71] G,[72] H,[73] and I.[74]

The comparison with other theoretical and semiempirical curves shows that DPASS[54] is the closest to our results. It is worth mentioning that for Hf and W, the 4*f* electrons are close in energy to the 5*p* electrons. The SLPA includes the screening among electrons of similar binding energy. For these cases, inter-shell $5p$–$4f$ screening is included considering the small difference between the binding energies and the meantime for the ion to get through the subshells, as discussed in Ref. 24.

The results for Pb and Bi are different. As can be noted in Fig. 3, the closed $5d$ subshell is located just below the FEG and, consequently, it becomes the main contribution to the stopping power around the maximum. The theoretical structure calculation of such multielectronic targets is a difficult task; particularly for the outer electrons. We can observe in Figs. 8 and 9 that the NR, sR and R calculations are very similar, which follows the corresponding binding energy comparison in Fig. 3. Moreover, these values are close to other models, such as DPASS.[54] The double maximum shape of our R curves indicates that the bound electron contribution could be underestimated. The difference between theoretical (deeper) and experimental (smaller) binding energy for the $5d$ subshell may explain this. In fact, the sR*, which agrees the $5d$ experimental binding energy, describes better the data for impact energies below 500 keV/amu. It should be noted that the latest data on Bi and Pb, for energies around the stopping maximum and below, were measured 30 years ago. Bi and Pb are exceptional cases and new experimental efforts at impact energies below 200 keV are encouraged to clarify these differences.

4. Conclusions

In this contribution, we advanced in the theoretical description of the target's electronic structure by using the best relativistic descriptions available. The combination of these results with the dielectric formalism, the SLPA and the nonperturbative model based on the screened potential allowed us to predict stopping values in an extent energy region. Hf and W were examined to illustrate postlanthanides transition metals, while Pb and Bi exemplified the posttransition metals of the group *p*. The present stopping power results are compared with the experimental data available, and other theoretical and semiempirical models. The influence of the $4f$ electrons (of Hf and W) and the $5d$ electrons (of Pb and Bi) is discussed. From the stopping cross sections analysis, we show that even though there are limitations in the atomic structure approximations, doubts rise over certain experimental values around the stopping maximum. It is obvious from the present analysis that the disagreement among data sets makes it imperative to count on independent theoretical models. Of course, semiempirical models will adapt to the existing data, and this is valid, as far as the existing measurements are correct. However, for the targets considered here, our model provides one of the best fully theoretical description available.

Acknowledgments

The authors acknowledge the financial support of Consejo Nacional de Investigaciones Científicas y Tecnológicas, Agencia Nacional de Promoción de la Investigación, el Desarrollo Tecnológico y la Innovación, and Universidad de Buenos Aires. The authors thanks Jorge Miraglia for previous discussions on this subject.

References

1. Rosenblum, S. Recherches Expérimentales sur le Passage des Rayons Travers la Matière. *Ann. Phys.* **1928**, *10* (10), 408–471. https://doi.org/10.1051/anphys/192810100408.
2. Bethe, H. Zur Theorie des Durchgangs schneller Korpuskularstrahlen durch Materie. *Ann. Phys.* **1930**, *397* (3), 325–400. https://doi.org/10.1002/andp.19303970303.
3. Bohr, N. *The Penetration of Atomic Particles Through Matter*; Munksgaard Copenhagen, 1948.
4. Oddershede, J.; Sabin, J. R. Orbital and Whole-Atom Proton Stopping Power and Shell Corrections for Atoms With Z $\leq$ 36. *At. Data Nucl. Data Tables* **1984**, *31* (2), 275–297. https://doi.org/10.1016/0092-640X(84)90024-X.
5. Oddershede, J.; Sabin, J. R. Bragg Rule Additivity of Bond Stopping Cross Sections. *Nucl. Instrum. Methods Phys. Res. Sect. B* **1989**, *42* (1), 7–10. https://doi.org/10.1016/0168-583X(89)90003-7.
6. Cabrera-Trujillo, R.; Sabin, J. R.; Deumens, E.; Öhrn, Y. Dynamical Processes in Stopping Cross Sections. In *Theory of the Interaction of Swift Ions With Matter. Part 1*; Advances in Quantum Chemistry, Vol. 45; Academic Press, 2004; pp. 99–124. https://doi.org/10.1016/S0065-3276(04)45005-9.
7. Quashie, E. E.; Andrade, X. A. A. C. Directional Dependency of Electronic Stopping in Nickel, Projectile'S Excited Charge State and Momentum Transfer. *Eur. Phys. J. D* **2021**, *75*, 280. https://doi.org/10.1140/epjd/s10053-021-00266-6.
8. Schiwietz, G.; Grande, P. Stopping of Protons – Improved Accuracy of the UCA Model. *Nucl. Instrum. Methods Phys. Res. Sect. B* **2012**, *273*, 1–5. https://doi.org/10.1016/j.nimb.2011.07.023 (20th International Conference on Ion Beam Analysis).
9. Sigmund, P.; Schinner, A. Progress in Understanding Heavy-Ion Stopping. *Nucl. Instrum. Methods Phys. Res. Sect. B* **2016**, *382*, 15–25. https://doi.org/10.1016/j.nimb.2015.12.041 (The 21st International workshop on Inelastic Ion Surface Collisions (IISC-21)).
10. Arista, N. R. Energy Loss of Ions In Solids: Non-Linear Calculations for Slow and Swift Ions. *Nucl. Instrum. Methods Phys. Res. Sect. B* **2002**, *195* (1), 91–105. https://doi.org/10.1016/S0168-583X(02)00687-0.
11. Lindhard, J. On the Properties of a Gas of Charged Particles. *K. Dan. Vidensk. Selsk. Mat.-Fys. Medd.* **1954**, *28* (8).
12. Ritchie, R. H. Interaction of Charged Particles With a Degenerate Fermi-Dirac Electron Gas. *Phys. Rev.* **1959**, *114* (3), 644–654. https://doi.org/10.1103/PhysRev.114.644.
13. Echenique, P. M.; Brandt, W.; Ritchie, R. H. Self-Consistent Wake Binding Energies. *Phys. Rev. B* **1986**, *33* (1), 43–48. https://doi.org/10.1103/PhysRevB.33.43.
14. Echenique, P. M.; Flores, F.; Ritchie, R. H. Dynamic Screening of Ions in Condensed Matter. In *Solid State Physics*; Ehrenreich, H., Turnbull, D., Eds.; Vol. 43; Academic Press, 1990; pp. 229–308. https://doi.org/10.1016/S0081-1947(08)60325-2.

15. Lindhard, J.; Scharff, M. Energy Loss in Matter by Fast Particles of Low Charge. *K. Dan. Vidensk. Selsk. Mat.-Fys. Medd.* **1953**, *27* (15), 1–31.
16. Chu, W. K.; Powerd, D. Calculation of Mean Excitation Energy for All Elements. *Phys. Lett. A* **1972**, *40* (1), 23–24. https://doi.org/10.1016/0375-9601(72)90181-8.
17. Meltzer, D. E.; Sabin, J. R.; Trickey, S. B. Calculation of Mean Excitation Energy and Stopping Cross Section in the Orbital Local Plasma Approximation. *Phys. Rev. A* **1990**, *41* (1), 220–232. https://doi.org/10.1103/PhysRevA.41.220.
18. Montanari, C. C.; Miraglia, J. E. The Dielectric Formalism for Inelastic Processes in High-Energy Ion-Matter Collisions. In *Advances in Quantum Chemistry: Theory of Heavy Ion Collision Physics in Hadron Therapy*; Belkic, D., Ed.; Vol. 2; Elsevier: New York, 2013; pp. 165–201 (Chapter 7).
19. de Vera, P.; Abril, I.; Garcia-Molina, R. Energy Spectra of Protons and Generated Secondary Electrons Around the Bragg Peak in Materials of Interest in Proton Therapy. *Radiat. Res.* **2018**, *190* (3), 282–297. https://doi.org/10.1667/RR14988.1.
20. Montanari, C.; Dimitriou, P. The IAEA Stopping Power Database, Following the Trends in Stopping Power of Ions in Matter. *Nucl. Instrum. Methods Phys. Res. Sect. B* **2017**, *408*, 50–55. https://doi.org/10.1016/j.nimb.2017.03.138 (Proceedings of the 18th International Conference on the Physics of Highly Charged Ions (HCI-2016), Kielce, Poland, 11–16 September 2016).
21. Roth, D.; Bruckner, B.; Moro, M. V.; Gruber, S.; Goebl, D.; Juaristi, J. I.; Alducin, M.; Steinberger, R.; Duchoslav, J.; Primetzhofer, D.; Bauer, P. Electronic Stopping of Slow Protons in Transition and Rare Earth Metals: Breakdown of the Free Electron Gas Concept. *Phys. Rev. Lett.* **2017**, *118* (10), 103401. https://doi.org/10.1103/PhysRevLett.118.103401.
22. Selau, F. F.; Trombini, H.; Marmitt, G. G.; de Andrade, A. M. H.; Morais, J.; Grande, P. L.; Alencar, I.; Vos, M.; Heller, R. Stopping and Straggling of 60-250-keV Backscattered Protons on Nanometric Pt Films. *Phys. Rev. A* **2020**, *102* (3), 032812. https://doi.org/10.1103/PhysRevA.102.032812.
23. Moro, M. V.; Wolf, P. M.; Bruckner, B.; Munnik, F.; Heller, R.; Bauer, P.; Primetzhofer, D. Experimental Electronic Stopping Cross Section of Tungsten for Light Ions in a Large Energy Interval. *Nucl. Instrum. Methods Phys. Res. Sect. B* **2021**, *498*, 1–8. https://doi.org/10.1016/j.nimb.2021.04.010.
24. Montanari, C. C.; Miranda, P. A.; Alves, E.; Mendez, A. M. P.; Mitnik, D. M.; Miraglia, J. E.; Correa, R.; Wachter, J.; Aguilera, M.; Catarino, N.; da Silva, R. C. Stopping Power of Hydrogen in Hafnium and the Importance of Relativistic 4*f* Electrons. *Phys. Rev. A* **2020**, *101* (6), 062701. https://doi.org/10.1103/PhysRevA.101.062701.
25. Montanari, C. C.; Miraglia, J. E. Low- and Intermediate-Energy Stopping Power of Protons and Antiprotons in Solid Targets. *Phys. Rev. A* **2017**, *96* (1), 012707. https://doi.org/10.1103/PhysRevA.96.012707.
26. Levine, Z. H.; Louie, S. G. New Model Dielectric Function and Exchange-Correlation Potential for Semiconductors and Insulators. *Phys. Rev. B* **1982**, *25* (10), 6310–6316. https://doi.org/10.1103/PhysRevB.25.6310.
27. Peralta, J. P.; Fiori, M.; Mendez, A. M. P.; et al. Stopping Power Calculations and the Levine-Mermin Dielectric Function for Inner Shells. *Phys. Rev. A* **2022**, *105*, 062814. https://doi.org/10.1103/PhysRevA.105.062814.
28. Mermin, N. D. Lindhard Dielectric Function in the Relaxation-Time Approximation. *Phys. Rev. B* **1970**, *1* (5), 2362–2363. https://doi.org/10.1103/PhysRevB.1.2362.
29. Werner, W. S. M.; Glantschnig, K.; Ambrosch-Draxl, C. Optical Constants and Inelastic Electron-Scattering Data for 17 Elemental Metals. *J. Phys. Chem. Ref. Data* **2009**, *38* (4), 1013–1092. https://doi.org/10.1063/1.3243762.

30. Abril, I.; Garcia-Molina, R.; Denton, C. D.; Pérez-Pérez, F. J.; Arista, N. R. Dielectric Description of Wakes and Stopping Powers in Solids. *Phys. Rev. A* **1998**, *58* (1), 357–366. https://doi.org/10.1103/PhysRevA.58.357.
31. Grant, I. P. Relativistic Atomic Structure Calculations. *Comput. Phys. Commun.* **1979**, *17* (1–2), 149–161. https://doi.org/10.1016/0010-4655(79)90077-8.
32. Fischer, C. F.; Godefroid, M.; Brage, T.; Jönsson, P.; Gaigalas, G. Advanced Multiconfiguration Methods for Complex Atoms: I. Energies and Wave Functions. *J. Phys. B At. Mol. Opt. Phys.* **2016**, *49* (18), 182004. https://doi.org/10.1088/0953-4075/49/18/182004.
33. Eissner, W.; Jones, M.; Nussbaumer, H. Techniques for the Calculation of Atomic Structures and Radiative Data Including Relativistic Corrections. *Comput. Phys. Commun.* **1974**, *8* (4), 270–306. https://doi.org/10.1016/0010-4655(74)90019-8.
34. Badnell, N. R. On the Effects of the Two-Body Non-Fine-Structure Operators of the Breit - Pauli Hamiltonian. *J. Phys. B: At. Mol. Opt. Phys.* **1997**, *30* (1), 1. https://doi.org/10.1088/0953-4075/30/1/005.
35. Klapisch, M.; Schwob, J. L.; Fraenkel, B. S.; Oreg, J. The 1s-3p Kb-Like X-Ray Spectrum of Highly Ionized Iron. *J. Opt. Soc. Am.* **1977**, *67* (2), 148–155.
36. Desclaux, J. P. Relativistic Dirac-Fock Expectation Values for Atoms With Z = 1 to Z = 120. *At. Data Nucl. Data Tables* **1973**, *12* (4), 311–406. https://doi.org/10.1016/0092-640X(73)90020-X.
37. Parpia, F. A.; Froese Fischer, C.; Grant, I. P. GRASP92: A Package for Large-Scale Relativistic Atomic Structure Calculations. *Comput. Phys. Commun.* **1996**, *94* (2–3), 249–271. https://doi.org/10.1016/0010-4655(95)00136-0.
38. Klapisch, M. A Program for Atomic Wavefunction Computations by the Parametric Potential Method. *Comput. Phys. Commun.* **1971**, *2* (5), 239–260. https://doi.org/10.1016/0010-4655(71)90001-4.
39. Bar-Shalom, A.; Klapisch, M.; Oreg, J. HULLAC, an Integrated Computer Package for Atomic Processes in Plasmas. *J. Quant. Spectrosc. Radiat. Transf.* **2001**, *71* (2–6), 169–188. https://doi.org/10.1016/S0022-4073(01)00066-8.
40. Badnell, N. R. A Breit-Pauli Distorted Wave Implementation for Autostructure. *Comput. Phys. Commun.* **2011**, *182* (7), 1528–1535. https://doi.org/10.1016/J.CPC.2011.03.023.
41. Bautista, M. A. Electronic Correlations and Polarizability of the Thomas-Fermi-Dirac-Amaldi Potential: Applications to the Singly Ionized Iron-Peak Species. *J. Phys. B At. Mol. Opt. Phys.* **2008**, *41* (6), 065701. https://doi.org/10.1088/0953-4075/41/6/065701.
42. Gombás, P. Statistische Behandlung des Atoms. In Atoms II/Atome II; Springer: Berlin, Heidelberg, 1956; pp. 109–231. https://doi.org/10.1007/978-3-642-85687-7_2.
43. Froese Fischer, C. General Hartree-Fock Program. *Comput. Phys. Commun.* **1987**, *43* (3), 355–365. https://doi.org/10.1016/0010-4655(87)90053-1.
44. Williams, G. *DPASS21.06*; 2020. https://www.sdu.dk/en/dpass.
45. Montanari, C. C.; Archubi, C. D.; Mitnik, D. M.; Miraglia, J. E. Energy Loss of Protons in Au, Pb, and Bi Using Relativistic Wave Functions. *Phys. Rev. A* **2009**, *79* (3), 032903. https://doi.org/10.1103/PhysRevA.79.032903.
46. Montanari, C. C.; Mitnik, D. M.; Archubi, C. D.; Miraglia, J. E. Energy Loss of Protons in W Using Fully Relativistic Calculations and Mean Excitation Energies of W, Au, Pb, and Bi. *Phys. Rev. A* **2009**, *80* (1), 012901. https://doi.org/10.1103/PhysRevA.80.012901.
47. Montanari, C. C.; Miraglia, J. E.; Arista, N. R. Suppression of Projectile-Electron Excitations in Collisions With a Free-Electron Gas of Metals. *Phys. Rev. A* **2000**, *62* (5), 052902. https://doi.org/10.1103/PhysRevA.62.052902.
48. Ziegler, J. F. *SRIM*; 2013. http://www.srim.org/.

49. Ziegler, J. F.; Ziegler, M. D.; Biersack, J. P. SRIM—The Stopping and Range of Ions in Matter (2010). *Nucl. Instrum. Methods Phys. Res. Sect. B* **2010**, *268* (11), 1818–1823. https://doi.org/10.1016/j.nimb.2010.02.091 (19th International Conference on Ion Beam Analysis).
50. IAEA.; Paul, H.; Montanari, C. C. *Electronic Stopping Power of Matter for Ions Graphs, Data, Comments and Programs*; 1928–2021. https://www-nds.iaea.org/stopping/.
51. Sirotinin, E. I.; Tulinov, A. F.; Khodyrev, V. A.; Mizgulin, V. N. *Nucl. Instrum. Methods Phys. Res. Sect. B* **1984**, *4* (3), 337–345.
52. Grande, P. L.; Schiwietz, G. *CasP version 6.0*; 2021. http://www.casp-program.org/.
53. Matias, F.; Fadanelli, R. C.; Grande, P. L.; Arista, N. R.; Koval, N. E.; Schiwietz, G. Stopping Power of Cluster Ions in a Free-Electron Gas From Partial-Wave Analysis. *Phys. Rev. A* **2018**, *98* (6), 062716. https://doi.org/10.1103/PhysRevA.98.062716.
54. Schinner, A.; Sigmund, P. *DPASS21.06*; 2020. https://www.sdu.dk/en/dpass.
55. Schinner, A.; Sigmund, P. Expanded PASS Stopping Code. *Nucl. Instrum. Methods Phys. Res. Sect. B* **2019**, *460*, 19–26. https://doi.org/10.1016/j.nimb.2018.10.047.
56. Luomajarvi, M. Stopping Powers of Some Metals for 0.3–1.5 MeV Protons. *Radiat. Effects* **1979**, *40* (3), 173–179. https://doi.org/10.1080/00337577908237920.
57. Chumanov, V. Y.; Izmailov, S. Z.; Pokhil, G. P.; Sirotinin, E. I.; Tulinov, A. F. *Phys. Status Solidi A* **1979**, *53* (1), 51–62. https://onlinelibrary.wiley.com/doi/epdf/10.1002/pssa.2210530104.
58. Sirotinin, E. I.; Tulinov, A. F.; Fiderkevich, A.; Shyshkin, K. S. The Determination of Energy Losses From the Spectrum of Particles Scattered by a Thick Target. *Radiat. Effects* **1972**, *15* (3–4), 149–152. https://doi.org/10.1080/00337577208234688.
59. Green, D. W.; Cooper, J. N.; Harris, J. C. Stopping Cross Sections of Metals for Protons of Energies From 400 to 1000 kev. *Phys. Rev.* **1955**, *98* (2), 466–470. https://doi.org/10.1103/PhysRev.98.466.
60. Ogino, K.; Kiyosawa, T.; Kiuchi, T. Stopping Powers for MeV Tritons in Solids. *Nucl. Instrum. Methods Phys. Res. Sect. B* **1988**, *33* (1), 155–157. https://doi.org/10.1016/0168-583X(88)90535-6.
61. Sörensen, H.; Andersen, H. H. Stopping Power of Al, Cu, Ag, Au, Pb, and U for 5–18-MeV Protons and Deuterons. *Phys. Rev. B* **1973**, *8* (5), 1854–1863. https://doi.org/10.1103/PhysRevB.8.1854.
62. Bader, M.; Pixley, R. E.; Mozer, F. S.; Whaling, W. Stopping Cross Section of Solids for Protons, 50–600 kev. *Phys. Rev.* **1956**, *103* (1), 32–38. https://doi.org/10.1103/PhysRev.103.32.
63. Eppacher, C.; Semrad, D. Dependence of Proton and Helium Energy Loss in Solids Upon Plasma Properties. *Nucl. Instrum. Methods Phys. Res. Sect. B* **1992**, *69* (1), 33–38. https://doi.org/10.1016/0168-583X(92)95735-A.
64. Sakamoto, N.; Ogawa, H.; Mannami, M.; Kimura, K.; Susuki, Y.; Hasegawa, M.; Katayama, I.; Noro, T.; Ikegami, H. Stopping Powers of Metallic Elements for High Energy Ions. *Radiat. Effects Defects Solids* **1991**, *117* (1–3), 193–195. https://doi.org/10.1080/10420159108220612.
65. Sakamoto, N.; Shiomi, N.; Ogawa, H.; Ishiwari, R. Stopping Powers of Sn and Pb for 3.0–8.5 MeV Protons. *Nucl. Instrum. Methods Phys. Res. Sect. B* **1986**, *13* (1), 115–118. https://doi.org/10.1016/0168-583X(86)90485-4.
66. Ishiwari, R.; Shiomi, N.; Sakamoto, N. Stopping Powers of Zr, Pd, Cd, in and Pb for 6.5 MeV Protons and Mean Excitation Energies. *Nucl. Instrum. Methods Phys. Res. Sect. B* **1984**, *2* (1), 195–198. https://doi.org/10.1016/0168-583X(84)90187-3.
67. Bischel, H.; Hiraoka, T. Energy Loss of 70 MeV Protons in Elements. *Nucl. Instrum. Methods Phys. Res. Sect. B* **1992**, *66* (3), 345–351. https://doi.org/10.1016/0168-583X(92)95995-4.

68. Eckardt, J. C. Energy Loss and Straggling of Protons and Helium Ions Traversing Some Thin Solid Foils. *Phys. Rev. A* **1978**, *18* (2), 426–433. https://doi.org/10.1103/PhysRevA.18.426.
69. Krist, T.; Mertens, P. Proton Energies at the Maximum of the Electronic Stopping Cross Section in Materials With 57≤Z≤83. *Nucl. Instrum. Methods Phys. Res.* **1983**, *218* (1), 790–794. https://doi.org/10.1016/0167-5087(83)91084-0.
70. Krist, T.; Mertens, P. Stopping Ratios for 30–330 keV Light Ions in Materials With 57≤Z≤83. *Nucl. Instrum. Methods Phys. Res.* **1983**, *218* (1), 821–826. https://doi.org/10.1016/0167-5087(83)91090-6.
71. Arkhipov, E. P.; Gott, Y. V. Slowing Down of 0.5-30 keV Protons in Some Materials. *Soviet J. Exp. Theor. Phys.* **1969**, *29*, 615.
72. Eppacher, C. *Die "effektive Ladung" von schnellen Heliumionen in Metallen* (Ph.D. thesis), Schriften der Johannes-Kepler-Universität Linz, Universitätsverlag Rudolf Trauner, 1995.
73. Valdés, J.; Tamayo, G.; Lantschner, G.; Eckardt, J.; Arista, N. Electronic Energy Loss of Low Velocity H+ Beams in Al, Ag, Sb, Au and Bi. *Nucl. Instrum. Methods Phys. Res. Sect. B* **1993**, *73* (3), 313–318. https://doi.org/10.1016/0168-583X(93)95744-P.
74. Knudsen, H.; Andersen, H. H.; Martini, V. Hydrogen and Helium Stopping Powers of Rare-Earth Metals. *Nucl. Instrum. Methods* **1980**, *168* (1), 41–50. https://doi.org/10.1016/0029-554X(80)91229-X.

CHAPTER NINE

Density-functional methods for extended helical systems

J.W. Mintmire*
Department of Physics, Oklahoma State University, Stillwater, OK, United States
*Corresponding author: e-mail address: john.mintmire@okstate.edu

Contents

Abstract

Over the past several decades, we have made progress in developing an approach for density-functional electronic structure calculations on quasi-one-dimensional nanostructures with helical symmetry, an approach that builds on early work by the author in collaboration with Prof. Sabin. In this chapter we review the development and details of these first-principles methods using Gaussian basis sets for calculating the electronic band structure of periodic graphitic nanostructures such as carbon nanotubes and graphene nanoribbons. An application of this method to recent experimental results for ortho-connected polythiophenes is also presented.

1. Introduction

I am pleased to write an article in this volume of *Advances in Quantum Chemistry* which honors Prof. John R. Sabin for his achievements in the field and his service as Serial Editor of this book series. I first met Jack when I was an undergraduate student at the University of Florida in 1974 and he was the instructor of one of my introductory general physics courses. I liked Jack's teaching methods very much, and when I was looking for a mentor for

Advances in Quantum Chemistry, Volume 85
ISSN 0065-3276
https://doi.org/10.1016/bs.aiq.2022.04.002
Copyright © 2022 Elsevier Inc.
All rights reserved.

undergraduate research as part of my major, I approached Jack to be my advisor and started working with him on a project calculating total energies of small molecules.[1] After beginning graduate school at Florida in 1976, I continued working with Jack as his graduate student. Brett Dunlap had arrived around 1976 as a postdoctoral associate working with John Connolly and Jack, with the development of nonmuffin-tin-based density-functional molecular calculations using Gaussian basis sets as his project. As an introductory project, Jack asked me to carry out some related calculations on the F_2 and carbon monoxide molecules using Brett's Gaussian basis set method vs standard Hartree–Fock methods with the same basis set, and also compare these results with current muffin-tin and with any other density-functional results.[2,3] With that project completed, Jack and I discussed carrying out similar density-functional calculations on surfaces, which resulted in my doctoral dissertation work applying these Gaussian basis set density-functional methods to extended systems with two-dimensional periodicity.[4–6]

After graduating from the University of Florida in 1980, I went to the US Naval Research Laboratory as a postdoctoral associate in the Theoretical Chemistry Section working with Carter White. Brett Dunlap had recently been hired as a new staff researcher in that group, and I was looking for opportunities to apply density-functional methods to new extended systems working with Brett and Carter. The Chemistry Division had active research programs in electroactive polymers such as polyacetylene at that time, and I adapted the computational approach I had developed at Florida while working with Jack to handle systems with one-dimensional translational periodicity, with applications to a range of polymeric materials such as polyacetylene[7,8] and polysilane[9,10] systems. In addition to pure translational symmetry, many chain polymers exhibit helical conformations. Using helical symmetry to calculate electronic structures of such systems was first proposed for semiempirical systems[11,12] with an ab initio scheme later reported by Blumen and Merkel.[13] At the same time that our group at the Naval Research Laboratory was studying systems with translational symmetry, Gaussian basis sets were being applied to a range of realistic chain polymer systems with helical symmetry.[14–16] When I was on sabbatical during 1989–90 at the Ohio State University in John Wilkins' group, my main effort was to extend our then-current computational framework to treat systems with helical symmetry as well.[17] Since then, this approach has been used to study a broad range of quasi-one-dimensional extended systems including carbon nanotubes, graphene nanoribbons, and other nanowires.[18–23]

In the following, we present a summary of this approach for helical chain polymers. The basics of the method are presented, with a focus on the methods used to evaluate energy gradients and the long-range lattice sums needed for the Coulomb matrix elements. Evaluation of these lattice sums has been a major challenge to improving computational performance of this method, and herein we discuss the use of polylogarithms to help evaluate long-range sums of multipole terms. Finally, a brief example will be presented with application to recent experimental results.

2. Helical band structure methods

Here, we discuss our algorithms for calculating the local-density-functional total energies for systems that have a quasi-one-dimensional screw symmetry. By "quasi-one-dimensional," we mean the system has periodicity in one dimension but is finite in the other two dimensions. Such systems are full three-dimensional systems, and the phrase "quasi-one-dimensional" is used to distinguish this system from model systems that are strictly defined with a single coordinate dimension. Quasi-one-dimensional systems are one example of a family of nanostructured systems denoted as reduced dimensional systems.

We can define a helical chain polymer as a nuclear lattice constructed from a finite basis of atoms and a screw operation $\mathcal{S}(a, \phi)$. For mathematical convenience, we define the screw operation in terms of a translation a units down the z-axis in conjunction with a right-handed rotation ϕ about the z-axis. That is,

$$\mathcal{S}(a, \phi)\,\mathbf{r} = \begin{pmatrix} x\cos\phi - y\sin\phi \\ x\sin\phi + y\cos\phi \\ z + a \end{pmatrix} \tag{1}$$

with the screw operation henceforth denoted as $\mathcal{S}$, with arguments a and ϕ implicitly understood. Because the symmetry group generated by the screw operation $\mathcal{S}$ is isomorphic with the one-dimensional translation group, Bloch's theorem can be generalized so that the one-electron wavefunctions will transform under $\mathcal{S}$ according to

$$\mathcal{S}^m \psi_i(\mathbf{r}; \underline{\kappa}) = e^{i\underline{\kappa} m}\, \psi_i(\mathbf{r}; \underline{\kappa}) \tag{2}$$

The quantity $\underline{\kappa}$ is a dimensionless quantity which is conventionally restricted to a range of $-\pi < \underline{\kappa} \leq \pi$, a central Brillouin zone. For the case $\phi = 0$ (i.e., $\mathcal{S}$ a pure translation), $\underline{\kappa}$ corresponds to a normalized quasimomentum; i.e., $\underline{\kappa} = ka$, where k is the traditional wavevector from Bloch's theorem in solid-state band-structure theory.

The one-electron wavefunctions ψ_i are constructed from a linear combination of Bloch functions φ_j, which are in turn constructed from a linear combination of nuclear-centered Gaussian-type orbitals $\chi_j(\mathbf{r})$ (in this case, products of Gaussians and the real solid spherical harmonics),

$$\psi_i(\mathbf{r}; \underline{\kappa}) = \sum_j c_{ji}(\underline{\kappa})\, \varphi_j(\mathbf{r}; \underline{\kappa}) \tag{3}$$

$$\varphi_j(\mathbf{r}; \underline{\kappa}) = \sum_m e^{-i\underline{\kappa} m}\, \mathcal{S}^m \chi_j(\mathbf{r}) \tag{4}$$

The total energy for the polymer system is given by

$$E = E_{kin} + E_c + E_{xc} \tag{5}$$

where E_{kin} is the kinetic energy

$$E_{kin} = -\frac{1}{2} \sum_{ij} \sum_m P_{ij}^m \langle \chi_i^m | \nabla^2 | \chi_j^0 \rangle, \tag{6}$$

E_c is the total Coulomb energy given by

$$E_c = \iint d^3 r_1 d^3 r_2 [\rho(\mathbf{r}_1) - \rho_N(\mathbf{r}_1)] \frac{1}{r_{12}} [\rho(\mathbf{r}_2) - \rho_N(\mathbf{r}_2)], \tag{7}$$

and E_{xc} is the exchange-correlation energy given by

$$E_{xc} = \int d^3 r\, \rho(\mathbf{r})\, \varepsilon_{xc}(\mathbf{r}). \tag{8}$$

These terms can be written explicitly in terms of matrix elements of the basis functions using

$$
\begin{aligned}
E =& \sum_{ij}\sum_{m} P_{ij}^{m}\left\{-\frac{1}{2}\langle\chi_i^m|\nabla^2|\chi_j^0\rangle + \langle\chi_i^m|\varepsilon_{xc}(\mathbf{r})|\chi_j^0\rangle\right\} \\
&+\frac{1}{2}\sum_{m''}\left\{\left(\sum_{n}{}'\sum_{n'}\frac{Z_nZ_{n'}}{|\mathbf{R}_n^0-\mathbf{R}_{n'}^{m''}|}\right)\right. \\
&+\sum_{ij}\sum_{m}P_{ij}^{m}\left(\sum_{i'j'}\sum_{m'}P_{i'j'}^{m'}\left[\chi_i^m\chi_j^0|\chi_{i'}^{m'+m''}\chi_{j'}^{m''}\right]\right. \\
&\left.\left.-2\sum_{n}\langle\chi_i^m|\frac{Z_n}{|\mathbf{r}-\mathbf{R}_n^{m''}|}|\chi_j^0\rangle\right)\right\}
\end{aligned}
\tag{9}
$$

where Z_n and $\mathbf{R}_n$ denote the nuclear charges and coordinates within a single unit cell, $\mathbf{R}_n^m$ denotes the nuclear coordinates in unit cell m ($\mathbf{R}_n^m = \mathcal{S}^m\mathbf{R}_n$), and $[\rho_1|\rho_2]$ denotes an electrostatic interaction integral,

$$
[\rho_1|\rho_2] = \int d^3r_1 \int d^3r_2\, \frac{\rho_1(\mathbf{r}_1)\rho_2(\mathbf{r}_2)}{r_{12}} \tag{10}
$$

Rather than solve for the total energy directly as expressed in Eq. (8), we follow the approach of earlier workers to fit the charge density (in the Coulomb potential) to a linear combination of Gaussian-type functions.[24,25] The methods used are essentially a straightforward extension of fitting methods developed for molecular calculations;[26–28] the primary innovations necessary for treating an extending system occur for fitting the charge density and will be discussed in detail in this work.

We do not have analytic forms of the integrals needed to evaluate the exchange-correlation energy in Eq. (8). We can either use least-squares fitting methods conceptually analogous to those used in the charge fitting, or we can use direct numerical integration of the matrix elements needed. Historically, we have used least-squares fitting methods in the polymer code, but most recent molecular codes use direct numerical integration methods. Both are included in the code implementing our approach.

3. Geometry optimization

In practical computations, we assume that integrations over the Brillouin zone are handled by summations over discrete sets of points in the Brillouin zone, and the occupation numbers $n_\alpha(\kappa)$ include appropriate weight factors for the chosen grid. Thus, the one-electron density matrix used in the computations can be expressed

$$\rho(\mathbf{r};\mathbf{r}') = \sum_{\alpha}\sum_{\kappa} n_\alpha(\kappa)\psi^*_\alpha(\mathbf{r}';\kappa)\,\psi_\alpha(\mathbf{r};\kappa) \tag{11}$$

$$= \sum_{ij}\sum_{m} P^m_{ij}\sum_{m'}\chi_i(\mathbf{r}';m+m')\,\chi_j(\mathbf{r};m') \tag{12}$$

where P^m_{ij} are the coefficients of the real lattice expansion of the density matrix given by

$$P^m_{ij} = \sum_{\alpha}\sum_{\kappa} n_\alpha(\kappa)c^*_{i\alpha}(\kappa)c_{j\alpha}(\kappa)\,e^{im\kappa} \tag{13}$$

We typically want to evaluate the "forces" on the nuclear centers, which are just the negative values of the gradients of the total energy with respect to the nuclear coordinates. We use an approach analogous to the approach we used earlier for molecular calculations.[29] The total energy is given by

$$E = E_{kin} + E_{xc} + E_c \tag{14}$$

where

$$E_{kin} = -\frac{1}{2}\sum_{ij}\sum_{m} P^m_{ij}\,\langle\chi^m_i|\nabla^2|\chi^0_j\rangle \tag{15}$$

$$E_{xc} = \sum_{ij}\sum_{m} P^m_{ij}\,\langle\chi^m_i|\varepsilon_{xc}(\mathbf{r})|\chi^0_j\rangle \tag{16}$$

$$E_c = \frac{1}{2}\sum_{m''}\left\{\sum_{ij}\sum_{m} P^m_{ij}\left(\sum_{i'j'}\sum_{m'} P^{m'}_{i'j'}\left[\chi^m_i\chi^0_j|\chi^{m'+m''}_{i'}\chi^{m''}_{j'}\right] -2\sum_{n}\langle\chi^m_i|\frac{Z_n}{|\mathbf{r}-\mathbf{R}^{m''}_n|}|\chi^0_j\rangle\right) + \left(\sum_{n}{}'\sum_{n'}\frac{Z_nZ_{n'}}{|\mathbf{R}^0_n-\mathbf{R}^{m''}_{n'}|}\right)\right\} \tag{17}$$

We are interested in finding the derivative dE/dX with respect to some general coordinate X. Here, X could be one of the Cartesian components of a given nucleus n in the origin unit cell, $\mathbf{R}^0_n$. This derivative will usually be expressed in terms of the partial derivatives with respect to the variables that directly define the energy expression: the nuclear coordinates, $\mathbf{R}^m_n$, the band orbital coefficients, $c_{ij}(\kappa)$, and the basis function values at each point in space, $\chi_j(\mathbf{r};m)$. When we use a fitting procedure, we will also have a dependence on the fitting coefficients, f_m.

We assume that both the nuclear charges, Z_n, and the occupation numbers, $n_i(\kappa)$, remain constant with respect to any deformations. Thus, we restrict consideration to insulators with filled bands in this immediate analysis, and we get the following expression:

$$\begin{aligned}\frac{dE}{dX} &= \sum_{i,m} \frac{d\mathbf{R}_i^m}{dX} \cdot \nabla_{\mathbf{R}_i^m}(E) + \sum_{j,m} \frac{\delta E}{\delta \chi_j} \frac{\delta \chi_j}{\delta X} \\ &+ \sum_{j,i,\kappa} \left[\frac{\partial E}{\partial c_{ji}(\kappa)} \frac{\partial c_{ji}(\kappa)}{\partial X} + \frac{\partial E}{\partial c_{ji}^*(\kappa)} \frac{\partial c_{ji}^*(\kappa)}{\partial X} \right]\end{aligned} \tag{18}$$

The first set of terms generate what are frequently denoted as the Hellmann–Feynman terms for the gradients of the energy. We will explain the second set of terms further on in our discussion.

The coefficients for the molecular one-electron wavefunctions, $c_{ji}(\kappa)$, satisfy

$$\begin{aligned}\frac{\partial E}{\partial c_{ji}(\kappa)} &= \varepsilon_i(\kappa) \frac{\partial N_i(\kappa)}{\partial c_{ji}(\kappa)} \\ \varepsilon_i(\kappa) &= \frac{\partial E}{\partial N_i(\kappa)}\end{aligned} \tag{19}$$

where

$$\begin{aligned}N_i(\kappa) &= n_i(\kappa) \sum_{j,j'} c_{ji}^*(\kappa)\, c_{j'i}(\kappa) \int d^3 r\, \chi_j^*(\mathbf{r};\kappa)\, \chi_{j'}(\mathbf{r};\kappa) \\ &= n_i(\kappa) \sum_{j,j'} c_{ji}^*(\kappa)\, c_{j'i}(\kappa)\, S_{jj'}(\kappa)\end{aligned} \tag{20}$$

Normalization will require that the quantity $N_i(\kappa)$ must remain invariant under variations of the coordinates and satisfy

$$N_i(\kappa) = n_i(\kappa) \tag{21}$$

The normalization condition, along with the assumption that the occupation numbers are constant, leads to the relationship

$$\begin{aligned}\frac{dN_i(\kappa)}{dX} &= \sum_{j,j'} \left[\frac{\partial c_{ji}^*(\kappa)}{\partial X} c_{j'i}(\kappa) + c_{ji}^*(\kappa) \frac{\partial c_{j'i}(\kappa)}{\partial X} \right] S_{jj'}(\kappa) \\ &+ \sum_{j,j'} c_{ji}^*(\kappa)\, c_{j'i}(\kappa)\, \frac{S_{jj'}}{\partial X} = 0\end{aligned} \tag{22}$$

which implies

$$\sum_{j,j'}\left[\frac{\partial c_{ji}^*(\kappa)}{\partial X}c_{j'i}(\kappa)+c_{ji}^*(\kappa)\frac{\partial c_{j'i}(\kappa)}{\partial X}\right]S_{jj'}(\kappa)=-\sum_{j,j'}c_{ji}^*(\kappa)\,c_{j'i}(\kappa)\,\frac{\partial S_{jj'}}{\partial X}\tag{23}$$

Also, we see that

$$\frac{\partial N_i(\kappa)}{\partial c_{ji}^*}=n_i(\kappa)\sum_{j'}c_{j'i}(\kappa)S_{jj'}(\kappa)\tag{24}$$

$$\frac{\partial N_i(\kappa)}{\partial c_{ji}}=n_i(\kappa)\sum_{j'}c_{j'i}^*(\kappa)S_{j'j}(\kappa)\tag{25}$$

The Pulay contributions[30] from the derivatives of the total energy with respect to the orbital coefficients can then be expressed as

$$\begin{aligned}&\sum_{j,i,\kappa}\left[\frac{\partial E}{\partial c_{ji}(\kappa)}\frac{\partial c_{ji}(\kappa)}{\partial X}+\frac{\partial E}{\partial c_{ji}^*(\kappa)}\frac{\partial c_{ji}^*(\kappa)}{\partial X}\right]\\&=\sum_{j,i,\kappa}\varepsilon_i(\kappa)\left[\frac{\partial N_i(\kappa)}{\partial c_{ji}(\kappa)}\frac{\partial c_{ji}(\kappa)}{\partial X}+\frac{\partial N_i(\kappa)}{\partial c_{ji}^*(\kappa)}\frac{\partial c_{ji}^*(\kappa)}{\partial X}\right]\\&=-\sum_{i,\kappa}n_i(\kappa)\,\varepsilon_i(\kappa)\sum_{j,j'}c_{ji}^*(\kappa)\,c_{j'i}(\kappa)\,\frac{\partial S_{jj'}}{\partial X}\end{aligned}\tag{26}$$

How do the variational derivatives behave that involve derivatives of the basis functions with respect to the deformation? Let us examine the derivatives for the individual component, E_{kin}, with respect to our deformation coordinate, X.

$$\begin{aligned}\sum_{j,m}\frac{\delta E_{kin}}{\delta\chi_j}\frac{\delta\chi_j}{\delta X}&=-\frac{1}{2}\sum_{ij}\sum_{m}P_{ij}^m\left(\langle\frac{\partial\chi_i^m}{\partial X}|\nabla^2|\chi_j^0\rangle+\langle\chi_i^m|\nabla^2|\frac{\partial\chi_j^0}{\partial X}\rangle\right)\\&=-\sum_{ij}\sum_{m}P_{ij}^m\,\langle\frac{\partial\chi_i^m}{\partial X}|\nabla^2|\chi_j^0\rangle\end{aligned}\tag{27}$$

$$\sum_{j,m}\frac{\delta E_c}{\delta\chi_j}\frac{\delta\chi_j}{\delta X} = \frac{1}{2}\sum_{m''}\left\{\sum_{ij}\sum_{m}P_{ij}^{m}\left(\sum_{i'j'}\sum_{m'}P_{i'j'}^{m'}\left[\chi_i^m\chi_j^0|\chi_{i'}^{m'+m''}\chi_{j'}^{m''}\right]\right.\right.$$
$$\left.\left.-2\sum_{n}\langle\chi_i^m|\frac{Z_n}{|\mathbf{r}-\mathbf{R}_n^{m''}|}|\chi_j^0\rangle\right)+\left(\sum_{n}{}'\sum_{n'}\frac{Z_nZ_{n'}}{|\mathbf{R}_n^0-\mathbf{R}_{n'}^{m''}|}\right)\right\} \tag{28}$$

The Coulomb energy is a bit more complicated because we are using fitting methods to reduce the computational complexity of our simulations. These methods are frequently denoted as a resolution of the identify (ROI) approach. First, we note that we also need to consider terms of the form

$$\sum_{i}\frac{\partial\widetilde{E}_c}{\partial f_i}\frac{df_i}{dX}, \tag{29}$$

which will contribute to the total derivative of the Coulomb energy with respect to the deformation. Here we take advantage of the equations that the fitting algorithm requires be satisfied under any deformation,

$$\begin{aligned}\frac{\partial\widetilde{E}_c}{\partial f_i} &= \lambda\eta_i, \text{ and}\\ \sum_{i} f_i\eta_i &= 1.\end{aligned} \tag{30}$$

This yields the useful relationship that

$$\begin{aligned}\sum_{i}\frac{\partial\widetilde{E}_c}{\partial f_i}\frac{df_i}{dX} &= \lambda\sum_{i}\eta_i\frac{df_i}{dX}\\ &= -\lambda\sum_{i}f_i\frac{d\eta_i}{dX} = 0,\end{aligned} \tag{31}$$

which tells us that the contributions resulting from the fitting coefficients will sum to zero in the total energy derivative.

Let us go back to our earlier definition of the fitted charge density, $\widetilde{\rho}(\mathbf{r})$,

$$\widetilde{\rho}(\mathbf{r}) = \sum_{i} f_i\,\mathcal{F}_i(\mathbf{r}). \tag{32}$$

The approximate Coulomb energy, $\widetilde{E}_c$, used in our calculations can then be expressed

$$\widetilde{E}_c = [\rho - \rho_N|\widetilde{\rho} - \rho_N] - \frac{1}{2}[\widetilde{\rho} - \rho_N|\widetilde{\rho} - \rho_N]. \tag{33}$$

We have already examined what the contributions will be to the derivatives of the total energy, E, with respect to various deformations, X, that arise from variations in the orbital coefficients and the charge fitting coefficients. The remaining contributions we get arise from variations of the orbital and fitting functions themselves. With the restriction to orbital and fitting functions in mind, we can write the variation of the total energy in the shorthand

$$\begin{aligned}\delta\widetilde{E}_c &= [\delta\rho - \delta\rho_N|\widetilde{\rho} - \rho_N] + [\delta\widetilde{\rho} - \delta\rho_N|\rho - \rho_N] - [\delta\widetilde{\rho} - \delta\rho_N|\widetilde{\rho} - \rho_N] \\ &= [\delta\rho - \delta\widetilde{\rho}|\widetilde{\rho} - \rho_N] + [\delta\widetilde{\rho}|\widetilde{\rho} - \rho_N] - [\delta\rho_N|\widetilde{\rho} - \rho_N] \\ &= [\delta\rho|\widetilde{\rho} - \rho_N] - [\delta\widetilde{\rho}|\widetilde{\rho} - \rho_N] + [\delta\widetilde{\rho}|\widetilde{\rho} - \rho_N] - [\delta\rho_N|\widetilde{\rho} - \rho_N]\end{aligned} \tag{34}$$

4. Multipole moment expansions

Most of the Coulomb integrals we need involve interactions between charges at two points, $\mathbf{r}_1$ and $\mathbf{r}_2$. For charge distributions localized about two different atomic centers separated by a vector $\mathbf{R}$, we can frequently evaluate the long-range Coulomb interactions with a multipole expansion. For two charges at $\mathbf{r}_1$ and $\mathbf{R} + \mathbf{r}_2$, we can use the modified regular and irregular solid spherical harmonics to expand Coulomb interactions between non-overlapping charge distributions by expanding the inverse of the inter-electronic separation in a multipole expansion.

$$\begin{aligned}\frac{1}{|\mathbf{r}_1 - \mathbf{r}_2 - \mathbf{R}|} &= \sum_{\ell,m}\sum_{\ell',m'}(-1)^{\ell'+m'+m}\, \mathcal{Y}_\ell^m(\mathbf{r}_1)\mathcal{Y}_{\ell'}^{m'}(\mathbf{r}_2)\, \mathcal{Z}_{\ell+\ell'}^{-(m+m')}(\mathbf{R}) \\ &= \sum_{\ell,m}\sum_{\ell',m'}(-1)^{\ell'}\, \mathcal{Y}_\ell^m(\mathbf{r}_1)\mathcal{Y}_{\ell'}^{m'}(\mathbf{r}_2)\left[\mathcal{Z}_{\ell+\ell'}^{m+m'}(\mathbf{R})\right]^* \\ \mathcal{Y}_\ell^m(\mathbf{r}) &= r^\ell\, Y_\ell^m(\theta,\phi)\sqrt{\frac{4\pi}{(2\ell+1)(\ell-m)!(\ell+m)!}} \\ \mathcal{Z}_\ell^m(\mathbf{r}) &= r^{-(\ell+1)}\, Y_\ell^m(\theta,\phi)\sqrt{\frac{4\pi(\ell-m)!(\ell+m)!}{(2\ell+1)}}\end{aligned}$$

$$\iint d\mathbf{r}_1 d\mathbf{r}_2 \frac{\rho_1(\mathbf{r}_1)\rho_2(\mathbf{r}_2)}{|\mathbf{r}_1 - \mathbf{r}_2 - \mathbf{R}|} = \sum_{\ell,m}\sum_{\ell',m'}(-1)^{\ell'} M_{\ell,m}(\rho_1) M_{\ell',m'}(\rho_2)\left[\mathcal{Z}_{\ell+\ell'}^{m+m'}(\mathbf{R})\right]^*$$
$$M_{\ell,m}(\rho) = \int d^3\mathbf{r}\,\rho(\mathbf{r})\,\mathcal{Y}_\ell^m(\mathbf{r}) \tag{35}$$

The above sums can be truncated to finite length using

$$\sum_{\ell=0}^{\infty}\sum_{\ell'=0}^{\infty}\sum_{m=-\ell}^{\ell}\sum_{m'=-\ell'}^{\ell'}(-1)^{\ell'}\mathcal{Y}_\ell^m(\mathbf{r}_1)\mathcal{Y}_{\ell'}^{m'}(\mathbf{r}_2)\left[\mathcal{Z}_{\ell+\ell'}^{m+m'}(\mathbf{R})\right]^*$$
$$= \sum_{L=0}^{\infty}\sum_{\ell=0}^{L}\sum_{m=-\ell}^{\ell}\sum_{m'\le(L-\ell)}(-1)^{(L-\ell)}\mathcal{Y}_\ell^m(\mathbf{r}_1)\mathcal{Y}_{L-\ell}^{m'}(\mathbf{r}_2)\left[\mathcal{Z}_L^{m+m'}(\mathbf{R})\right]^*$$

If we truncate at a finite value of L, the finite expansion is invariant to rotation and shift of origin. We can rearrange the order of summations to a form where we can precalculate partial summations.

$$\sum_{L=0}^{\mathcal{L}}\sum_{\ell=0}^{L}\sum_{m=-\ell}^{\ell}\sum_{m'=\ell-L}^{L-\ell}(-1)^{(L-\ell)}\mathcal{Y}_\ell^m(\mathbf{r}_1)\mathcal{Y}_{L-\ell}^{m'}(\mathbf{r}_2)\left[\mathcal{Z}_L^{m+m'}(\mathbf{R})\right]^*$$
$$=\sum_{\ell=0}^{\mathcal{L}}\sum_{m}\mathcal{Y}_\ell^m(\mathbf{r}_1)\sum_{\ell'=0}^{\mathcal{L}-\ell}\sum_{m'}(-1)^{\ell'}\mathcal{Y}_{\ell'}^{m'}(\mathbf{r}_2)\left[\mathcal{Z}_{\ell+\ell'}^{m+m'}(\mathbf{R})\right]^*$$
$$=\sum_{\ell=0}^{\mathcal{L}}\sum_{m}\left[\mathcal{X}_\ell^m\right]^*\mathcal{Y}_\ell^m(\mathbf{r}_1)$$
$$\mathcal{X}_\ell^m = \sum_{\ell'=0}^{\mathcal{L}-\ell}\sum_{m'}(-1)^{\ell'+m'}\mathcal{Y}_{\ell'}^{-m'}(\mathbf{r}_2)\mathcal{Z}_{\ell+\ell'}^{m+m'}(\mathbf{R})$$

5. Long-range axial multipole moment expansions

For long-range Coulomb interactions using a multipole expansion, we consider the multipole moments centered about a point on the helical axis of symmetry. In particular, we consider multipole interaction expansions for multipole moments of various charge distributions centered about the origin and other points along the helical axis of symmetry translated by integer multiples of the helical step size. This corresponds to $\mathbf{R}$ pointing along the z-axis, which restricts the irregular solid spherical harmonics to $m = 0$.

$$\mathbf{R} = R\,\mathbf{e}_z$$
$$\mathcal{Z}_L^M(\mathbf{R}) = \frac{L!}{R^{L+1}}\,\delta_{M,0}$$
$$\mathcal{Z}_L^M(-\mathbf{R}) = (-1)^L \frac{L!}{R^{L+1}}\,\delta_{M,0}$$
$$\mathcal{Z}_L(\mathbf{R}) \equiv \mathcal{Z}_L^0(\mathbf{r})$$

Using a bipolar expansion[31,32] of the Coulomb potential, we can express the long-range interaction (i.e., at sufficient distance to neglect any overlap between the charge distributions) between two charge distributions localized in the origin unit cell and unit cell m by

$$[\rho_1^n|\rho_2^0] = \sum_{L=0}^{\infty} \mathcal{Z}_L(na) \sum_{\ell=0}^{L} (-1)^{L-\ell} \sum_{|m|\le min[\ell, L-\ell]} M_1^n(\ell, m)\, M_2^0(L-\ell, -m).$$

All of the summations for which we use the multipole expansion contain contributions from both the "n" and "-n" unit cells. Taking advantage of the symmetries of the spherical harmonics, we find

$$[\rho_1^n + \rho_1^{-n}|\rho_2^0] = 2\sum_{L=0}^{\infty}\sum_{m=0}^{[L/2]} \mathcal{Z}_L(na) g_L(mn\phi) Q_{Lm} \tag{36}$$
$$g_L(\theta) = \begin{cases} \cos(\theta) & L\ \text{even}, \\ \sin(\theta) & L\ \text{odd}, \end{cases}$$

and

$$Q_{Lm} = \begin{cases} \sum_{\ell=0}^{L} (-1)^{L-\ell} M_1(\ell,0)\, M_2(L-\ell,0) & m=0, \text{else} \\ 2\sum_{\ell=m}^{L-m} (-1)^{\ell-m}\, \Re e[M_1(\ell,m)\, M_2^*(L-\ell,m)] & L\ \text{even}, \\ 2\sum_{\ell=m}^{L-m} (-1)^{\ell-m}\, \Im m[M_1(\ell,m)\, M_2^*(L-\ell,m)] & L\ \text{odd}, \end{cases} \tag{37}$$

where $M_i(\ell, m) \equiv M_i^0(\ell, m)$. In our computational algorithm for calculating summations such as those in Eq. (35), we used a truncated expansion up to $L = 8$ in Eq. (36).[17] For a given summation, we first calculate the required values of $M_i(\ell, m)$ and generate the values of Q_{Lm}. The sum in Eq. (35), for example, is then evaluated by starting with the exact expression in Eq. (26)

for $m = 0$, then summing the differences between the exact expression for a given value of m and the value that is evaluated from the truncated multipole expansion in Eq. (36). The truncation of this series is then determined by the convergence of these differences. The sum to infinity of the multipole interaction terms is then added to extrapolate the series in Eq. (35) for finite order in L. Evaluation of this term,

$$\sum_{n=1}^{\infty} Z_L(na)\, g_L(mn\phi) = L!\, \zeta_{L+1}(m\phi)/a^{(L+1)}$$

requires the evaluation of the quantity $\zeta_L(\theta)$, which has the general form

$$\zeta_\ell(\theta) = \begin{cases} \sum_{n=1}^{\infty} \cos(n\theta)/n^\ell & \ell \text{ odd}, \\ \sum_{n=1}^{\infty} \sin(n\theta)/n^\ell & \ell \text{ even}. \end{cases} \tag{38}$$

As noted by other workers,[33] closed analytic expressions are known for particular values of the arguments; in particular, $\zeta_L(0)$ is the Riemann zeta function of integer argument, $\zeta(L)$,[34] and values are also tabulated in this reference for $\theta = \pi/2$ and π. For computational convenience and precision, we have chosen to use the identity,[35]

$$\begin{aligned} \zeta_1(\theta) &= \sum_{n=1}^{\infty} \frac{\cos n\theta}{n} \\ &= -\frac{1}{2} \ln[2(1 - \cos\theta)]. \end{aligned} \tag{39}$$

We used the *Mathematica* software package to evaluate the coefficients for the series expansion to Eq. (36),

$$\zeta_1(\theta) = -\ln\theta + \frac{\theta^2}{24} + \frac{\theta^4}{2880} + \cdots, \tag{40}$$

Using the relation $\zeta_\ell'(\theta) = (-1)^\ell\, \zeta_{\ell-1}(\theta)$, we find that using the series expansion to 40th order and explicitly integrating each term to obtain ζ_ℓ for $\ell > 1$ yields a relative precision of $\sim 10^{-14}$ for $0 \leq \theta \leq \pi$.

6. Polylogarithm evaluation methods

The multipole expansion is not accurate for close lattice spacings, so we frequently need to start the long-range Coulomb summation at some $N > 1$. This polynomial expansion has precision problems in evaluating the multipole interaction sums starting at larger distances of the form:

$$\sum_{n=N}^{\infty} Z_L(na)\, g_L(mn\phi) = L!\, \zeta_{L+1}(m\phi)/a^{(L+1)} - \sum_{n=1}^{N-1} Z_L(na)\, g_L(mn\phi).$$

Because the differences between the first and second expressions in the right-hand side of the equation above are frequently several orders of magnitude smaller than the individual expressions, we will often need the evaluation of the $\zeta_L(\theta)$ functions to more precision than we can obtain with the polynomial approach described above using Eq. (40). Instead, we can relate the $\zeta_L(\theta)$ functions for nonzero θ to polylogarithm functions[36–38] using extended precision algorithms in our code.[39]

$$Li_L(z) = \sum_{k=1}^{\infty} \frac{z^k}{k^L}. \tag{41}$$

We can choose z to be the complex quantity, $z = e^{i\theta}$, leading to the relationship

$$\zeta_L(\theta) = \begin{cases} \Re\left[Li_L(e^{i\theta})\right], & L \text{ even}, \\ \Im\left[Li_L(e^{i\theta})\right], & L \text{ odd}. \end{cases} \tag{42}$$

From Crandall,[36] we find we can compute Eq. (41) using the expression given below:

$$Li_L(z) = \frac{\ln^{L-1} z}{(L-1)!}\left(H_{L-1} - \ln(-\ln z)\right) + \sum_{m=0}^{\infty}{}' \left[\frac{\zeta(L-m)}{m!} \ln^m z\right], \tag{43}$$

and

$$H_n = \sum_{k=1}^{n} \frac{1}{k} = H_{n-1} + \frac{1}{n}, \; H_0 = 0,$$

where the primed summation indicates that we exclude any singular term for which $L - m = 1$. We can obtain D decimal digits of precision for the truncated summation with an upper limit on the primed summation so that $m \leq N$, with the limit N given by:

$$N = [D \log_2 10] \approx [3.322D].$$

The Riemann zeta functions of integer argument, $\zeta(m)$, must also be evaluated to high precision. For $m < 1$, this is straightforward:

$$\begin{aligned} \zeta(0) &= -\frac{1}{2} \\ \zeta(1-2n) &= -\frac{B_{2n}}{2n} \\ \zeta(-2n) &= 0, \end{aligned} \tag{44}$$

where n is a positive integer, and B_{2n} are the Bernoulli numbers. For $m > 1$, we can use the algorithm[37,38]:

$$\begin{aligned} \zeta(s) &= \frac{-1}{d_{0n}(1 - 2^{1-s})} \sum_{k=1}^{n} \frac{(-1)^k d_{kn}}{k^s} + \gamma_n(s) \\ d_{kn} &= n \sum_{j=k}^{n} \frac{(n+j-1)!4^j}{(n-j)!\,(2j)!} \end{aligned} \tag{45}$$

The error term, $\gamma_n(s)$, given by truncating the above expression can be estimated using:

$$|\gamma_n(s)| \leq \frac{3}{(3+\sqrt{8})^n} \frac{1}{|1-2^{1-s}|}$$

We see that by estimating

$$10^{-D} \approx |\gamma_n(s)| \approx \frac{6}{(3+\sqrt{8})^n},$$

we can get roughly D decimal digits of precision if the cutoff limit n is given by

$$n \approx 1.3D + 1.$$

7. Ortho-connected polythiophenes

Polythiophenes have shown great potential for device applications as a result of their high stability, ease of structural modification, and appropriate optical gap.[40] The most commonly studied polythiophenes are the poly(3-alkyl-2,5-thiophene)s (P3AT), with a chain structure depicted in the inset of Fig. 1. In the equilibrium conformation, all atoms in the heteroatomic rings in the backbone lie in the same plane, and the polymer consists of monomers of individual heteroatomic rings with an attached alkyl substituent generated with a 180 degrees helical twist. Researchers at the Oak Ridge National Laboratory (ORNL) synthesized the ortho-linked isomers of the thiophenes, the poly(5-alkyl-2,3-thiophene)s (P5AT),[41] in particular the hexyl-substituted P5AT. As seen in Fig. 2, the adjacent monomers in the polymer backbone connect via bonds between the 2- and 3-position carbons in the heteratomic ring. These polymers were reported to have a helical conformation with twist angle of roughly 120 degrees and step length of 0.65–0.68 nm (corresponding to that suggested from scanning tunneling microscopy results). These experimental results provide us an opportunity to show how the helical density-functional approach can provide a useful tool for understanding the experimental materials. As a summary of this review, we will present some computational results comparing the energetics as a function of twist

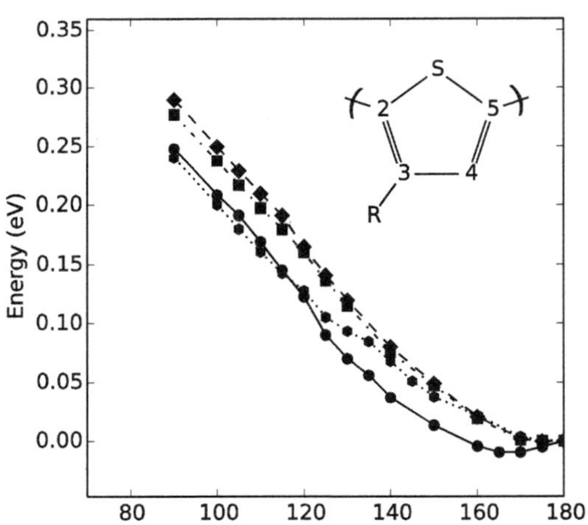

Fig. 1 Total energy per monomer unit cell of the optimized geometry at each value of the helical twist angle for the poly(3-alkyl-2,5-thiophene)s. Line markers denote substituents on each chain: hydrogen (*circles*), methyl (*hexagons*), ethyl (*diamonds*), and propyl (*squares*). Inset shows bonding structure of the monomer unit well, with chain-backbone bonds connecting monomers attached to the 2,5 carbon sites in the thiophene ring. Total energies are shown relative to the energy at the planar, 180 degrees twist angle for the backbone.

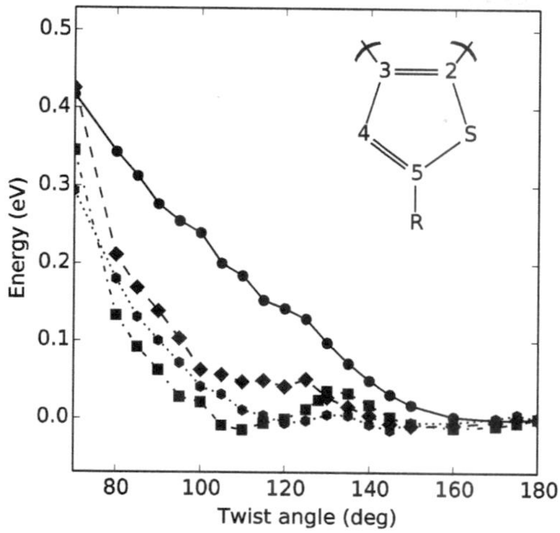

Fig. 2 Total energy per monomer unit cell of the optimized geometry at each value of the helical twist angle for the poly(5-alkyl-2,3-thiophene)s. Line markers denote substituents on each chain: hydrogen (*circles*), methyl (*hexagons*), ethyl (*diamonds*), and propyl (*squares*). Inset shows bonding structure of the monomer unit well, with chain-backbone bonds connecting monomers attached to the 2,3 carbon sites in the thiophene ring. Total energies are shown relative to the energy at the planar, 180 degrees twist angle for the backbone.

angle for the regioregular P3AT and P5AT polymers with hydrogen, methyl, ethyl, and propyl substituents. These calculations were carried out with a 6-31G* basis set, including six Cartesian *d* polarization functions for all nonhydrogen atoms,[42] and a full geometry optimization was carried out at each evaluated twist angle. The largest substituent for which we were able to complete simulations was the propyl group.

We see that for the P3AT systems, the hydrogen-substituted polymer has a weak minimum in energy slightly away from the planar conformation, but the addition of any of the alkyl substituents results in a minimum at the 180 degrees twist angle giving the planar conformation. For the P5AT systems, however, we find that only the H-substituted system has a minimum at 180 degrees. The propyl-substituted polymer shows a pronounced energy minimum at 110 degrees with a calculated repeat length of 1.9 Å, which would correspond to a helical pitch of 5.8 Å for the assumed structure with three monomers per turn in the helix, in good agreement with the reported experimental results.

Acknowledgments

The approach presented herein represents the culmination of a line of research over several decades that began with Jack Sabin working with a young graduate student in the 1970s. Jack was a wonderful person to work with and helpful mentor to those working with him, and has remained a good friend over all these years. In addition to Jack, Sam Trickey, and others associated with the Quantum Theory Project at the University of Florida who helped me progress, many others along the way have contributed to this development including Carter White and Brett Dunlap at the Naval Research Laboratory. This work was supported in part by a sabbatical period at the ORNL Center for Nanophase Materials Science funded through the ORAU ORNL-HERE program.

References

1. Mintmire, J. W.; Sabin, J. R. Intermolecular Potential Studies of Hydrogen Interactions With Rare-Gas Atoms. *Int. J. Quantum Chem. Symp.* **1976**, *10*, 213.
2. Mintmire, J. W. LCAO Methods in the Xα Formalis A Comparison of Results for the F_2 Molecule. *Int. J. Quantum Chem. Symp.* **1979**, *13*, 163.
3. Mintmire, J. W.; Sabin, J. R. A Comparison of the LCAO-Xα Method With the Hartree-Fock and Multiple-Scattering-Xα Methods on Carbon Monoxide. *Chem. Phys.* **1980**, *51*, 91.
4. Mintmire, J. W. LCAO Local Density Functional Approach to Surface Electronic Structure Calculations (Ph.D. thesis), University of Florida 1980.
5. Mintmire, J. W.; Sabin, J. R. Local Density Functional Methods in Two-Dimensionally Periodic Systems: The Atomic Hydrogen Monolayer. *Int. J. Quantum Chem. Symp.* **1980**, *14*, 707.
6. Mintmire, J. W.; Sabin, J. R.; Trickey, S. B. Local Density Functional Methods in Two-Dimensionally Periodic Systems. Hydrogen and Beryllium Monolayers. *Phys. Rev. B* **1982**, *26*, 1743.
7. Mintmire, J. W.; White, C. T. Xα Approach for the Determination of Electronic and Geometric Structure of Polyacetylene and Other Chain Polymers. *Phys. Rev. Lett.* **1983**, *50*, 101.
8. Mintmire, J. W.; White, C. T. Local-Density Functional Approach to All-Trans Polyacetylene. *Phys. Rev. B* **1983**, *28*, 3283.
9. Mintmire, J. W. First-Principles Geometry Optimization of Polysilane,. In Tersoff, J., Vanderbilt, D., Vitek, V., Eds.; Atomic Scale Calculations in Materials Science, Vol. 141 of MRS Symposia Proceedings; Materials Research Society. Pittsburgh, 1988; pp. 235–239.
10. Mintmire, J. W. Conformational Effects in Organopolysilanes: A First-Principles Approach. *Phys. Rev. B* **1989**, *39*, 13350.
11. Imamura, A. Electronic Structures of Polymers Using the Tight-Binding Approximation. Polyethylene by the Extended Hückel Method. *J. Chem. Phys.* **1970**, *52*, 3168–3175.
12. Fujita, H.; Imamura, A. Electronic Structures of Polymers Using the Tight-Binding Approximation. Polyethylene and Polyglycine by the CNDO Method. *J. Chem. Phys.* **1970**, *53*, 4555–4566.
13. Blumen, A.; Merkel, C. Energy Band Calculations on Helical Systems. *Phys. Stat. Solidi B* **1977**, *83*, 425–431.
14. Karpfen, A.; Beyer, A. Ab Initio Studies on Polymers. Torsional Potential in Regular Polyethylene Chains. *J. Comput. Chem.* **1984**, *5*, 11–18.
15. André, J. M.; Vercauteren, D. P.; Bodart, J. P.; Fripiat, J. G. Ab Initio Calculations of the Electronic Structure of Helical Polymers. *J. Comp. Chem.* **1984**, *5*, 536–547.

16. Teramae, H.; Takeda, K. Ab Initio Studies on Silicon Compounds. The Gauche Structure of the Parent Polysilane. *J. Am. Chem. Soc.* **1989**, *111*, 1281–1285.
17. Mintmire, J. W. Local-Density Functional Electronic Structure of Helical Chain Polymers. In *Density Functional Theory Approaches to Chemistry*; Labanowski, J. K., Andzelm, J. W., Eds.; Springer-Verlag: New York, 1991; pp. 125–137.
18. Mintmire, J. W.; Dunlap, B. I.; White, C. T. Are Fullerene Tubules Metallic? *Phys. Rev. Lett.* **1992**, *68*, 631–635.
19. Mintmire, J. W.; Robertson, D. H.; White, C. T. Properties of Fullerene Nanotubules. *J. Phys. Chem. Solids* **1993**, *54*, 1835.
20. White, C. T.; Mintmire, J. W. Fundamental Properties of Single-Wall Carbon Nanotubes. *J. Phys. Chem. B* **2005**, *109*, 52–65.
21. Zhang, M. L.; Miao, M. S.; VanDoren, V. E.; Ladik, J. J.; Mintmire, J. W. Local Density Approximation Calculation of the Conformation and Electronic Structure of Poly(Fluoroethylene). *J. Chem. Phys.* **1999**, *111*, 8696–8700.
22. Elizondo, S. L.; Mintmire, J. W. First-Principles Study of the Optical Properties of ZnO Single-Wall Nanotubes. *J. Phys. Chem. C* **2007**, *111*, 17821–17826.
23. Li, J. W.; Mintmire, J. W. Surface Passivation Effects in Silicon Nanowires. *Mol. Phys.* **2015**, *113*, 274–281.
24. Whitten, J. L. Coulombic Potential Energy Integrals and Approximations. *J. Chem. Phys.* **1973**, *58*, 4496–4501.
25. Sambe, H.; Felton, R. H. A New Computational Approach to Slater's SCF Xα Equation. *J. Chem. Phys.* **1975**, *62*, 1122–1126.
26. Dunlap, B. I.; Connolly, J. W. D.; Sabin, J. R. On Some Approximations in Applications of Xα Theory. *J. Chem. Phys.* **1979**, *71*, 3396.
27. Dunlap, B. I.; Connolly, J. W. D.; Sabin, J. R. On First-Row Diatomic Molecular and Local Density Models. *J. Chem. Phys.* **1979**, *71*, 4993.
28. Mintmire, J. W.; Dunlap, B. I. Fitting the Coulomb Potential Variationally in Linear Combination of Atomic Orbitals Density Functional Calculations. *Phys. Rev. A* **1982**, *25*, 88.
29. Dunlap, B. I.; Andzelm, J.; Mintmire, J. W. Local-Density-Functional Total Energy Gradients in the Linear Combination of Gaussian-Type Orbitals Method. *Phys. Rev. A* **1990**, *42*, 6354–6359.
30. Pulay, P. Ab Initio Calculation of Force Constants and Equilibrium Geometries in Polyatomic Molecules: Theory. *Mol. Phys.* **1969**, *17*, 197–204.
31. Steinborn, E. O.; Ruedenberg, K. Rotation and Translation of Regular and Irregular Solid Spherical Harmonics. *Adv. Quantum Chem.* **1973**, *7*, 1.
32. Weniger, E. J.; Steinborn, E. O. A Simple Derivation of the Addition Theorems of the Irregular Solid Harmonics, the Helmholtz Harmonics, and the Modified Helmholtz Harmonics. *J. Math. Phys.* **1985**, *26*, 664–670.
33. Piela, L.; Delhalle, J. Efficient Procedure to Evaluate Long-Range Coulombic Interactions Within Framework of LCAO-CO Method for Infinite Polymers. *Int. J. Quantum Chem.* **1978**, *13*, 605–617.
34. Abramowitz, M.; Stegun, I. *Handbook of Mathematical Functions With Formulas, Graphs, and Mathematical Tables*; Dover Publications: New York, 1965.
35. Gradshteyn, I. S.; Ryzhik, I. M. *Table of Integrals, Series, and Products*, 7th ed.; Elsevier/Academic Press: Amsterdam, 2007.
36. Crandall, R. E. *Note on Fast Polylogarithm Computation*; 2006. http://academic.reed.edu/physics/faculty/crandall/papers/Polylog.pdf (accessed 25 January 2015).
37. Borwein, P. An Efficient Algorithm for the Riemann Zeta Function. In *Constructive, Experimental, and Nonlinear Analysis*; Ra, M. T., Ed.; Vol. 27; Canadian Mathematical Society/American Mathematical Society: Providence, RI, 2000; pp. 29–34.

38. Gourdon, X.; Sebah, P. *Numerical Evaluation of the Riemann Zeta-Function*; 2003. http://numbers.computation.free.fr/Constants/Miscellaneous/zeta.pdf (accessed 25 January 2015).
39. Bailey, D. H. A Fortran 90 Based Multiprecision System. *ACM Trans. Math. Softw.* **1995**, *21*, 453–458. https://doi.org/10.1006/jcph.1998.5909.
40. Kaloni, T. P.; Giesbrecht, P. K.; Schreckenbach, G.; Freund, M. S. Polythiophene: From Fundamental Perspectives to Applications. *Chem. Mater.* **2017**, *29*, 10248–10283.
41. Zhang, H. H.; Ma, C.; Bonnesen, P. V.; Zhu, J.; Sumpter, B. G.; Carrillo, J. M. Y.; Yin, P.; Li, A. P.; Hong, K. Helical Poly(5-Alkyl-2,3-Thiophene)s: Controlled Synthesis and Structure Characterization. *Macromolecules* **2016**, *49*, 4691–4698.
42. Francl, M. M.; Pietro, W. J.; Hehre, W. J.; Binkley, J. S.; Gordon, M. S.; DeFrees, D. J.; Pople, J. A. Self-Consistent Molecular Orbital Methods. XXIII. A Polarization-Type Basis Set for Second-Row Elements. *J. Chem. Phys.* **1982**, 77, 3654–3665.

CHAPTER TEN

Atomic ionization, capture, and stopping cross sections by ion impact examined with the Benford law

Jorge E. Miraglia* **and Maria Silvia Gravielle**
Instituto de Astronomía y Física del Espacio (UBA-CONICET), Buenos Aires, Argentina
*Corresponding author: e-mail address: miraglia@iafe.uba.ar

Contents

Abstract

Cross-section databases for atomic ionization, capture, and stopping by ion impact are intensively used in different areas of basic research, as well as in technological developments. We propose examining these databases by using the Benford law. In this chapter, the applicability of the Benford law for given data sets of theoretical and experimental atomic cross sections by ion impact is studied. We found that the data sets corresponding to theoretical ionization, capture, and stopping cross sections satisfy quite well the Benford law, not only for the first digits but also within a given order of magnitude. Experimental stopping power values from the International Atomic Energy Agency database were also scrutinized, but in this case, the Benford conformity was not so satisfactory due to its small rank width. In all the cases, densities of prime

Advances in Quantum Chemistry, Volume 85
ISSN 0065-3276
https://doi.org/10.1016/bs.aiq.2022.04.006
Copyright © 2022 Elsevier Inc.
All rights reserved.

numbers and theorems of Nigrini and Pinkham are evaluated and discussed. Furthermore, we introduce different relative errors which represent useful tools to test atomic databases in general.

Abbreviations

BL Benford law
CDW-EIS continuum distorted wave-eikonal initial state
EI eikonal impulse
IAEA International Atomic Energy Agency
MAD mean absolute deviation

1. Introduction

Atomic ionization, capture, and stopping power cross sections are the basic pillars of atomic physics, being involved in most collisions processes. In particular, atomic cross sections by ion impact play a fundamental role in a wide variety of fields, ranging from plasma physics to medical treatments. Concerning atomic ionization by fast multicharged ions, this process has been extensively studied due to its importance not only to fully understand the interactions involved in the few-body problem[1–3] but also to describe the propagation of ions through matter,[4] including biological tissues.[5–7] Single-electron capture cross sections are also important to study fundamental problems,[8] as well as to investigate the effect of ion irradiation on biological targets,[9] to unravel the mechanisms involved in astrophysical processes, such as the X-ray emission triggered by the solar wind,[10–12] or to model plasma for fusion developments,[13–15] among other uses.[16] Moreover, stopping cross sections are essential in many applications involving biological matter,[17] like hadron therapy,[18] or in materials science.[19,20] Prof. J. Sabin has developed extensive research in many of these fields. Much of his pioneering work in atomic and molecular physics has served as input for different applications or has been used as a basis for new developments. In this chapter, however, we will focus only on one article from his long list of publications, namely the data table for proton stopping power for atoms from H to Kr published in Ref. 19, which has been widely used over time. We will use this data set as an example of results able to be assessed with a very useful statistics method: the Benford law.

In all the above mentioned areas, the state of the art of scientific research is commonly based on large databases containing theoretical cross sections, which are used as a primary input in complex computational codes.[21,22]

However, the accuracy of such databases was recently questioned, prompting the determination of the uncertainties inherent to each of these calculations.[23–25] In addition, some of the numerical codes employed to model ion-matter interactions make use of experimental data sets,[26] for which experimental biases should be also analyzed.[27,28]

In this chapter, we examine a novel aspect of big atomic cross-section databases: the performance within the Benford law (BL). This law states the probability of occurrence of a given digit, or even within a given order of magnitude, in a big data set.[29,30] There are an enormous number of uses of the BL: fight against tax fraud by detecting manipulation as anomalies, stock exchange data, corporate disbursements, analysis of sales figures, demographics and scientific data, etc.[31,32] Hence, the application of the BL crosses a huge range of disciplines, including physical sciences and also astrophysics, where several astrophysical data, like exoplanet masses, pulsar rotation frequencies, γ-ray source fluxes, and fundamental physics constants, satisfy the BL.[33–35] For the interested reader, a complete list of articles on the BL can be found in Ref. 36.

The applicability of the BL to big data sets corresponding to ionization and capture cross sections by multicharged ions is here analyzed considering accurate theoretical models for the intermediate- and high-energy ranges. In the case of atomic ionization, we make use of the well-known continuum distorted wave-eikonal initial state (CDW-EIS) approximation[37,38] to calculate partial ionization cross sections from different subshells of atomic targets. The CDW-EIS theory represents one of the most reliable methods for ion-atom ionization at intermediate and high velocities, which has been extensively applied for the description of different ionization processes.[39–42] Note that results derived within the CDW-EIS approach are frequently used as input data in Monte Carlo codes.[22] For capture, we scrutinize the conformity level of the BL of atomic cross sections for state-selective electron capture to different subshells of ionic projectiles. These cross sections were evaluated within the eikonal impulse (EI) approximation,[43] which has been successfully applied to different charge-exchange processes.[44–47] As mentioned, theoretical stopping power cross sections were extracted from the detailed table published by Oddershede and Sabin,[19] whose results are based on Sigmund's kinetic theory.[48] Finally, experimental stopping cross sections extracted from the large database of the International Atomic Energy Agency (IAEA),[49] corresponding to more than 50 different experiments, are analyzed within the BL.

The chapter is organized as follows. The BL is briefly summarized in Section 2, while the considered atomic data sets are specified in Section 3. In Section 4, we define the overall features of a given data set to verify the BL and the different levels of conformity are quantified. Results of the BL for the four atomic data sets—ionization, capture, and theoretical and experimental stopping—are presented and discussed in Section 5, and in Section 6 we outline our conclusions.

2. The Benford law

The BL was originally proposed by the astronomer Newcomb in 1881[50] and formulated much later by Benford in 1938.[51] For a given big data set $\{\eta\}$, both man-made and from nature, the Benford distribution states that the frequency of the first significant digit η_1 follows the following law:

$$P^B_{\eta_1} = \log_{10}(1 + 1/\eta_1), \tag{1}$$

where $P^B_{\eta_1}$ is the Benford probability of occurring η_1 (not considering the number of zeros on the left of the numbers).

From a mathematical point of view,[52] to identify the digits of a given real number, one can define an operator corresponding to the first ($j = 1$), second ($j = 2$), and higher digits ($j > 2$) of a real number η as $\widehat{D}_j(\eta) = \eta_j$. For example, if $\eta^* = 0.0234$, then $\widehat{D}_1(\eta^*) = \eta_1 = 2$, $\widehat{D}_2(\eta^*) = \eta_2 = 3$, and $\widehat{D}_3(\eta^*) = \eta_3 = 4$. The numbers formed by the first two digits are here denoted as $\{\eta_1, \eta_2\}$ (in our example, $\{\eta_1, \eta_2\} = 23$), and so on.

The BL can be extended for all decimal digits. For example, for the first two digits it holds

$$P^B_{\{\eta_1,\eta_2\}} = \log_{10}\left(1 + \frac{1}{10\eta_1 + \eta_2}\right). \tag{2}$$

Like any probability, the closure relation is normalized, that is,

$$\sum_{\eta_1=1}^{9} P^B_{\eta_1} = \sum_{\{\eta_1,\eta_2\}=10}^{99} P^B_{\{\eta_1,\eta_2\}} = 1. \tag{3}$$

In addition, if we are interested in the order of magnitude M, we need to use the operator $\widehat{M}$ so that $\widehat{M}\ (\eta) = M$. In our example, $\widehat{M}(\eta^*) = -2$. Furthermore, if we are working with a nondecimal system, the BL still applies replacing $\log_{10}$ with $\log_b$, where b is the working base.

From Eq. (1), for the first digit, $\eta_1 \in [1, 9]$, the BL states that the number $\eta_1 = 1$ appears in more than 30% of the cases, while $\eta_1 = 9$ in less than 5%. But the probability of appearance of the second digit,

$$P^B_{\eta_2} = \Sigma_{\eta_1} P^B_{\{\eta_1,\eta_2\}}, \tag{4}$$

varies from $P^B_{\eta_2=0} = 0.1196$ to $P^B_{\eta_2=9} = 0.0850$, tending to the *uniform* distribution $P^B_{\eta_m} \rightarrow 1/10$ as $m \rightarrow \infty$.

If we consider two-digit integer numbers, so $\{\eta_1, \eta_2\} \in [10, 99]$, the probability of appearing 10 is $P^B_{\{\eta_1,\eta_2\}=10} = 0.047$, while the one of 99 is $P^B_{\{\eta_1,\eta_2\}=99} = 0.004$. A list of occurring of a given digit in the first second and third positions can be found in several articles, for example, in Refs. 31, 32.

At first sight, these results are against the human perception, which would tend to assume a randomness. For a single digit, one would guess a *uniform* distribution, that is, $P^U_{\eta_1} = 1/9$ for any digit, and for two digits, $P^U_{\{\eta_1,\eta_2\}} = 1/90$, and so on. This is precisely what makes the BL interesting: if the numbers are naively forged by the human mind, for example, the BL will no longer be satisfied. However, it is important to note that not all data sets follow the BL, but just the ones resulting from the product of multiple independent factors that produce a probability covering several orders of magnitude when they are plotted in a lognormal scale.

3. The atomic data sets

The values η that we are going to scrutinize under the BL correspond to four different atomic-collision data sets, namely: ionization from the different *nl* subshells of neutral atoms by impact of multicharged ions, electron capture from hydrogen to the subshell *nl* of bare projectiles, stopping power cross sections for proton impact on atomic targets, and experimental stopping power cross sections by multicharged ions moving in gases. The numbers *n* and *l* denote the principal and orbital quantum numbers, respectively, of the atomic subshell. Details of these data sets will be provided in the next sections.

3.1 Ionization cross-section data set

During the last years, we have been calculating *nl*-ionization cross sections by multicharged ion impact on different neutral atoms. Calculations were carried out with the CDW-EIS method,[38] being numerically performed as accurate as possible. All the results are available in Ref. 53. Such atomic cross sections were satisfactorily used for the evaluation of molecular ionization within a stoichiometric model.[54]

In this work, CDW-EIS nl-ionization cross sections from Ref. 53 for antiprotons, H^+, He^{2+}, Be^{4+}, C^{6+}, and O^{8+} impinging on H, He, Li, Be, B, C, N, O, F, Ne, P, S, and Ar neutral atoms, for impact energies ranging from 100 to 10,000 keV/amu, were put together. We gather 2808 results with three significant figures, covering about 10 orders of magnitude. All these values were expressed in atomic units, forming a data set that will be noted, for short, as $\eta = I = I(nl)$. We also explore the reduced set of total ionization cross sections, $\eta = X = \Sigma_{nl} o_{nl} I(nl)$, where o_{nl} is the occupation number of the nl-subshell of the atomic target.

3.2 Electron capture cross-section data set

We consider capture cross sections from hydrogen to the nl-subshell of bare-ion projectiles calculated within the EI approximation.[43] Results for capture by H^+ and He^{2+} projectiles to the principal quantum numbers $n = 1, 2, 3$, and 4, and by Li^{3+}, Be^{4+}, B^{5+}, C^{6+}, N^{7+}, and O^{8+} ions to the principal quantum numbers $n = 1, \ldots, 9$, both for orbital momentum numbers $l = 0, \ldots, n-1$, are analyzed. In all the cases, 13 impact energies, ranging from 25 to 1220 keV/amu, were considered.[47] We totalize a set of 3275 values, expressed in atomic units, which will be here noted, for short, as $\eta = C = C(nl)$. Also the smaller set of capture cross sections to a given principal number n, defined as $\eta = T = T(n) = \sum_l C(nl)$, is explored.

3.3 Theoretical stopping power data set

Based on the momentum conservation, Sigmund[48] developed a general theory of stopping expressed in terms of the electron velocity in the initial atomic state. Oddershede and Sabin[19] implemented the Sigmund theory by using the quantum mechanical atomic velocity distribution state to state. They published a complete table of stopping cross sections by proton impact on neutral atoms, ranging from hydrogen to krypton. All the values were transformed to atomic units. This stopping data set will be noted as $\eta = S$. Furthermore, notice that from this data it is possible to derive, in a first-order perturbative approach, stopping cross sections for higher values of the projectile charge by using the Z^2 law.

3.4 Experimental stopping power data set

We use data of experimental stopping power as stored by the website of the IAEA.[49] We collected all the values corresponding to heavy ion impact on neutral gas targets, bringing together 4118 values from 52 files of the IAEA website for the following colliding systems: H^+, He^{2+}, Li^{3+}, N^{9+}, Cu^{29+}, and Kr^{36+} impinging on H_2, He, N_2, O_2, Ne, Ar, Kr, and Xe. All the values were again standardized to atomic units from the practical units 10^{-15} eVcm^2 and MeV cm/mg, as published in the site. Furthermore, the targets were described as they are found in nature, that is, for dimers, such as H_2, N_2, and O_2, the results of the IAEA were multiplied by 2 because they are normalized to the number of atoms. Note that this data set represents a great challenge for the BL because it contains several experiments differing each other, especially in the intermediate- and low-energy regions where we may have different first digits for the same collision parameters. This set will be noted as $\eta = E$.

On summarizing, the following sets will be considered

$$\eta = \begin{cases} I = I(nl) & \text{Ionization cross section from the } nl \text{ subshell.} \\ X = \sum_{nl} I(nl), & \text{Total ionization cross section.} \\ C = C(nl) & \text{Capture cross section to the } nl \text{ subshell.} \\ T = T(n) = \sum_{l} C(nl), & \text{Total capture cross section to the } n \text{ shell.} \\ S & \text{Theoretical stopping power cross section.} \\ E & \text{Experimental stopping power cross section.} \end{cases} \tag{5}$$

All of the atomic cross-section sets have no node and they are unimodal, containing only positive quantities. These features make them good candidates to check if the BL is satisfied.

In addition, to study any set of numerical results under the Benford scheme, one important point to bear in mind is the role of the last digit. In a numerical calculation, the last digit is generally worked through a rounding-up process, while the last digit of a Benford distribution is understood via a truncation process. Therefore, in our study of the BL the third (last) digit will be out of discussion.

4. Parameters of the Benford law

4.1 The width and the density of points

A first requirement to inspect the BL is that the data set is expected to be evenly distributed in a logarithmic scale. It means that the numbers η, when ranked from smallest to largest, can be approximated by a linear form in a logarithmic scale, that is,

$$log_{10}\eta \simeq a + bj, \tag{6}$$

where η represents a given value of the set and j indicates its order index in the data set, i.e., $j \in [1, J]$, with J being the total amount of values of the set. In all the cases, the actual values of $log_{10}\eta$ do not fall on a straight line from start to finish; it is reasonably straight in the middle, but it is curvy in the tails, as shown in Fig. 1. Following Nigrini,[32] the width Δ of a set of values can be estimated by the difference between the extremes of the linear fitting,

$$\Delta = (a + bJ) - (a + b) = b(J - 1) \simeq bJ. \tag{7}$$

The width Δ represents the effective number of orders of magnitude that the data set covers. For the five sets studied in this work, the values of Δ are given in Table 1 and displayed in Fig. 1. The BL requires $\Delta \gg 2$ to sample the first

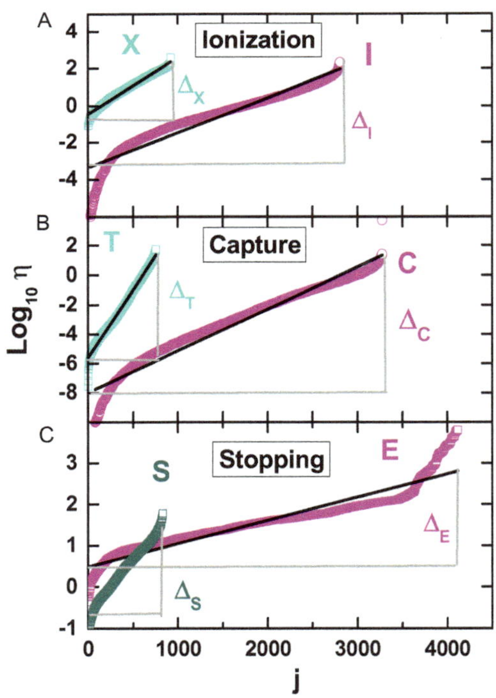

Fig. 1 Symbols, logarithm of population numbers as a function of the rank of the magnitude η under study (ordered from smallest to largest). Panels (A), (B), and (C) correspond to ionization, capture, and stopping power data sets, respectively. *Solid line* is the linear fitting as given by Eq. (6) and Δ indicates the corresponding width.

Table 1 Widths, densities, MAD errors, and ratios, as defined in the text, for the data sets given in Eq. (5).

η	*I*	*X*	*C*	*T*	*S*	*E*
J	2808	925	3275	755	828	4118
Δ	5.34	2.78	9.31	6.96	2.28	2.32
δ	526	331	351	108	363	1778
ε_1^B (%)	0.52	1.02	0.46	0.76	0.70	1.52
ε_1^U (%)	6.33	5.11	5.75	5.92	5.28	5.27
ε_1^N (%)	0.63	1.22	0.73	1.08	0.77	1.72
ρ_1^B	0.08	0.20	0.08	0.13	0.13	0.29
ρ_2^B	0.27	0.56	0.47	1.03	0.39	0.46
π_2	0.262	0.270	0.257	0.244	0.257	0.268

digits at least twice. It is expected that the larger the width Δ, the more robust the prediction of the BL.

For the I and C data sets, the Δ values are 5.34 and 9.31, respectively, being large enough to warrant a good spread of numbers. Instead for the stopping E, $\Delta = 2.32$, being a small value, which originates critical problems of borders, as we will see. A similar problem arises if we reduce the number of data. For example, by considering the total ionization cross section set X, the total number of data is reduced to $J = 925$, with values covering just two orders of magnitude.

In order to measure the effect of both Δ and J, we can define a new important parameter: the density of points $\delta = J/\Delta \simeq 1/b$. From our experience, we find that it is required $\delta \gg 200$ to have a reasonable occurrence of the first digit only. For total capture cross sections to different n-shells, represented by the set T, the resulting range Δ still covers seven orders of magnitude, but the amount of values reduces dramatically to $J = 755$, given rise to $\delta \sim 10^2$, which is a small density. Hence, we are in the presence of different situations: for X, Δ is substantially reduced, while for T, δ is too small. For all the cases, values of δ are also displayed in Table 1.

4.2 The degree of *"Benfordness"*

At this point, one should develop a tool to quantify how good the data conform the BL or, in other words, to quantify the *"Benfordness"* of a given data set under study. A traditional approach is to use the chi-squared statistic.

However, this test is not useful for very large data sets since for a large value of J, the calculated chi-square will generally be higher than the critical value, leading us to conclude that the data set does not conform the BL.[32] A specific method to measure the *Benfordness* of the data is the mean absolute deviation (MAD) test. For the first digit, the MAD test is defined in percentages as[32]

$$\varepsilon_1^B = \frac{1}{9}\sum_{\eta_1=1}^{9}\left|P_{\eta_1} - P_{\eta_1}^B\right| \times 100, \tag{8}$$

where P_{η_1} is the actual frequency of the first digit of the magnitude η and $P_{\eta_1}^B$ is the corresponding prediction of the BL, as given by Eq. (1). Nigrini[32] published some guidelines, based on his personal experience, to qualify the *Benfordness* according to the value of ε_1^B as it follows:

$$\begin{cases} 0.0\% \leqslant \varepsilon_1^B \leqslant 0.6\%, & \textit{close conformity}, \\ 0.6\% \leqslant \varepsilon_1^B \leqslant 1.2\%, & \textit{acceptable conformity}, \\ 1.2\% \leqslant \varepsilon_1^B \leqslant 1.5\%, & \textit{marginal conformity}. \end{cases} \tag{9}$$

If we rule out the rigor of the accountancy to detect frauds, we would be very satisfied if we can comply with these figures. Just for comparison, we could define equivalently a MAD error corresponding to the uniform distribution as

$$\varepsilon_1^U = \frac{1}{9}\sum_{\eta_1=1}^{9}\left|P_{\eta_1} - P_{\eta_1}^U\right| \times 100, \tag{10}$$

where $P_{\eta_1}^U = 1/9$ is the uniform (random) prediction. One would expect that the ratio

$$\rho_1^B = \frac{\varepsilon_1^B}{\varepsilon_1^U} \tag{11}$$

verifying $\rho_1^B \ll 1$ indicates that we are in the presence of a Benford distribution.

As far as the first digit is concerned, there is an interesting alternative test introduced by Nigrini[32] to diagnose a Benford distribution. Consider an any-digit set of a given magnitude $\eta = \{\eta_1, \eta_2, \ldots\}$ and calculate

$$N_{\mu_1} = \frac{\sum_{\eta}\{\eta_1, \eta_2, \ldots\}\delta_{\eta_1,\mu_1}}{\sum_{\eta}\{\eta_1, \eta_2, \ldots\}}, \tag{12}$$

with $\mu_1 \in [1, 9]$. The second theorem of Nigrini[32] states that the Benford distribution produces $N_{\mu_1} \rightarrow N^B_{\mu_1} = 1/9$. Therefore, this is another independent strategy to check whether the set under study satisfies the BL. This new parameter allows us to introduce an equivalent Nigrini MAD error, ε_1^N, to assert the *Benfordness* of the set under study, as

$$\varepsilon_1^N = \frac{1}{9}\sum_{\eta_1=1}^{9}\left|N_{\eta_1} - N^B_{\eta_1}\right| \times 100. \tag{13}$$

The analysis of the second digit demands a more fine attention because $P^B_{\eta_2}$ is close to the value for randomness, i.e., $P^U_{\eta_2} = 1/10$, and a differentiation is required. To this end, we find convenient to define the ratio of MAD errors

$$\rho_2^B = \frac{\varepsilon_2^B}{\varepsilon_2^U} = \frac{\frac{1}{10}\sum_{\eta_2=0}^{9}\left|P_{\eta_2} - P^B_{\eta_2}\right|}{\frac{1}{10}\sum_{\eta_2=0}^{9}\left|P_{\eta_2} - P^U_{\eta_2}\right|}, \tag{14}$$

which gives an indication of the goodness of the second digit to conform the Benford distribution. In similar fashion with ρ_1^B, we can state: if $\rho_2^B > 1$, the second digit of the set is ruled by randomness, while $\rho_2^B < 1$ indicates that the set is ruled by Benford. Then, the smaller ρ_2^B, the closer to the Benford distribution. For all the sets, values of ε_1^B, ε_1^U, ε_1^N, ρ_1^B, and ρ_2^B are shown in Table 1.

5. Results and discussion

5.1 Examination of the atomic data sets

5.1.1 Ionization cross sections

We start analyzing the probability of occurring the first figure of the *nl*-ionization data, denoted as P_{I_1}, in Fig. 2A. We compare P_{I_1} with the Benford prediction $P^B_{I_1}$ given by Eq. (1), which is indicated as a histogram in light gray. The agreement is very good and, in consequence, the MAD

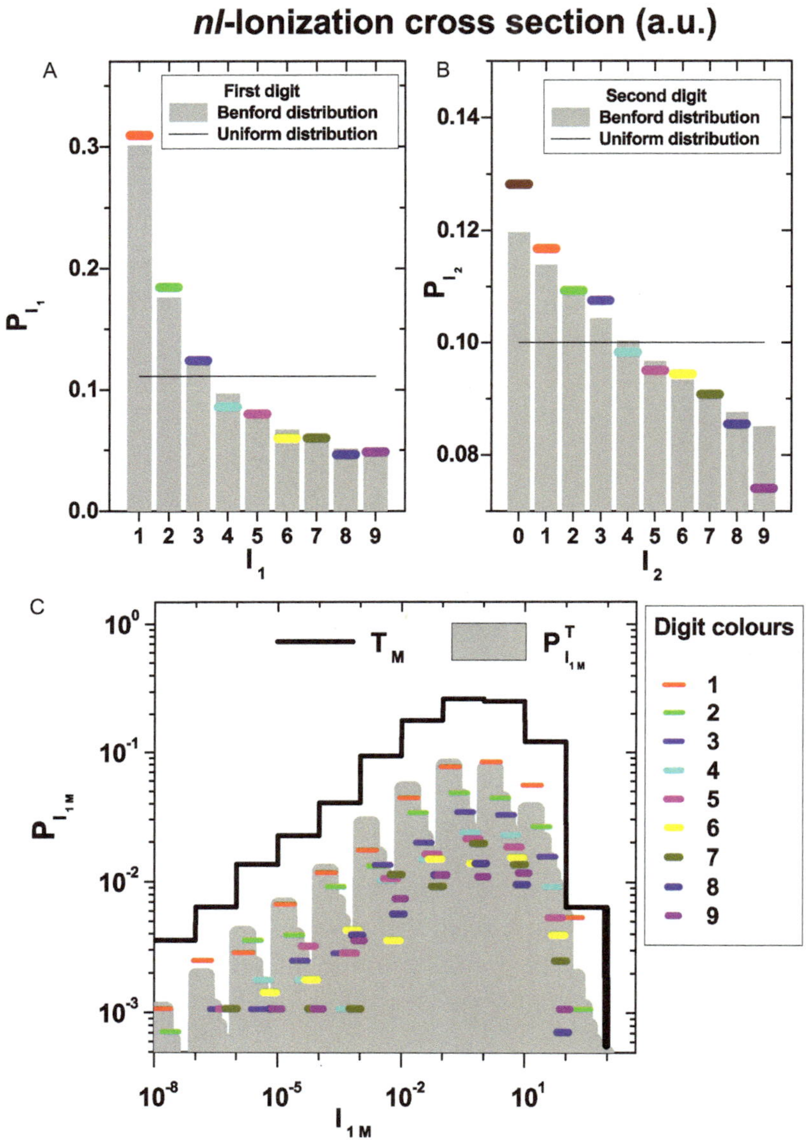

Fig. 2 *Color symbols* are the frequencies of (A) the first digits and (B) second digits from the data set of *nl*-ionization cross sections. The *light gray histogram* represents the Benford predictions and the *horizontal black solid line* denotes the uniform distribution. (C) *Color symbols* correspond to the distributions of the first digits within each decade: $l_{1M} = l_1 \times 10^M$. The *solid line histogram* represents the top function as defined in Eq. (15). The *light gray histogram* represents the *Benfordization* estimation given by Eq. (16).

error is very small, i.e., $\varepsilon_1^B = 0.52\%$ (see Table 1). It means that, according to (9), we can certify the agreement as *close conformity*. The second theorem of Nigrini produces a very similar error, $\varepsilon_1^N = 0.63\%$, which stands as an alternative criterion to assess the *Benfordness* using the same categorization of (9).

We can go further by studying the probability to have a given second digit $I_2 \in [0, 9]$, denoted as P_{I_2}, which is shown in Fig. 2B. Again, the agreement with the Benford prediction is quite good.

To deepen the analysis we should calculate the two-digit probability $P_{\{I_1,I_2\}}$. But instead, we prefer to introduce a novel criterion based on the prediction of the density of two-digit prime numbers π_2, which is a very sensitive and sharp value reduced to the range [10–99]. This represents a new parameter that again subverts human thinking. One would tend to think that π_2 equals that of the uniform distribution, $\pi_2^U = 21/90 = 0.233$. But this is not true for a Benford distribution. By using Eq. (2), one can easily obtain $\pi_2^B = 0.266$. Our $\{I_1, I_2\}$ data produce $\pi_2 = 0.262$, in close agreement with the BL prediction. Therefore, we can conclude categorically that our *nl* -ionization cross-section data set satisfies very well the BL.

We are also interested in studying the BL within a given order of magnitude. In Fig. 2C, we plot the data corresponding to the occurrence of a single digit within a given order of magnitude, that is, $I_{1M} = I_1 \times 10^M$. Even though the results for I_{1M} spread along 9 orders of magnitude, from 10^{-8} to 10^0, it becomes evident that the BL still applies within each order of magnitude. To visualize this behavior more clearly, we first define a top function for a given order of magnitude M as

$$T_M = \sum\nolimits_{I_1} P_{I_{1M}}, \tag{15}$$

which plays the role of a closure relation within the range of magnitude M. Hence, the probability $P_{I_{1M}}$ can be estimated by simply *Benfordizing* the top function, i.e.,

$$P_{I_{1M}}^T \simeq \log_{10}(1 + 1/I_1)T_M, \tag{16}$$

which is plotted in Fig. 2C in light gray. From this figure, one can observe that $P_{I_1 M}^T$ guides quite well the data values, indicating that all the substantial information can be reduced to M values of the top function T_M.

On the other hand, the total cross section set X is another matter because it is spread in a little more than two orders of magnitude (see Table 1).

But we can still verify that X follows the first-digit BL with an error of $\varepsilon_1^B = 1.02\%$, which means that we can certify the agreement as *acceptable conformity*.

5.1.2 Capture cross sections

Fig. 3A shows the probability of occurring the first figure of the *nl*-capture cross-section data set P_{C_1}, which is compared with the Benford prediction

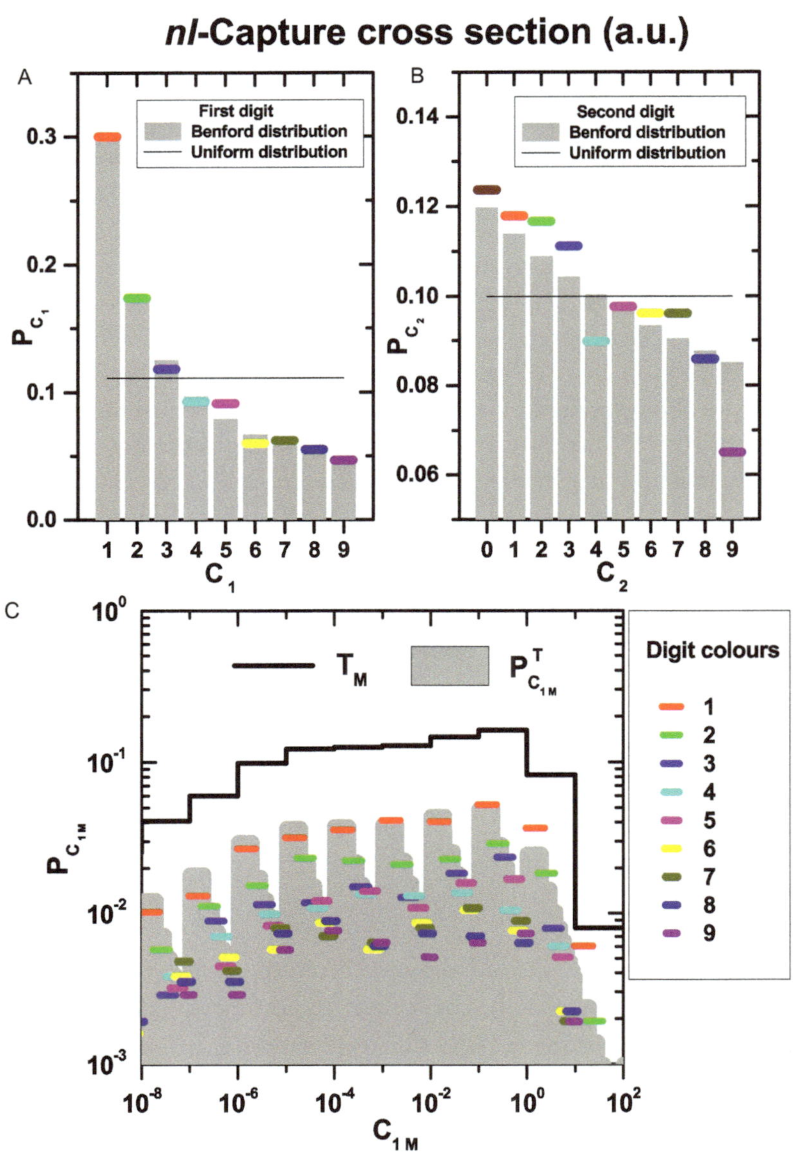

Fig. 3 Similar to Fig. 2 for *nl*-capture cross sections.

given by Eq. (1). The agreement is good and the MAD error is small, $\varepsilon_1^B = 0.46\%$, complying with the *close conformity* according with the categorization (9). The second digit does not look so well but it is better than the uniform distribution, with $\rho_2^B < 1$ (see Table 1). Moreover, the distribution along 10 orders of magnitude follows reasonably well the *"Benfordization"* of the top probability, shown in light gray in Fig. 3C.

Instead, for the capture cross sections to the n-state, corresponding to the data set T, the first digit probability degrades to *acceptable conformity* with $\varepsilon_1^B = 0.76\%$ (see Table 1), while its second digit does not present any bias to Benford nor to uniform because $\rho_2^B = 1.03$.

5.1.3 Theoretical stopping power cross sections

This data set presents very small values of both Δ and J. Even in this case, the MAD error is quite good, $\varepsilon_1^B = 0.7\%$ ($\varepsilon_1^N = 0.77\%$), and the set qualifies quite well within *acceptable conformity*. Fig. 4A and B shows the performance of the first and second digits, respectively, in the framework of the BL. The second digit is more close to the Benford distribution, but it should be noticed that such a conclusion is not definitive since more values and a larger width are required, as shown in Fig. 4C.

5.1.4 Experimental stopping power cross sections

This is a case in which not only the width Δ is small, but also we are dealing with a large variety of experimental data from different laboratories. As shown in Fig. 5A, the first digit probability P_{E_1} resembles the histogram of Benford but the MAD error is high, $\varepsilon_1^B = 1.52\%$, which hardly qualifies within the category of *marginal conformity*, as classified in (9). Nonetheless, the shape is definitively more close to the BL than to the uniform distribution ($\varepsilon_1^U = 5.3\%$). The second digit distribution (Fig. 5B) does not differ a lot from the one of Benford because $\rho_2^B = 0.46 < 1$. On the contrary, $P_{E_{1M}}$ does not follow the logarithmic structure within each order of magnitude as in the previous cases, as shown in Fig. 5C.

5.2 Two numerical experiences

5.2.1 Testing the universal scaling of Pinkham

One of the most extraordinary properties of the BL is its scale invariance, to the point that it is possible to obtain it mathematically by just invoking this property, as discovered by Pinkham.[55] From the point of view of physics, this is evident because the BL should be independent of the units that we

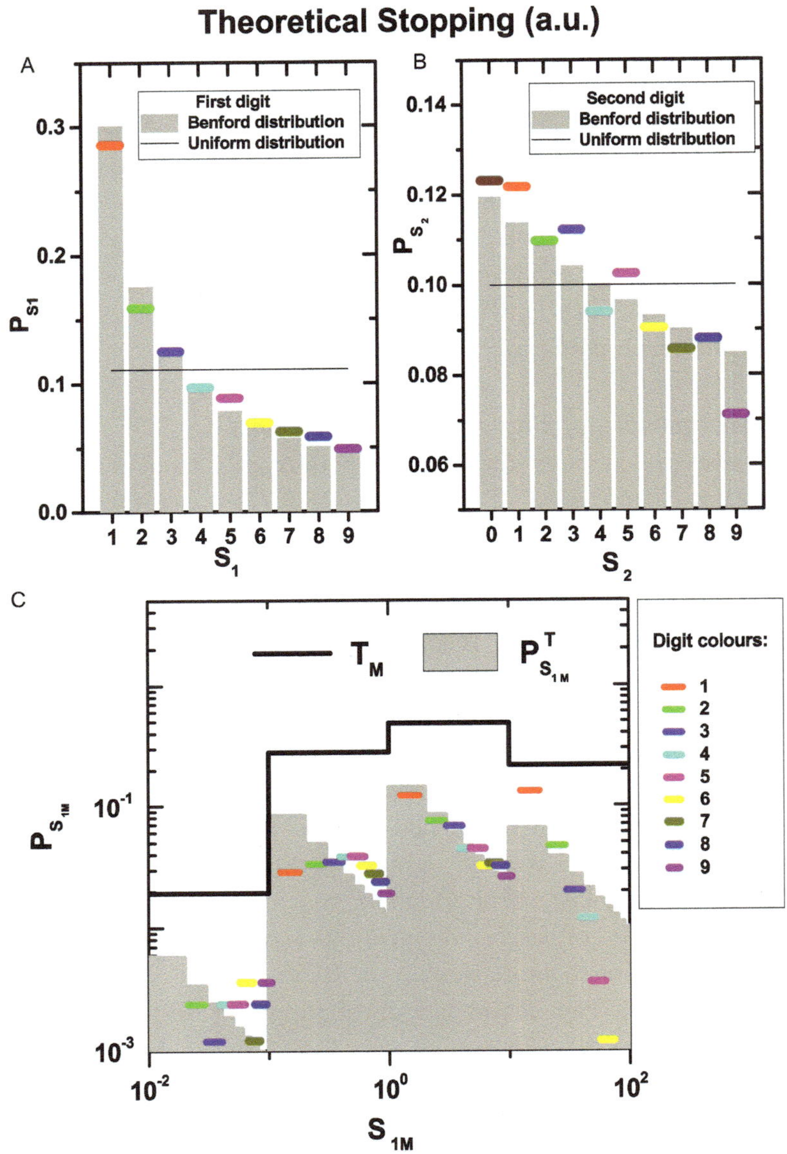

Fig. 4 Similar to Fig. 2 for theoretical stopping power cross sections from Ref. 19.

use. Ideally, as the Benford distribution is based on the logarithmic function, any multiplicative factor will just shift the distribution, keeping intact the occurrences of the digits. This is the case if we have a big width Δ. However, there is a limitation when the width Δ is small, like in the case

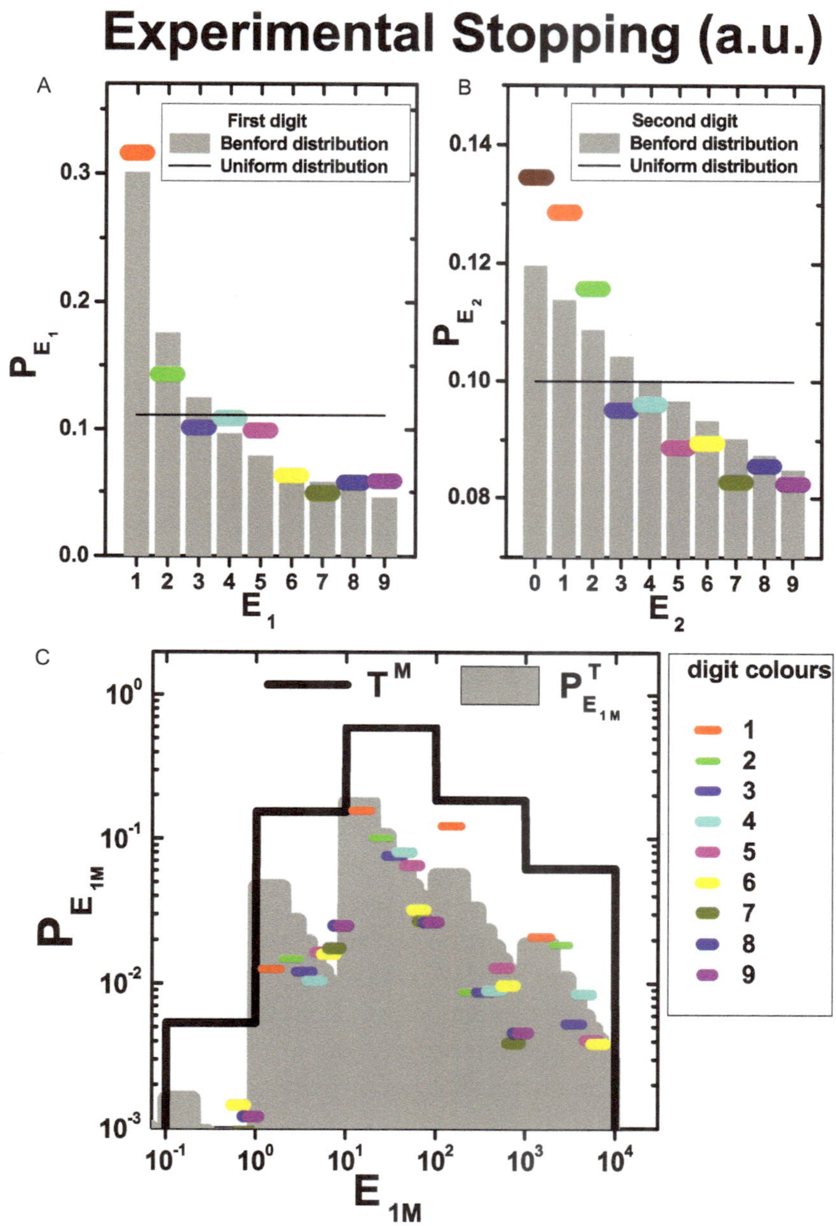

Fig. 5 Similar to Fig. 2 for experimental stopping power cross sections from IAEA database.[49]

of the stopping power cross sections. This limitation surfaces clearly due to the border effects in a limited width. As Δ is small, the reduction or enhancement of the probabilities corresponding to some digits will be transferred to other digits when different units are used.

In our analysis, we use atomic units for ionization cross sections, but we could have used any other unit system, like cm^2 or whatever we choose. For example, if we transform the set of *nl*-ionization values I in cm^2, we need to multiply by 2.800×10^{-17}. This change of scale produces $\varepsilon_1^B = 0.43\%$, which slightly differs from the error 0.52% obtained with atomic units, i.e., a difference of 0.1%. It is important to note that the order of magnitude of the units is irrelevant since it just replicates the same digits. But if we built the set of experimental stopping power cross sections in 10^{-15} eVcm^2, we should multiply the set by 1.213, finding $\varepsilon_1^B = 1.9\%$, which is substantially greater than the error 1.5% obtained by expressing these values in atomic units (that is, a difference of 0.4%). This difference is a consequence of the border effects due to the small width Δ. Note that in units of 10^{-15} eVcm^2, the experimental stopping power set will no longer classify as *marginal conformity*.

5.2.2 The whole data set

We essay a daring experience by gathering all the data, *nl*-ionization, *nl*-capture, and stopping power cross sections (all expressed in atomic units). We totalize 10,201 values in only one numerical universe. We found that this big set of values satisfies reasonably well the BL with $\varepsilon_1^B = 0.67\%$ and $\rho_2^B = 0.33$. However, it does not represent an overall improvement of the BL conformity. The explanation may be related to the stopping power set, which adds 4118 values in a short width, introducing a high density that disturbs the required even distribution in the logarithmic scale. This behavior demonstrates that a large number of values not necessarily improve the Benford distribution, but the stability of the density of points does.

6. Conclusion

We have studied six different sets of atomic data with the BL. We found three different qualifications according to the Nigrini scheme—*close, acceptable*, and *marginal conformities*—for the different sets of ionization, capture, and theoretical and experimental stopping power cross sections. This findings allows us to conclude that any atomic collision data set having the appropriate width Δ and density of points δ will satisfy the Benford distribution. Furthermore, the parameters here introduced to quantify the degree of conformity of the BL, like the MAD errors, could be used to check the quality of any atomic data set in order to detect systematic experimental errors or theoretical biases. For example, if an unnoticed experimental error or a bug in the numerical code introduces a spurious additive term in the cross sections of a given data set, this data set would be far from *close conformity* with the BL, regardless of the value of its width Δ and/or its density δ.

Acknowledgments

The authors acknowledge the financial support from the following institutions of Argentina: Consejo Nacional de Investigaciones Científicas y Técnicas (CONICET), Agencia Nacional de Promoción Científica y Técnológica (ANPCyT), and Universidad of Buenos Aires.

References

1. Schulz, M.; Moshammer, R.; Fischer, D.; Kollmus, H.; Madison, D. H.; Jones, S.; Ullrich, J. Three-Dimensional Imaging of Atomic Four-Body Processes. *Nature* **2003**, *422*, 48–50.
2. Bhattacharya, S.; Choudhury, K. B.; Deb, N. C.; Sinha, C.; Roy, K.; Msezane, A. Z. Ionization of Sodium by the Impact of Alpha Particle. *Eur. Phys. J. D* **2008**, *47*, 335–338.
3. Gassert, H.; et al. Agreement of Experiment and Theory on the Single Ionization of Helium by Fast Proton Impact. *Phys. Rev. Lett.* **2016**, *116*, 073201.
4. Kaganovich, I. D.; Startsev, E.; Davidson, R. C. Scaling and Formulary of Cross-Sections for Ion-Atom Impact Ionization. *New J. Phys.* **2006**, *8*, 278.
5. Tribedi, L. C.; Agnihotri, A. N.; Galassi, M. E.; Rivarola, R. D.; Champion, C. Ionization of Uracil in Collisions With Fast Bare Ions. *Eur. Phys. J. D* **2012**, *66*, 303.
6. Bhattacharjee, S.; Bagdia, C.; Chowdhury, M. R.; Mandal, A.; Monti, J. M.; Rivarola, R. D.; Tribedi, L. C. Bare-Carbon-Ion-Impact Electron Emission From Adenine Molecules: Differential and Total Cross-Section Measurements. *Phys. Rev. A* **2019**, *100*, 012703.
7. Bhattacharjee, S.; Mandal, A.; Chowdhury, M. R.; Bagdia, C.; Monti, J. M.; Rivarola, R. D.; Tribedi, L. C. Electron Emission in Ionization of Adenine Molecule Induced by 5 MeV/u Bare C Ions. *Eur. Phys. J. D* **2020**, *74*, 163.
8. Stolterfoht, N.; Cabrera-Trujillo, R.; Krstić, P. S.; Öhrn, Y.; Deumens, E.; Sabin, J. R. Stueckelberg Oscillations in the Charge Transfer into the $n = 2$ and $n = 3$ Shells of He^{2+} on Collision With H. *Int. J. Quantum Chem.* **2009**, *109*, 3063–3074.
9. Champion, C.; Weck, P. F.; Lekadir, H.; Galassi, M. E.; Fojón, O. A.; Abufager, P.; Rivarola, R. D.; Hanssen, J. Proton-Induced Single Electron Capture on DNA/RNA Bases. *Phys. Med. Biol.* **2012**, *57*, 3039–3049.
10. Cravens, T. E. X-Ray Emission From Comets. *Science* **2002**, *296*, 1042–1045.
11. Otranto, S.; Olson, R. E.; Beiersdorfer, P. X-Ray Emission Cross Sections Following Charge Exchange by Multiply Charged Ions of Astrophysical Interest. *Phys. Rev. A* **2006**, *73*, 022723.
12. Schultz, D. R.; Ovchinnikov, S. Y.; Stancil, P. C.; Zaman, T. Elastic, Charge Transfer, and Related Transport Cross Sections for Proton Impact of Atomic Hydrogen for Astrophysical and Laboratory Plasma Modeling. *J. Phys. B* **2016**, *49*, 084004.

13. Jorge, A.; Errea, L. F.; Illescas, C.; Méndez, L. Calculation of Ionization and Total and Partial Charge Exchange Cross Sections for Collisions of C^{6+} and N^{7+} With H. *Eur. Phys. J. D* **2014**, *68*, 227.
14. Cariatore, N. D.; Otranto, S.; Olson, R. E. Classical Description of H(ls) and H*(n= 2) for Cross-Section Calculations Relevant to Charge-Exchange Diagnostics. *Phys. Rev. A* **2015**, *91*, 042709.
15. International Thermonuclear Experimental Reactor (ITER), https://www.iter.org/.
16. Aumayr, F.; et al. Roadmap on Photonic, Electronic and Atomic Collision Physics: III. Heavy Particles: With Zero to Relativistic Speeds. *J. Phys. B* **2019**, *52*, 171003.
17. Sauer, S. P. A.; Oddershede, J.; Sabin, J. R. Mean Excitation Energies and Their Directional Characteristics for Energy Deposition by Swift Ions on the DNA and RNA Nucleobases. *J. Phys. Chem. C* **2010**, *114*, 20335–20341.
18. Schardt, D.; Elsasser, T.; Schulz-Ertner, D. Heavy-Ion Tumor Therapy: Physical and Radiobiological Benefits. *Rev. Mod. Phys.* **2010**, *82*, 383–425.
19. Oddershede, J.; Sabin, J. R. Orbital and Whole-Atom Proton Stopping Power and Shell Corrections for Atoms With $Z \leq 36$. *At. Data Nucl. Data Tables* **1984**, *31*, 275–297.
20. Montanari, C. C.; Miranda, P. A.; Alves, E.; Mendez, A. M. P.; Mitnik, D. M.; Miraglia, J. E.; Correa, R.; Wachter, J.; Aguilera, M.; Catarino, N.; da Silva, R. C. Stopping Power of Hydrogen in Hafnium and the Importance of Relativistic 4f Electrons. *Phys. Rev. A* **2020**, *101*, 062701.
21. Young, P. R.; Dere, K. P.; Landi, E.; Zanna, G. D.; Mason, H. E. The CHIANTI Atomic Database. *J. Phys. B* **2016**, *49*, 074009.
22. Quinto, M. A.; Monti, J. M.; Weck, P. F.; Fojón, O. A.; Hanssen, J.; Rivarola, R. D.; Senot, P.; Champion, C. Monte Carlo Simulation of Proton Track Structure in Biological Matter. *Eur. Phys. J. D* **2017**, *71*, 130.
23. Paul, H.; Sánchez-Parcerisa, D. A Critical Overview of Recent Stopping Power Programs for Positive Ions in Solid Element. *Nucl. Instrum. Methods Phys. Res. B* **2013**, *312*, 110–117.
24. Chung, H. K.; Braams, B. J.; Bartschat, K.; Császár, A. G.; Drake, G. W. F.; Kirchner, T.; Kokoouline, V.; Tennyson, J. Uncertainty Estimates for Theoretical Atomic and Molecular Data. *J. Phys. D* **2016**, *49*, 363002.
25. Drake, G. W. F. Accuracy in Atomic and Molecular Data. *J. Phys. B* **2020**, *53*, 223001.
26. Paul, H. *Theory of Heavy Ion Collision Physics in Hadron Therapy*; Elsevier, Academic Press: Amsterdam, 2013; pp. 39–61.
27. Paul, H. The Solid-Gas Difference in Stopping Powers, and Statistical Analysis of Stopping Power Data. *Nucl. Instrum. Methods Phys. Res. B* **2009**, *267*, 9–14.
28. Parfitt, W. A.; Jackman, R. B. Machine Learning for the Prediction of Stopping Powers. *Nucl. Instrum. Methods Phys. Res. B* **2020**, *478*, 21–33.
29. Hill, T. P. The First Digit Phenomenon. *Am. Sci.* **1998**, *86*, 358–363.
30. Berger, A.; Hill, T. P. The Mathematics of Benford's Law: A Primer. *Stat. Methods Appl.* **2021**, *30*, 779–795.
31. Nigrini, M. J. Fraud Detection: I've Got Your Number. *J. Accountancy* **1999**, *187*, 79–83.
32. Nigrini, M. J. *Benford's Law: Applications for Forensic Accounting, Auditing, and Fraud Detection*; John Wiley *& Sons: New York, 2012.*
33. Sambridge, M.; Tkalčić, H.; Jackson, A. Benford's Law in the Natural Sciences. *Geophys. Res. Lett.* **2010**, *37*, L22301.
34. Alexopoulos, T.; Leontsinis, S. J. Astrophys. Astron. *J. Astrophys. Astron.* **2014**, *35*, 639–648.
35. Melita, M. D.; Miraglia, J. E. On the Applicability of Benford Law to Exoplanetary and Asteroid Data. *New Astron.* **2021**, *89*, 101654.
36. https://www.benfordonline.net/list/alphabetical.

37. Crothers, D. S. F.; McCann, J. F. Ionisation of Atoms by Ion Impact. *J. Phys. B* **1983**, *16*, 3229–3242.
38. Fainstein, P. D.; Ponce, V. H.; Rivarola, R. D. Two-Centre Effects in Ionization By Ion Impact. *J. Phys. B* **1991**, *24*, 3091–3119.
39. Bhattacharya, S.; Choudhury, K. B.; Deb, N. C.; Sinha, C.; Roy, K.; Msezane, A. Z. Ionization of Sodium by the Impact of Alpha Particle. *Eur. Phys. J. D* **2008**, *47*, 335–338.
40. Gulyás, L.; Egri, S.; Kirchner, T. Differential Cross Sections for Single Ionization of Li in Collisions With Fast Protons and O^{8+} Ions. *Phys. Rev. A* **2014**, *90*, 062710.
41. Igarashi, A.; Gulyás, L. Fully Differential Cross Sections for Ionization of Li(2s) and Li(2p) by 16 MeV Li^{2+} Impact. *J. Phys. B* **2020**, *53*, 075203.
42. Rojas, M. F.; Quinto, M. A.; Rivarola, R. D.; Monti, J. M. Dynamic Effective Charge in the Target Continuum Within the CDW-EIS Model for Ionisation in Ion-Atom Collisions: Angular Dependence. *Eur. Phys. J. D* **2021**, *75*, 154.
43. Gravielle, M. S.; Miraglia, J. E. Eikonal Impulse Approximation in Electron-Capture Processes. *Phys. Rev. A* **1991**, *44*, 7299–7306.
44. Gravielle, M. S.; Miraglia, J. E. State-Selective Scaling in Electron Capture by Multicharged Ions on Light Atoms. *Phys. Rev. A* **1995**, *51*, 2131–2139.
45. Montanari, C. C.; Gravielle, M. S.; Miraglia, J. E. Propensity Rule for the Magnetic Substate Distribution in Electron Capture at High Impact Energies. *Phys. Rev. A* **1997**, *56*, 4314–4316.
46. Montanari, C. C.; Gravielle, M. S.; Miraglia, J. E. Capture From $n = 2$ States of Hydrogen by Heavy Fast Ions. *Phys. Scr.* **1997**, *56*, 279–283.
47. Jorge, A.; Illescas, C.; Miraglia, J. E.; Gravielle, M. S. Scaling for State-Selective Charge Exchange Due to Collisions of Multicharged Ions With Hydrogen. *J. Phys. B* **2015**, *48*, 235201.
48. Sigmund, P. Kinetic Theory of Particle Stopping in a Medium With Internal Motion. *Phys. Rev. A* **1982**, *26*, 2497–2517.
49. International Atomic Energy Agency (IAEA). *Stopping Power of Matter for Ions, Graphs, Data, Comments and Programs*; 2021. https://www-nds.iaea.org/stopping/.
50. Newcomb, S. Note on the Frequency of Use of the Different Digits in Natural Numbers. *Am. J. Math.* **1881**, *4*, 39–40.
51. Benford, F. The Law of Anomalous Numbers. *Proc. Am. Philos. Soc.* **1938**, *78*, 551–572.
52. Berger, A.; Hill, T. P. A Basic Theory of Benford's Law. *Prob. Surv.* **2011**, *8*, 1–126.
53. Miraglia, J. E. Shell-to-Shell Ionization Cross Sections of Antiprotons, H^+, He^{2+}, Be^{4+}, C^{6+} and O^{8+} on H, He, Li, Be, B, C, N, O, F, Ne, P, S and Ar Neutral Atoms. *arXiv:1909.13682v2 [physics.atom-ph]* **2019**, 1–18.
54. Mendez, A. M. P.; Montanari, C. C.; Miraglia, J. E. Ionization of Biological Molecules by Multicharged Ions Using the Stoichiometric Model. *J. Phys. B* **2020**, *53*, 055201.
55. Pinkham, R. S. On the Distribution of First Significant Digits. *Ann. Math. Stat.* **1961**, *32*, 1223–1230.

CHAPTER ELEVEN

Long-lived molecular dications: A selected probe for double ionization

Lucas Sigaud[a,*] and Eduardo C. Montenegro[b]
[a]Instituto de Física, Universidade Federal Fluminense, Niterói, Brazil
[b]Instituto de Física, Universidade Federal do Rio de Janeiro, Rio de Janeiro, Brazil
[*]Corresponding author: e-mail address: lsigaud@id.uff.br

Contents

Abstract

The energy loss of projectiles passing through a medium is highly dependent on the electronic orbitals of the molecules which compose this medium. This includes systems as diverse as, for example, material modification and characterization, solar radiation and cosmic rays penetrating planetary atmospheres, radiotherapy cancer treatments and plasma fusion reactors such as tokamaks. Double ionization is not only present, but may play an important role in the effects of ionizing radiation in these environments. In this study we present a short review of metastable molecular dication production measurements by electron impact. We further analyze which mechanisms contribute for the dication production of several simple molecules, including evenly symmetric ones (such as N_2, O_2 and C_2H_4), whose dication production cross sections were directly measured only recently by means of the DETOF technique.

1. Introduction

The passage of a charged particle through a material medium—solids, liquids, gases, plasmas, biologic media, etc.—leaves a trail of ions produced by the removal of electrons from various atomic or molecular states.[1]

Advances in Quantum Chemistry, Volume 85
ISSN 0065-3276
https://doi.org/10.1016/bs.aiq.2022.05.001
Copyright © 2022 Elsevier Inc.
All rights reserved.

This process as a whole involves an interdependence between the projectile and the target that is mainly tied to the exchange of energy. This exchange of energy comes down, in most cases, to the transfer of kinetic energy from the projectile to the target and leaves its imprint both on the projectile and on the state in which the target is left. John Sabin, along with his many collaborators, has devoted much of his scientific life to unraveling the interaction effects on these two partners.

Due to the completeness relation behind Bethe's sum rule,[2,3] the basic calculation of electronic energy loss from swift ($v \gtrsim 3v_{Bohr} = e^2/\hbar$) projectiles does not depend on much detail on the target's final state.[4] This does not mean, however, that the projectile energy loss is not highly sensitive to some average parameters of the target, such as average excitation energy and associated shell corrections.[5] In fact, the calculation of the average excitation energy is essential for the determination of the energy loss of charged particles in matter, and the quality of this calculation has important practical consequences, such as in tumor therapy using accelerators.[6–8] Within this scenario, high quality calculations of this parameter for water[9] and more complex biological molecules,[5,10,11] as well as experiments to feed these calculations, are required.[12]

One complement of the projectile's energy loss is the set of excited states in which the target molecules can be left. As a consequence of the non-selectivity of the interaction between the projectile and the various electrons of the target, the average excitation energy is significantly higher than the first ionization energy of the molecule.[4] Outer valence shell and less bound electrons are, as a rule, more likely to be removed. On the other hand, inner shell electrons, such as those in the K shell of the atoms that make up the molecule, are less likely to be removed, but, when they are, a large energy transfer is made. In this way, all electrons contribute to the average excitation energy.

As a rule, and except for the case of the outermost valence electrons when a stable molecular parent ion can be formed, the removal of one or more inner valence or core electrons results in the fragmentation of the molecular ion. In biological tissues,[13] planetary atmospheres[14] or combustion plasmas,[15] molecular fragmentation is one of the main traces of the projectile's passage and energy transfer. Highly reactive fragments of the parent molecule that can significantly affect the composition and the inventory of the medium can be produced. The branching ratios for the various resulting fragments depend on the selection of which electrons from the molecule are removed by the projectile, which, in turn, depends, among other factors, on its velocity, its net charge and the removal mechanism.

The removal of any two electrons by their interaction with the projectile completely destabilizes the molecule, which can fragment into two or several pieces. Despite this being the general rule, with the fragmentation channel dominant for the relaxation of molecular dications, various molecules, even some small ones, can be left in the form of stable or metastable dications. This is the case, for example, of N_2, O_2, NH_3, CO_2, NO and CO. On the other hand, stable dications have not been observed so far for CH_4 and H_2O. In the latter case, and due to its importance in tumor therapy and in planetary atmospheres, the fragmentation from electrons, protons, and heavy ions impact has been extensively explored. Various mechanisms that result in the removal of two electrons such as double impact,[16] single impact with Auger decay[16] or populating of satellite states,[17] double capture and transfer ionization,[18–21] have been studied.

The removal of a very specific pair of electrons, either by charged particles or photons, and which usually belong to the outermost valence shells, can result in a stable or metastable dication. The stability of these ions is easier to be visualized in the case of diatomic molecules, as depicted in Fig. 1. The repulsive potential energy curve corresponding to the two single charged fragments separated at large internuclear distances, becomes an attractive well at short internuclear distances due the covalent bonding contribution added to polarization effects.[22] The value of the minimum potential energy

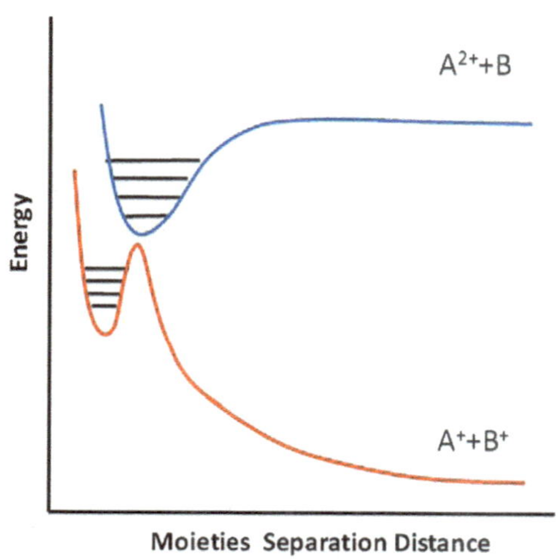

Fig. 1 Schematic representation of the potential energy curves for a diatomic dication, showing the repulsive or attractive behavior depending of the asymptotic charges. The $A^{2+}+B$ reaction is fast due to the absence of the activation barrier and the $A^{+}+B^{+}$ fragmentation delivers moieties with large kinetic energies to the medium.

of this well is larger than the asymptotic energy corresponding to the two charged fragments. Trapped in the well, the dication can tunnel through the barrier and its half-life depends essentially on the characteristics of that well and the vibrational state in which it is formed.[23] Although most dications formed by double ionization or double photo-ionization have a very short half-life (<1 ns), quickly fragmenting, many of them have a long half-life (>100 μs) and, in rarefied gaseous media, such as planetary ionospheres, this half-life can be longer than the mean collision time so they are mainly destroyed by collision with other gas molecules.[24]

The important role played by dications in some environments, for example in interstellar space or planetary atmospheres, is directly associated with these characteristics. In energized media, more stable dications can significantly survive the radiation in which they are submerged due to their larger ionization potential compared to monocations or neutral molecules, and live long enough to efficiently react with other gas molecules.[23] Unlike neutral-neutral reactions which, at very low temperatures, must overcome the activation barrier, the attractive potential of monocation or dication-neutral reactions results in very high reaction rates which play a key role in the physical-chemistry of these media.[25] For example, the formation of complex molecules on Titan is fueled by ion-neutral reactions in its upper atmosphere.[26]

Metastable molecular dications are reservoirs of potential energy and can release fragments with high kinetic energy, as illustrated in Fig. 1. These released monocations can interact with neutral gas molecules with kinetic energy far above that of thermal reactions, resulting in different products from those produced by the latter. Another consequence of these fragmentation products with high kinetic energy is that their velocities can be higher than the escape velocity of planetary atmospheres, contributing to their erosion.[25]

Stable molecular dications are produced by photoionization or by charged particles and, in this text, we will focus only on the latter. There is a high selectivity of the electrons to be removed for the production of stable dications, favoring the most external and non-binding electrons of the parent molecule. Electron transfer or transfer-ionization by thermal or slow ions ($v < v_{Bohr}$) are efficient for the production of stable dications, as these mechanisms favor the transfer of these outer shell electrons. For higher velocities ($v \gtrsim 3v_{Bohr}$), electron capture selects the inner electrons of the parent molecule and double capture (two electrons are removed from the target and captured by the projectile) or transfer-ionization (two electrons are removed from the target, one is ionized while the other is captured by the projectile)[27] tends to produce unstable dications that break up quickly. In this regime, ionization becomes the most important mechanism for stable dication production, whether it is done by heavy ions or electrons.

Removal of two electrons through the ionization channel can be done by double ionization or by single ionization followed by Auger decay. In the case of double ionization, the main mechanism is the double impact, known in the literature as TS-2 (two-step-2),[28] in which the projectile independently interacts with two electrons of the target in succession. For swift projectiles, this mechanism imprints a signature in the form $\sim 1/E^2$ on the projectile energy (E) dependence of the dications production cross section. On the other hand, the K-shell ionization of one of the atoms of the parent molecule gives rise to a post-collisional, fast Auger decay of the vacancy formed, which may eventually result in a stable double-vacancy configuration in the outermost molecular orbitals. In this case, the projectile ejects only one electron during the collisional process and its signature is a $\sim 1/E$ dependence in the dication production cross section for high energies. In any case the formation of a long lived dication requires a great discrimination of the molecular orbitals in which the two ejected electrons are located, otherwise the dication thus formed does not hold and breaks up rapidly.

This double vacancy production phase occurs in a time scale between tens of attoseconds (100 eV electron transit time) and a few femtoseconds (half-life of a K vacancy in C)[29] and is driven by the aforementioned mechanisms. Such phase precedes an equally fast and highly correlated rearrangement of electrons[30] which, in turn, will induce a rearrangement of the nuclei, either fragmenting the molecule or accommodating it in a stable dication. Here, we focus on this first phase, aiming to uncover some dynamical patterns among the double vacancy production mechanisms for several molecules.

Several issues must be addressed in order to comprehensively reveal dynamic features of the initial collisional phase of the process. Concerning the TS-2 mechanism, and since several electron pairs are energetically allowed to be ionized, it can be argued which pair results in a stable dication, as most of them result in molecular fragmentation. In this same line, it can be also argued whether these selected electrons need to belong to different molecular orbitals. Further, if there are any particular alignments of the pair of molecular orbitals with respect to the electron trajectory which inhibits or enhance dication production. Concerning the possibility of single ionization followed by Auger decay, the first question is if the Auger decay of a single K-shell vacancy can result in a stable dication. Indeed, this is not so straightforward as this decay should properly select some specific molecular orbitals where the two final vacancies are located. If Auger decay is energetically allowed, how likely it is compared to TS-2? For molecules with atoms heavier than O, Auger-like decay in these atoms can also occur if a single vacancy is produced in an inner valence shell, which has a significant larger ionization cross section compared to K-shell ionization. Does this primary single inner-valence shell vacancy contribute for a stable dication formation? For larger molecules, such as Benzene, for example, can the localized double charge produced by Auger decay be redistributed fast enough so as to result in a stable dication instead of fragmenting the molecule? And finally, without being exhaustive, does an isotopic substitution affect the rate of production of stable dications? These are some of the questions that will be addressed in this manuscript in light of the experimental results available.

Experimentally, the signature of these mechanisms, as well as their predominance for different electron velocities, can be obtained, for each molecule, by measuring the structure of the cross section producing stable dications over a wide range of projectile velocities. The reported measurements that allow such evaluation are not numerous and, as a rule, do not attempt to identify or analyze the mechanisms responsible for two-electron removal. Thus, many of the available experimental results were not conceptually explored in this regard so far.

Molecular dications have been observed since the early 1930s.[31,32] However, many important long-lived dications, such as N_2^{++}, O_2^{++}, $C_2H_4^{++}$, $C_6H_6^{++}$, etc., have an even symmetry regarding its components, meaning that the formation of two moieties with the same elements is possible in a fragmentation process. This presents an inherent difficulty for the dications to be identified with the widely used methods of mass spectrometry because its signature is embedded in the N^+, O^+, CH_2^+, or $C_3H_3^+$ fragments, following the previous examples, respectively. Since this disentanglement was only recently done for the above cases, and to be more comprehensive in the following sections, the next section will be devoted to a brief description of the DETOF (Delayed Extraction Time-of-Flight spectrometry) technique, which allows to disentangle the dication from the monocation peak in the mass spectra.

2. The DETOF technique

The DETOF technique was developed at the Universidade Federal do Rio de Janeiro (UFRJ) and has been employed in the past decade for several studies, which includes the identification and quantification of molecular dications for even molecules, such as N_2 and O_2. The presence of the dications of these two molecules, for instance, have been predicted in Earth's atmosphere for decades but were not yet measured due to the fact that the fragment ion of the break-up of the parent molecule has the same mass-to-charge (m/q) ratio of its corresponding dication.[33,34]

The technique itself is based in the well-known time-of-flight (TOF) mass spectrometer experimental setup, which can be seen in Fig. 2. The apparatus consists of an electron gun, operating in the 20–900 eV energy range, which is coupled to a target gas cell. The ions produced in an electron-molecule collision are afterwards guided by an electrostatic field of approximately 21 V/mm toward a TOF drift tube, to be finally detected

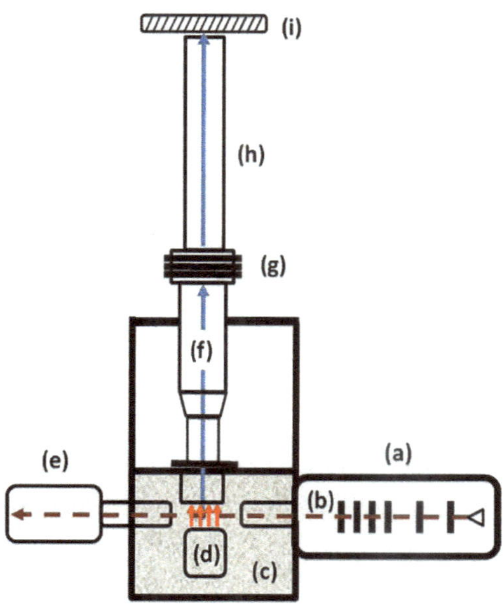

Fig. 2 Scheme of the experimental setup. (A) Electron gun; (B) Pulsed electron beam; (C) Gas cell; (D) Pulsed extractor (the red vertical arrows indicate the delayed electric field); (E) Faraday cup; (F) Recoil ions; (G) Einzel lens; (H) TOF drift tube; (i) Microchannel plate detector.

at its end by a Micro-Channel Plate (MCP) detector. Both the electron beam and the electrostatic field are pulsed and synchronized with repetition rate of 20 kHz so as the field is turned on only after the projectiles have already left the extraction region, limited by the TOF tube entrance. During the measurements, the incident electrons are collected by a Faraday Cup and integrated over the measurement time, and the pressure inside the gas cell is constantly monitored by an absolute capacitive manometer. Thus, the number of both projectiles and scattering centers is known and absolute cross sections for the different charged products of the collision can be determined. Further technical details of the apparatus itself are described elsewhere[35] and we will focus on the technique itself from now on.

The main principle of the DETOF technique relies on the synchronism between the electron beam and the electrostatic extraction field pulses. If this synchronization is implemented as described above, all positively-charged products of the electron-molecule collision inside the extraction region are collected. On the other hand, if there is a delay time between the pulses, the recoil ions are allowed a corresponding free-flight time that, depending on their post-collisional velocity, may cause them to not enter the TOF drift tube collimator. This collimator is conical, with a diameter of 6 mm at its base and of 7 mm at its top, and is positioned at the very entrance of the TOF drift tube. This means that there is a correlation between the number of collected ions and the delay time between electron beam and extraction field pulses via the velocity of the recoil ion after ionization. If different TOF spectra are taken while gradually and systematically increasing this delay time, a profiling of the different kinetic energy distributions of a certain resultant moiety of the collision is possible. Therefore, not only different fragmentation processes can be identified by their signatures in the different kinetic energy distributions present, but also species with the same mass-to-charge ratio, comprising a single peak in the TOF spectrum, can be identified if at least one of their kinetic energy distributions is well known.[36,37]

That is indeed the case for dications of molecules with even symmetry, such as described above, which are embedded in the peak for the fragment with half the molecule's original mass. Since molecular dications do not undergo fragmentation, they do not alter their initial Maxwell-Boltzmann energy distribution—which is the same as the unfragmented molecular monocation. Conversely, the fragment with the same m/q of the dication, which is the molecule broken in half, always acquires additional kinetic energy due to momentum conservation during fragmentation. Thus, by identifying the Maxwell-Boltzmann distribution contribution to the number of ions collected for that peak, one can disentangle it from the fragment and determine its formation absolute cross section.[36–41] The minimum half-life of metastable molecular dications for them to be measured in this manner before fragmentation depends on its mass, which affects their time-of-flight inside the drift tube, but can be estimated at a minimum value of 100 μs.

TOF spectra of the ethylene molecule is shown in Fig. 3 for 70 eV electron impact with different delay times between the beam and the extraction pulses. It can be clearly seen, by the narrowing of the peaks, that the faster fragments leave the extraction region much quicker than the slower fragments. In particular, fragments where the C—C bond is severed, since they have been produced by breaking the molecule into species with roughly the same mass, tend to have higher velocities than the larger moieties. The peak with the combined moieties CH_2^+ and $C_2H_4^{++}$ is the exception, due to the contribution from the dication, which gains no kinetic energy in the ionization process and therefore remains in the collection region for longer time delays. In particular, it can be seen in the inset of Fig. 3 that the peak for these combined moieties narrows considerably for 2 μs delay time, showing the

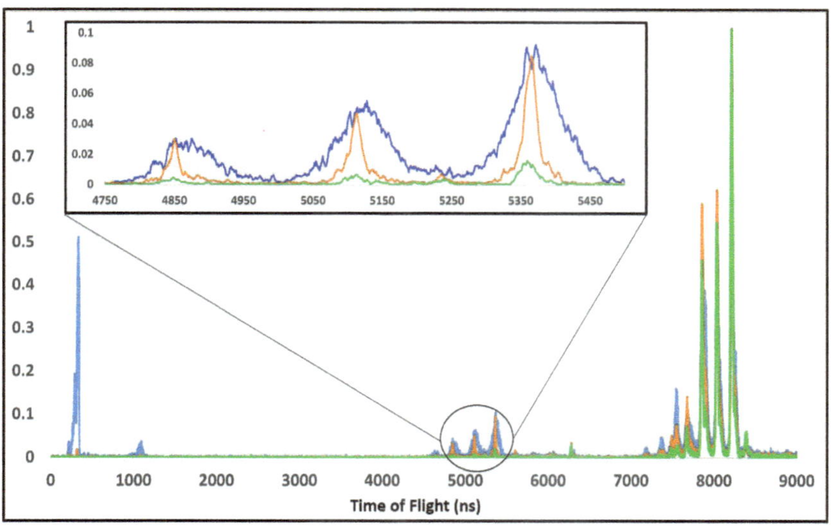

Fig. 3 Time-of-flight spectra for ethylene ionization by 70 eV electron impact for 0 μs (blue line), 2 μs (red line) and 6 μs (green line) delay times, normalized by the parent ion production. In the inset, the behavior with the delay time for the CH_n^+ ($n = 0,1,2$) can be seen in more detail zooming in on the corresponding region of the spectra. See text for a thorough description.

presence of CH_2^+ ions with a high kinetic energy distribution, besides the slow CH_2^+ and the thermal $C_2H_4^{++}$, which remains in the extraction region after 6 μs delay time, while almost all of the remaining of the CH_n^+ ($n = 0,1,2$) fragments do not. Also, it can be seen that collection of moieties that lose one or two hydrogen atoms are much less affected by the delay time, since they do not gain a significant contribution to their kinetic energy distribution by the much lighter hydrogen(s). On the other hand, hydrogen ions themselves leave the collection region almost immediately after we increase the delay time, making H^+ measurements unreliable with this apparatus.

The number of ions recorded by the detector, n_{Rec}, is assumed to be the same as the number of ions that enter the TOF drift tube. A conical collimator and an Einzel-like lens are present to ensure optimal focalization of the ions that pass through the TOF entrance collimator and avoid losses via collisions on the tube's walls. The electrostatic potentials applied at different points of the mass spectrometer are chosen by means of ion trajectories simulations, resulting in an average of 2% ion losses in the flight time path.[35] Assuming cylindrical symmetry with respect to the electron beam axis, n_{Rec} can then be written as a function of the time delay t:

$$n_{Rec}(t) = \int_0^\infty n_T(v_\perp, t) f(v_\perp) dv_\perp \tag{1}$$

where $v_\perp$ is the transverse velocity of the recoil ion with respect to the electron beam axis, $f(v_\perp)$ is the velocity distribution of the recoil ions and $n_T(v_\perp, t)$ is a transmission efficiency function that depends not only on the transverse velocity and the time delay but also on the geometrical properties of the interaction region.[36,37] A schematic representation of the collection of ions for different time delays is illustrated in Fig. 4. Nevertheless, the transmission efficiency function is analytical and well-known, while the number of recorded ions is measured experimentally; thus, Eq. (1), above, is a Fredholm-type equation, in which the integrand is unknown, and it can only be resolved numerically assuming trial functions for the velocity distributions.

Three different classes of velocity distributions are employed in order to analyze the experimental data, which are converted from the following kinetic energy distributions: (i) Gaussian ($G(E_G)$), representing fragmentations in which the recoil ion is left with an average kinetic energy of E_G;

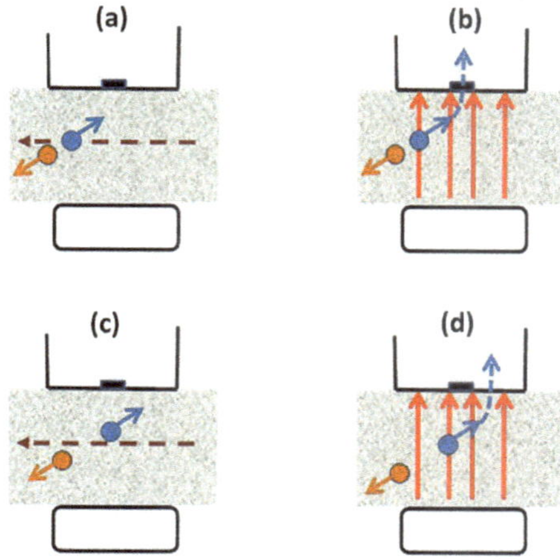

Fig. 4 Illustration of the time delay synchronism working as a velocity selector. Charged fragments are represented in blue, while neutral fragments are pictured in orange; as an example, sample velocity vectors are drawn for each fragments, while the charged fragment's trajectory is shown as a dashed curve. There are four situations represented: (A) for a short time delay t_1, the fragments produced are still close to the collision site; (B) once the extraction field is turned on for delay time t_1, the charged fragment is guided toward the TOF tube entrance, defined by its collimator diameter; (C) a longer time delay $t_2 > t_1$ allows the fragments more free flight time; and (D) when the extraction field is turned on for t_2, the charged fragment is no longer collected because it hits outside of the collimator leading to the TOF drift tube. The fragments sizes and velocities are overestimated in order to help visualization.

(ii) Exponential (*Expo*), accounting for processes where the recoil ion is left with a supra-thermal kinetic energy, e.g., fragmentations via vibrational continuum; and (iii) Maxwell-Boltzmann (*MB*), corresponding to unfragmented products, which acquire no kinetic energy in the collision process (neglecting the momentum produced by the ejection of the ionized electron). It must be observed here that the *MB* distribution has no free adjustment parameters and is the same for both the dication and monocation of a molecule. Thus, the parent ion peak analysis is always a consistent validation of the procedure.

An example of peak comprising a fragment and a molecular dication (for instance, CH_2^+ and $C_2H_4^{++}$[39]) would have its total kinetic energy distribution described by a function such as:

$$f(E) = f_0\left[2\sqrt{E/\pi}(kT)^{-3/2}e^{-E/kT}\right] + f_1\left(\alpha_1 e^{-\alpha_1 E}\right) + \sum_{i=2}^{N} f_i\left[\frac{2\alpha_i}{\sqrt{\pi}(1+\mathrm{erf}(\alpha_i E_{Gi}))}e^{-\alpha_i^2(E-E_{Gi})^2}\right] \quad (2)$$

where f_i are the coefficients for the different energy distributions present for that particular peak and both α_i and E_{Gi} are adjustable parameters. f_0 is the relative weight of the *MB* distribution, and represents the fraction of the total peak that corresponds to the molecular dication, while f_1 refers to the *Expo* distribution and the remaining coefficients are ascribed to the Gaussian distributions that may compose the total fragment energy distribution. Therefore, knowing the absolute cross section for the fragment + dication group and determining f_0, the absolute cross section for the dication production for a molecule with even symmetry is obtained.

The analysis is performed by optimizing the coefficients and free parameters listed above in order to adjust the velocity distribution (correspondent to the energy distribution shown in Eq. (2)) to the measured experimental data. The criteria defined for the analysis are that the overall adjustment has to have a R-squared fitting correspondence parameter of at least 0.99 and that uncertainties in the kinetic energy distributions and their relative weights are determined by keeping the R-squared parameter above 0.97.

Thus, the molecular dication with the same mass-to-charge ratio of a fragment ion can be disentangled and have its absolute production cross section directly measured. Now we turn to the molecular dication production, and its dynamical mechanisms, by electron impact.

3. Discussion of molecular dications data

Molecular dication data is scattered throughout the literature, and a sample collection will be presented here, including some molecules of astrophysical, biological and industrial interest. We will focus on small molecules for two reasons: (i) they are by far the most thoroughly studied; and (ii) the effects of molecular stability are much more pronounced for small molecules with respect to the removal of two electrons, i.e., there is a great variance among these compounds of dication production rate. In order to compare them, we will henceforth present the dication yields relative to the parent ionic molecule. This is also the key method to analyze their production mechanisms.

3.1 Metastable molecular dications

We will discuss dications coming primarily from diatomic (N_2, O_2, NO, CO and Cl_2) and triatomic (NO_2 and CO_2) molecules, as well as ammonia (NH_3) and a couple of basic compounds (ethylene and benzene, respectively C_2H_4 and C_6H_6) that work as building blocks for larger hydrocarbons (including polycyclic aromatic hydrocarbons, or PAHs). It must be noted that there is a group of small molecules that present no metastable dications, at least with lifetimes within the range of detection time of the experimental setups from the authors that studied them. These include, besides the obvious case of H_2, molecules such as H_2O,[42,43] N_2O,[44] H_2S,[44] and CH_4,[45] though it must be noted that sometimes controversies arise: Rao et al.[46] have reported the presence of H_2O^{++} at 200 eV electron impact energy three orders of magnitude below the parent ion yield, which would be enough to be detected by Montenegro et al.[42] and Straub et al.[43] The latter argue that Rao et al.[46] probably attributed the O^{++} yield to H_2O^{++}. On the other hand, Lindsay et al.[47] reported no presence of NO_2 dication by electron impact, while Stephan et al.[48] measured its production for many impact energies from threshold up to 180 eV, albeit very low ($\sim 1.1 \times 10^{-3}$ dication/parent ion ratio at its peak). First, this shows that NO_2 after the removal of two electrons is very unstable, mostly fragmenting instead of remaining in a metastable configuration. Second, the reason for this difference in detection by the two different groups may rely on their different times between production and detection of recoil ions and the lifetime of NO_2^{++}, as argued by Song et al.[49] in a recent review.

With these aspects in mind, and in order to provide the reader with a clearer picture of the behavior of different molecular dications, we divide the data exhibited here in three groups. In Fig. 5 we present data for more stable dications among the molecules listed above, meaning those that have a maximum relative yield to its parent ion above 2%. Those are N_2, Cl_2, CS_2 and C_6H_6 (which we present here in its "pure" form and with one isotopic substitution in one of the atoms of the aromatic ring, namely $^{13}CC_5H_6$).

Looking at Fig. 5, some key general aspects must be pointed out. First, benzene has a much larger dication/parent ion ratio than all other presented molecules (at least a factor of 3 at its maximum). Second, it can be clearly seen that the behavior of the ratio with respect to the impact energy is different for these molecules—all of them exhibit a peak in the vicinity of 100 eV impact energy, both N_2 and Cl_2 ratios decrease steadily with the increase of the impact energy, while benzene and CS_2 reach a plateau for the higher end of the impact energy range. This is an indication of different production mechanisms present for dication production, and will be discussed in detail below.

A particular feature present in Fig. 5 that is noteworthy is the difference between the two isotopical "versions" of benzene shown: $^{12}C_6H_6$ and

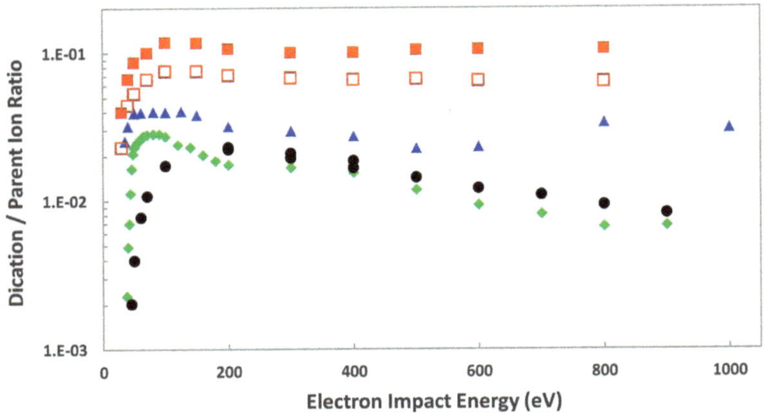

Fig. 5 Ratio between molecular dication and parent ion production with respect to the electron projectile impact energy. Maximum uncertainties estimated by the respective authors are provided in parentheses. Black circles—N_2 (10%)[36,40]; green lozenges—Cl_2 (15%)[50]; blue triangles—CS_2 (17%)[44]; and red squares—benzene (12%) (full symbols: C_6H_6; open symbols: $^{13}CC_5H_6$—see text).[51]

$^{13}C^{12}C_5H_6$ (or, for simplicity C_6H_6 and $^{13}CC_5H_6$, respectively). By the simple substitution of a single ^{13}C in the aromatic ring, the stability of the molecule after the removal of two electrons is so compromised as to reduce the dication production by ~40%.[51] This rather unexpected result is attributed to a geometrical rearrangement of the molecule's nuclei after electronic removal and subsequent redistribution. The final geometry of the benzene dication is a pyramidal one, instead of the original ring structure.[52] The conformational change of one of the atoms leaving the planar geometry seems to be very sensitive to the mass asymmetry introduced by the carbon isotope.[51] This feature had not been observed until very recently because of the symmetric even nature of the C_6H_6 molecule—its dication had only been measured before the DETOF measurements of Sigaud et al.[51] using the isotopic analogue $^{13}CC_5H_6$. The results exhibited in Fig. 5 show that the use of isotopic analogues is not a reliable practice when the molecule undergoes geometrical conformational change after the removal of two electrons.

This is not the case for the other two symmetric even molecules presented in Fig. 5, since they are diatomic ones: N_2 and Cl_2. The data presented here for N_2 was obtained with the aid of the DETOF technique[36,40]; before that, besides a measurement for the single impact energy of 70 eV and using a complex cryogenic detector,[53] molecular nitrogen dication production had been obtained only using the hetero-isotopic $^{14}N^{15}N$ in order to separate it from the N^+ fragment.[54,55] Despite difficulties in the latter normalization[55] due to the relative low abundance of ^{15}N, comparison between the homo-isotopic and hetero-isotopic species data agree very well with each other.[36]

For Cl_2, on the other hand, there are to date no reported results for the homo-isotopic $^{35}Cl_2$ and $^{37}Cl_2$ dications—the only data found in the literature are the $^{35}Cl^{37}Cl$ measurements of Basner and Becker.[50] Since there is no geometry change in the molecule after ionization and since the natural abundance of the ^{37}Cl isotope is ~24% and they are separated by two mass units, there is no reason to believe that homo-isotopic molecular chlorine dication would result in different yields than the ones reported; nevertheless, that was not directly observed yet.

In Fig. 6, we turn to the second group of molecules to be presented here: those whose relative dication-to-parent-ion ratio maximum lies between 1% and 2%. These are: O_2, C_2H_4, NO and CO_2. We also include the data of N_2 that was already shown in Fig. 5 as a frame of reference to the reader. In Fig. 7 the third group is presented—molecules with dication yields below 1% in comparison to its parent ion yield: CO, NH_3, ND_3 and NO_2.

We start by looking once again to the behavior of the dication-to-parent-ion ratio of different molecules with respect to the impact energy. Like N_2 (shown in Figs. 5 and 6)[36,40] and Cl_2 (Fig. 5),[50] CO_2 (Fig. 6)[56] also seems to have a smooth but steady decrease with the increase of the impact energy. Meanwhile, C_2H_4 (Fig. 6)[39] exhibits the same behavior as benzene

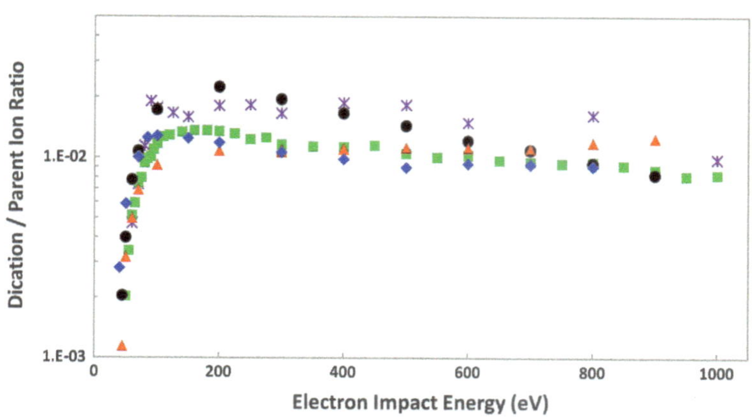

Fig. 6 Ratio between molecular dication and parent ion production with respect to the electron projectile impact energy. Maximum uncertainties estimated by the respective authors are provided in parentheses. Black circles—N_2 (10%)[36,40]; red triangles—O_2 (8%)[38,40]; blue lozenges—C_2H_4 (10%)[39]; green squares—CO_2 (5.5%)[56]; and purple stars—NO (30%).[47]

and CS_2 (both in Fig. 5)[44,51]; CO[45] also appears to follow the same trend, but since the impact energy covered by Tian and Vidal in their experiment[45] does not go above 600 eV, it is hard to be certain of that. On the other hand, O_2 (Fig. 6)[38,40] and both NH_3 and ND_3 (both in Fig. 7)[57,58] show a different behavior, in the sense that they do not exhibit a peak structure—the dication-to-parent-ion ratio for all three molecules appear to reach a maximum value around 150–200 eV impact energy and then retain a constant value for higher impact energies. One could argue that NO (Fig. 6)[47] seems to follow one of these three trends, but the high discrepancies (~30% for the dication alone, as reported by Lindsay et al.[47]) and fluctuating data points allow no definite evaluation. Likewise, the narrow impact energy range of the NO_2[48] data (limited at 180 eV), like the CO case, does not allow any inference on the behavior of the ratio for a broader range of the impact energy. These different trends will be discussed in the Dication Production Mechanisms subsection, below.

Two particular features should also be observed here. First, NH_3 and ND_3 also have different dication-to-parent-ion ratios, as reported independently by Rao and Srivastava for NH_3[57] and Rejoub et al. for ND_3.[58] Although the latter have also measured electron-impact ionization of NH_3, their spectrometer was not sensitive enough in order to separate the parent ion from its fragments, presenting, thus, absolute cross sections for the production of the whole (NH_3^+ + NH_2^+ + NH^+ + N^+) group.

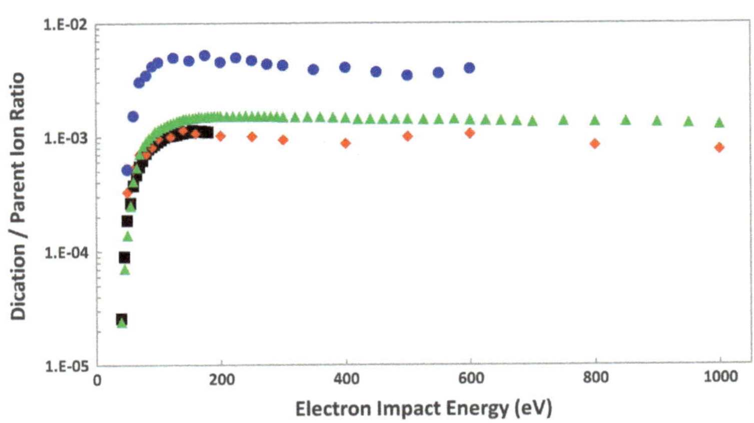

Fig. 7 Ratio between molecular dication and parent ion production with respect to the electron projectile impact energy. Maximum uncertainties estimated by the respective authors are provided in parentheses. Black squares—NO_2 (20%)[48]; green triangles—NH_3 (11%)[57]; red lozenges—ND_3 (20%)[58]; blue circles—CO (15%).[45]

Their absolute cross section values for NH_3^{++} have approximately the same values as the ones for ND_3^{++} in their measurements; nonetheless, data from Rao and Srivastava for NH_3^{++} and NH_3^{+} production[57] result in dication-to-parent-ion ratios above the ones for ND_3^{++} and ND_3^{++} presented by Rejoub et al.[58] More data seem to be needed in order to resolve this discrepancy.

Second, the DETOF technique was again employed in order to directly observe dication production for another two symmetric even molecules, namely C_2H_4[39] and O_2.[38,40] The latter had also been measured before using its hetero-isotopic analogue, $^{16}O^{17}O$.[55] However, and contrary to the N_2 case, the data presented here using the DETOF technique[38,40] lies above the data for the hetero-isotopic correspondent presented by Märk[55] by a factor of at least 60% throughout the whole common impact energy range. Once again, no conformational changes in the geometry of the molecule can be present for a diatomic molecule. This difference was attributed[38] mainly to the difficult normalization procedures and low statistics due to very low (0.04%) abundance of ^{17}O in Märk's measurements. Nevertheless, Bull et al.[59] reported, using an electron-molecule crossed beam experiment, velocity map imaging and a data analysis basis set of 11 Gaussian-like functions, oxygen molecular dications cross sections values between the data presented here[38,40] and data from Märk.[55] Their measurements for the nitrogen molecular dication agree with all the other reported ones, though.

And, finally, there is also controversy regarding the CO data. We presented here data from Tian and Vidal,[45] although there are also measurements by Mangan et al.[60] and Hille and Märk.[61] All data agree on the limits of the relative uncertainties, but that is also because they are relatively large—we chose the one where the error bars were smaller (~15% on the dication data),[45] while the other measurements have uncertainties of the order of 30%.[60,61] Besides, the data from Hille and Märk lie above the data from Tian and Vidal, but below the ones from Mangan et al.[60] The latter argue that this may be related to the open issue of the lifetime of CO^{++}, estimated by different groups to be anything from 1 to 30 μs. Since the three experiments have different times between ionization and detection, the question regarding the absolute values for the metastable CO^{++} production is yet to be answered.

3.2 Dication production mechanisms

Let us now take a deeper look at the different behaviors exhibited by the dication-to-parent-ion ratios of the different molecules with respect to the electron impact energy shown in Figs. 5–7. As described above, these can be of three different types, namely: (i) Peaked—with the increase of the impact energy, the ratio increases up to a maximum, then steadily decreases; (ii) Plateau—with the increase of the impact energy, the ratio increases up to a maximum, and retains that value constantly throughout the remainder of the impact energy range; and (iii) Combined, meaning a combination of the previous two—with the increase of the impact energy, the ratio increases up to a maximum, decreases down but stabilizes eventually in a constant value, exhibiting both the peaked structure of case (i) but also the plateau characteristic of case (ii). N_2, Cl_2 and CO_2 fall in case (i), O_2, NH_3 and ND_3 are representatives of case (ii), while C_6H_6 (and $^{13}CC_5H_6$), CS_2, C_2H_4 and maybe CO show a combination of both tendencies, leaving them in case (iii). As stated above, the available data allow one to draw no conclusions regarding NO and NO_2 yet.

Since we are observing the behavior of the dication-to-parent-ion ratio with the impact energy, these differences are consequence of the different dependencies that the dication production can have with respect not only to the impact energy but also with the dependency of the parent ion production with the impact energy. In other words, how dication and parent ion productions depend on the impact energy is the key aspect that define the shape of the data shown in Figs. 5-7.

The difference between the behaviors observed for cases (i) and (ii) rely on the nature of the collisional process itself. As stated above, in the Introduction section, two different mechanisms can produce molecular stable dications: the interaction of the projectile resulting in the simultaneous emission of two electrons from the target, which constitutes a double impact process known as TS-2, or the ionization during the projectile-target interaction of an inner electron (either core or inner valence shell) followed by an Auger decay, which as far as the collision itself is concerned is a single ionization process.

It is known that, in the higher electron impact velocities regime, the ionization cross section has a $\sim 1/E$ dependency with the impact energy E.[62,63] Therefore, the probability of a single projectile interacting with the target in order to simultaneously remove two electrons from it results in a $\sim 1/E^2$ dependency with the impact energy E. The parent ion production results from a single ionization process but the dication, albeit a doubly-ionized species, can be the result of a double vacancy production or a single vacancy production during the collision (TS-2 or inner shell ionization followed by Auger decay, respectively). The latter case should result in a $\sim 1/E$ tendency, as the parent ion, while the former should present a $\sim 1/E^2$ dependency with the impact energy E. Therefore, when we analyze the behavior of the dication-to-parent-ion ratio with the impact energy, if the molecular dication is produced via TS-2 the ratio should follow a $\sim 1/E$ tendency for higher energies, resulting in the peaked structure of case (i). On the other hand, if it is produced by single inner shell ionization followed by Auger decay, then the ratio should tend to a constant value for higher impact energies—the plateau characteristic of case (ii). And if the resulting dication is produced by both competing mechanisms, then the ratio dependency with the impact energy should exhibit both structures, which can be seen for the molecules of case (iii).

As an example, we will turn to the dication-to-parent-ion ratio of ethylene, a case (iii) representative, with respect to the electron impact energy. Fig. 8 shows the data for ethylene that can be seen in Fig. 6, but now on a log-log scale. A rough estimate for the contributions of both case (i) and (ii) processes can be attained by combining a $\sim 1/E$ dependency with weighted carbon K-shell ionization cross sections, obtained from Talukder et al.,[64] representing the contribution coming from single K-shell ionization followed by Auger decay. The weight for the $\sim 1/E$

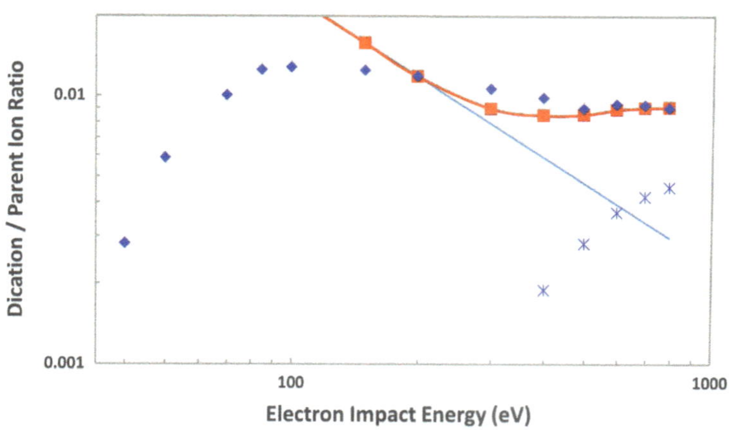

Fig. 8 Ratio between ethylene dication and parent ion production with respect to the electron projectile impact energy (blue lozenges).[39] The red squares represent the adjustment combining a $\sim 1/E$ dependency (blue line) from direct double ionization (TS-2) and the carbon K-Shell ionization cross section data (blue stars) from Talukder et al.[64] (see text). The red line is only to guide the eye.

dependency was set at 2.368 in order to normalize it for the ethylene ratio value at 200 eV impact energy. The weight for the carbons K-shell ionization was determined to be 0.67 in order to get the best possible fit for the higher end of the impact energy range. It must be noted that the purpose of this adjustment is to investigate qualitatively the contribution of the two competing dication production mechanisms.

Now the question that remains is why molecules that fall either in case (i) (TS-2) or (ii) (inner shell ionization followed by Auger decay) do not present the other mechanism (Auger decay or TS-2, respectively) as a pathway to metastable molecular dication production.

Analyzing case (i) first, let us use molecular nitrogen as a case study. It exhibits the $\sim 1/E$ tendency in its dication-to-parent-ion ratio with respect to impact energy for the high end of its range. Therefore, direct double impact ionization (TS-2) is present, i.e., the removal of two electrons concomitantly by the projectile is a possible pathway to stable dication production. K-shell ionization of nitrogen atoms has too low a cross section to be relevant for molecular dication production in the impact energy range presented here.[64] However, inner valence shell ionization followed by Auger decay is not an energetically allowed mechanism. The reason for that lies in the difference in the energy states of the molecular nitrogen ion: auto-ionization is energetically forbidden for vacancies produced in its $2\sigma_g$ inner molecular orbital.[40] In short, both K-shell ionization is not quantitatively relevant in comparison with the TS-2 process and inner valence shell ionization does not lead to Auger decay to a metastable molecular dication—thus, only the TS-2 tendency is observed in Figs. 5 and 6 for N_2. This reasoning must be valid in the analysis of all case (i) molecules.

Case (ii) molecules, on the other hand, pose the need for a more complex explanation, since they present only the mechanism of single inner valence shell ionization followed by Auger decay—the K-shell ionization cross section of oxygen is also too low to be a factor here.[64] The question becomes, as a consequence, why does direct double ionization not lead to a metastable molecular dication. Taking the O_2 molecule as a case study, it can be observed that the electronic configuration of its stable dication is without its two outermost electrons, which populate unpaired and geometrically orthogonal orbitals.[40] Therefore, the probability that a projectile assess only these two orbitals in its trajectory during the collision must be so small as to almost completely suppress the TS-2 mechanism in comparison to inner shell ionization followed by Auger decay,[40,41] since direct double ionization leading to the removal of other two electrons that are not these two outermost must lead to molecular fragmentation. This limitation due to the geometrical disposition of the electronic orbitals of the molecule with respect to the projectile's trajectory is illustrated in Fig. 9.

Other case (ii) molecules must have such limitations as well. Ammonia, for example, has an electronic configuration of $(1a_1)^2(2a_1)^2(1e)^4(3a_1)^2$, and only the removal of the two $3a_1$ electrons allows the molecule to go to a metastable state that will eventually dissociate after inter-system crossing.[65] Any other direct double ionization possibility leads to the fragmentation of the molecule, mainly because of the myriad of vibrational states of ammonia, leading to the stretching of one or more hydrogen-carbon bonds. Once again, the unlikelihood of a projectile assessing exclusively those two electrons in the collision process considerably suppress the presence of TS-2 as an active mechanism for stable dication production.

These features illustrate how highly selective dication production can be for different molecules, and its knowledge is of utmost importance for determining the contribution of double ionization in different molecular collisions.

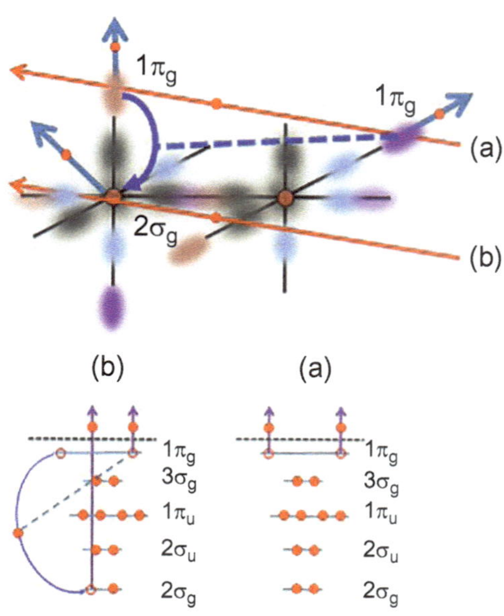

Fig. 9 Pictorial representation of the double ionization of O_2. Direct double ionization, represented by trajectory (A), involves two spatially displaced $1\pi_g$ electrons which, due to the constraint of geometric orientation of the molecule with respect to the projectile trajectory, have a small probability of occurring. Pathway (B) involves the ionization of an inner valence electron, $2\sigma_g$, which relaxes through Auger emission of a $1\pi_g$ electron. This latter mechanism is dominant at high impact velocities.

4. Final remarks

In this work, a review of molecular dication production by electron impact was presented, focusing mainly on the mechanisms that lead to its final doubly-ionized metastable state—either by removing two electrons during the collision, i.e., when the projectile interacts with the target simultaneously producing two vacancies during its transit time through the molecule, or by producing a vacancy in a core or inner valence shell, i.e., a single ionization followed by Auger decay.

The energy loss of ionizing radiation through a molecular medium, as stated above, is highly dependent on the characteristics of the electronic orbitals of the molecules present. Although for many cases the removal of two electrons leaves the original molecule in a fragmented state, in many cases metastable doubly-ionized states can be a relevant depository of energy of the impinging particle, producing moieties which are capable of being exotic participants in secondary reactions of energized media, due to their property of dividing into fragments which, due to Coulombic repulsion, can reach up to tens of eV.[23] Therefore, molecular dications may play important roles in the physical-chemistry of planetary atmospheres, including local temperature increases, planetary escape and the overall inventory of molecules, ions and atoms in these environments.[25,34,40]

Since the presence of dications of simple (and abundant) molecules such as nitrogen and oxygen in Earth's atmosphere have been predicted but never been measured,[33] information on their production cross section is needed when calculating their mean excitation energies[11] and modeling the penetration of ionizing radiation through our atmosphere and its consequences.

Molecular fragmentation is also an important aspect for many different other environments. Biological effects of ionizing radiation penetrating the human body in radiotherapy treatment procedures is highly dependent on the fragmentation patterns of not only water molecules[9] but also of DNA chains and DNA precursor molecules.[10] As simple hydrocarbons, such as ethylene, are the building blocks of larger organic molecules,[39] the knowledge of their dication production—stable or otherwise—may be a relevant aspect of this modeling. In fact, cyclic hydrocarbons, as shown here for the benzene case, may be a quantitatively relevant product for the projectile energy loss, since its metastable dication alone amounts for around 11% of the parent molecular ion production. Larger organic chains must be investigated in order to estimate their contribution.

Furthermore, dications, being highly reactive species, have also very particular roles in other environments besides the aforementioned planetary atmospheres and radiotherapy treatments. For example, emission spectra of polycyclic aromatic hydrocarbons have been shown to be markers, in combustion chambers, for the determination of soot inventories,[15] which are dependent of isotopic substitutions to whose presence in the dication production of benzene has been shown to be remarkably sensitive. Benzene dications, meanwhile, have also been shown recently to be able to rapidly react with oxygen anions, producing an oxidized agent in the formation of more complex molecules.[66]

Finally, it must be pointed out that, although dications are usually thought of as second-order process as far as ionization at high impact velocities are concerned, the discussion of dication production mechanisms above show that, when single inner shell ionization followed by Auger decay resulting in stable dications is present, this is not the case. In fact, for a large class of molecules (cases (ii) and (iii), as classified above) dication production retain a constant ratio comparing with the products of single ionization, meaning that they never become negligible to the overall ionization picture of highly energized media.

Acknowledgments

The authors gratefully acknowledge the Brazilian funding agencies CNPq, CAPES and FAPERJ for their financial support.

References

1. Betz, H. D. Charge States and Charge-Changing Cross Sections of Fast Heavy Ions Penetrating through Gaseous and Solid Media. *Rev. Mod. Phys.* **1972**, *44*, 465.
2. Bethe, H. A.; Jackiw, R. W. *Intermediate Quantum Mechanics*; The Benjamin/Cummings Publishing Company: Menlo Park, California, 1986.
3. Montenegro, E. C.; Meyerhof, W. E. Sum Rules and Electron-Electron Interaction in Two-Center Scattering. *Phys. Rev. A* **1991**, *43*, 2289.
4. Arista, N. R. Energy Loss of Ions in Solids: Non-Linear Calculations for Slow and Swift Ions. *Nucl. Instrum. Methods Phys. Res. B* **2002**, *195*, 91.
5. Sabin, J. R.; Oddershede, J.; Cabrera-Trujillo, R.; Sauer, S. P. A.; Deumens, E.; Öhrn, Y. Stopping Power of Molecules for Fast Ions. *Mol. Phys.* **2010**, *108*, 2891.
6. Smith, A. R. Proton Therapy. *Phys. Med. Biol.* **2006**, *51*, R491.
7. Levin, W. P.; Kooy, H.; Loeffler, J. S.; DeLaney, T. F. Proton Beam Therapy. *Br. J. Cancer* **2005**, *93*, 849.
8. Schardt, D.; Elsässer, T.; Schulz-Ertner, D. Heavy-Ion Tumor Therapy: Physical and Radiobiological Benefits. *Rev. Mod. Phys.* **2010**, *82*, 383.
9. Sabin, J. R.; Oddershede, J.; Sauer, S. P. A. On the Determination of the Mean Excitation Energy of Water. In *Advances in Quantum Chemistry*; Academic Press, 2013; p. 63.

10. Sauer, S. P. A.; Oddershede, J.; Sabin, J. R. Mean Excitation Energies for Biomolecules: Glycine to DNA. In *Advances in Quantum Chemistry*; Academic Press, 2013; p. 215.
11. Sauer, S. P. A.; Sabin, J. R.; Oddershede, J. Calculation of Mean Excitation Energies. In *Advances in Quantum Chemistry*; Academic Press, 2019; p. 225.
12. Nikjoo, H.; Emfietzoglou, D.; Liamsuwan, T.; Taleei, R.; Liljequist, D.; Uehara, S. Radiation Track, DNA Damage and Response—A Review. *Rep. Prog. Phys.* **2016**, *79*, 116601.
13. Ferreira, N.; Sigaud, L.; Montenegro, E. C. Kinetic Energy Distribution of OH + From Water Fragmentation by Electron Impact. *Phys. Rev. A* **2017**, *96*, 012705.
14. Westlake, J. H.; Waite, J. H., Jr.; Carrasco, N.; Richard, M.; Cravens, T. The Role of Ion-Molecule Reactions in the Growth of Heavy Ions in Titan's Ionosphere. *J. Geophys. Res. Space Phys.* **2014**, *119*, 5951.
15. Mueller, L.; Jakobi, G.; Orasche, J.; Karg, E.; Sklorz, M.; Abbaszade, G.; Weggler, B.; Jing, L.; Schnelle-Kreis, J.; Zimmermann, R. Online Determination of Polycyclic Aromatic Hydrocarbon Formation From a Flame Soot Generator. *Anal. Bioanal. Chem.* **2015**, *407*, 5911.
16. Tavares, A. C.; Luna, H.; Wolff, W.; Montenegro, E. C. Double Ionization of Water Molecules Induced by Swift Protons. *Phys. Rev. A* **2015**, *92*, 032714.
17. Ferreira, N.; Sigaud, L.; Montenegro, E. C. Three-Body Fragmentation From Single Ionization of Water by Electron Impact: The Role of Satellite States. *J. Phys. Chem. A* **2017**, *121*, 3234.
18. Stolterfoht, N.; Cabrera-Trujillo, R.; Hellhammer, R.; Pesić, Z.; Deumens, E.; Öhrn, Y.; Sabin, J. R. Charge Exchange and Fragmentation in Slow Collisions of He^{2+} With Water Molecules. *Adv. Quantum Chem.* **2007**, *52*, 149.
19. Cabrerra-Trujillo, R.; Deumens, E.; Öhrn, Y.; Quinet, O.; Sabin, J. R.; Stolterfoht, N. Water-Molecule Fragmentation Induced by Charge Exchange in Slow Collisions With He^{+} and He^{2+} Ions in the keV-Energy Region. *Phys. Rev. A* **2007**, *75*, 052702.
20. Sabin, J. R.; Cabrerra-Trujillo, R.; Stolterfoht, N.; Deumens, E.; Öhrn, Y. Fragmentation of Water on Swift $^3He^{2+}$ Ion Impact. *Nucl. Instrum. Meth. B* **2009**, *267*, 196.
21. Montenegro, E. C.; Shah, M. B.; Luna, H.; Scully, S. W. J.; de Barros, A. L. F.; Wyer, J. A.; Lecointre, J. Water Fragmentation and Energy Loss by Carbon Ions at the Distal Region of the Bragg Peak. *Phys. Rev. Lett.* **2007**, *99*, 213201.
22. Fantuzzi, F.; Cardozo, T. M.; Nascimento, M. A. C. On the Metastability of Doubly Charged Homonuclear Diatomics. *Phys. Chem. Chem. Phys.* **2017**, *19*, 19352.
23. Price, S. D.; Fletcher, J. D.; Gossan, F. E.; Parkes, M. A. Bimolecular Reactions of the Dications and Trications of Atoms and Small Molecules in the Gas-Phase. *Int. Rev. Phys. Chem.* **2017**, *36*, 145.
24. Falcinelli, S.; Pirani, F.; Alagia, M.; Schio, L.; Richter, R.; Stranges, S.; Balucani, N.; Vecchiocattivi, F. Molecular Dications in Planetary Atmospheric Escape. *Atmosphere* **2016**, 7, 112.
25. Thissen, R.; Witasse, O.; Dutuit, O.; Wedlund, C. S.; Gronoffe, G.; Lilensten, J. Doubly-Charged Ions in the Planetary Ionospheres: A Review. *Phys. Chem. Chem. Phys.* **2011**, *13*, 18264.
26. Waite, J. H., Jr.; Young, D. T.; Cravens, T. E.; Coates, A. J.; Crary, F. J.; Magee, B.; Westlake, J. The Process of Tholin Formation in Titan's Upper Atmosphere. *Science* **2007**, *316*, 870.
27. Montenegro, E. C.; Wong, K. L.; Wu, W.; Richard, P.; Ben-Itzhak, I.; Cocke, C. L.; Moshammer, R.; Giese, J. P.; Wang, Y. D.; Lin, C. D. Projectile Charge-State Dependence of Transfer Ionization to Single Capture Ratio in Collisions of Multiply Charged Ions With He. *Phys. Rev. A* **1997**, *55*, 209.

28. McGuire, J. H. Perturbation Expansions in Electron Correlation Dynamics In Atomic Collisions. In *Cambridge Monographs on Atomic, Molecular and Chemical Physics*; Cambridge University Press, 1997; pp. 144–173.
29. Campbell, J. L.; Papp, T. Widths of the Atomic k–n7 Levels. *At. Data Nucl. Data Tables* **2001**, *77*, 1.
30. Cederbaum, L. S.; Zobeley, J. Ultrafast Charge Migration by electron Correlation. *Chem. Phys. Lett.* **1999**, *307*, 205.
31. Conrad, R. The Appearance of Doubly Positively Charged Molecules in a Beam of Canal Rays. *Phys. Z.* **1930**, *31*, 888.
32. Vaughan, A. L. Mass Spectrograph Analyses, and Critical Potentials for the Production of Ions by electron Impact, in Nitrogen and Carbon Monoxide. *Phys. Rev.* **1931**, *38* (9), 1687.
33. Simon, C.; Lilensten, J.; Dutuit, O.; Thissen, R.; Witasse, O.; Alcaraz, C.; Soldi-Lose, H. Prediction and Modelling of Doubly-Charged Ions in the Earth's Upper Atmosphere. *Ann. Geophys.* **2005**, *23*, 781 (– and references therein).
34. Avakyan, S. V. The Role of Processes with a High Threshold Energy in the Physics of the Upper Atmospheres of the Planets. *J. Opt. Technol.* **2005**, *72*, 602.
35. Sigaud, L.; de Jesus, V. L. B.; Ferreira, N.; Montenegro, E. C. A Novel Double-Focusing Time-of-Flight Mass Spectrometer for Absolute Recoil Ion Cross Sections Measurements. *Rev. Sci. Instrum.* **2016**, *87*, 083112.
36. Ferreira, N.; Sigaud, L.; de Jesus, V. L. B.; Rocha, A. B.; Coutinho, L. H.; Montenegro, E. C. Fragmentation of $^{14,15}N_2$ by Electron Impact Investigated Using a Time-Delayed Spectroscopic Technique. *Phys. Rev. A* **2012**, *86*, 012702.
37. Ferreira, N.; Sigaud, L.; Montenegro, E. C. Molecular Fragmentation by Electron Impact Investigated Using a Time Delayed Spectroscopic Technique. *J. Phys.: Conf. Ser.* **2014**, *488*, 012042.
38. Sigaud, L.; Ferreira, N.; Montenegro, E. C. Absolute Cross Sections for O_2 Dication Production by Electron Impact. *J. Chem. Phys.* **2013**, *139*, 024302.
39. Sigaud, L.; Montenegro, E. C. Absolute Cross Sections for Ethylene-Dication Production by electron Impact. *Phys. Rev. A* **2017**, *95*, 012704.
40. Sigaud, L.; Montenegro, E. C. Highly Selective Mechanisms for the Production of N_2 and O_2 Dications by Electron Impact. *Phys. Rev. A* **2018**, *98*, 052701.
41. Sigaud, L.; Montenegro, E. C. Absolute Cross Sections for Production of Molecular Dications by Electron Impact. *J. Phys.: Conf. Ser.* **2020**, *1412*, 052007.
42. Montenegro, E. C.; Scully, S. W. J.; Wyer, J. A.; Senthil, V.; Shah, M. B. Evaporation, Fission and Auto-Dissociation of Doubly Charged Water. *J. Electron Spectrosc. Relat. Phenom.* **2007**, *155*, 81.
43. Straub, H. C.; Lindsay, B. G.; Smith, K. A.; Stebbings, R. F. Absolute Partial Cross Sections for Electron-Impact Ionization of H_2O and D_2O From Threshold to 1000 eV. *J. Chem. Phys.* **1998**, *108*, 109.
44. Lindsay, B. G.; Rejoub, R.; Stebbings, R. F. Absolute Cross Sections for electron-Impact Ionization of N_2O, H_2S, and CS_2 From Threshold to 1000 eV. *J. Chem. Phys.* **2003**, *118*, 5894.
45. Tian, C.; Vidal, C. R. Cross Sections of the Electron Impact Dissociative Ionization of CO, CH_4 and C_2H_2. *J. Phys. B: At., Mol. Opt. Phys.* **1998**, *31*, 895.
46. Rao, M. V. V. S.; Iga, I.; Srivastava, S. K. Ionization Cross-Sections for the Production of Positive Ions From H2O by Electron Impact. *J. Geophys. Res.* **1995**, *100*, 26421.
47. Lindsay, B. G.; Mangan, M. A.; Straub, H. C.; Stebbings, R. F. Absolute Partial Cross Sections for Electron-Impact Ionization of NO and NO_2 From Threshold to 1000 eV. *J. Chem. Phys.* **2000**, *112*, 9404.
48. Stephan, K.; Helm, H.; Kim, Y. B.; Seykora, G.; Ramler, J.; Grössl, M.; Märk, E.; Märk, T. D. Single and Double Ionization of Nitrogen Dioxide by electron Impact From Threshold up to 180 eV. *J. Chem. Phys.* **1980**, *73*, 303.

49. Song, M.-Y.; Yoon, J.-S.; Cho, H.; Karwasz, G. P.; Kokoouline, V.; Nakamura, Y.; Tennyson, J. Cross Sections for Electron Collisions with NO, N_2O, and NO_2. *J. Phys. Chem. Ref. Data* **2019**, *48*, 043104.
50. Basner, R.; Becker, K. Experimental Absolute Electron Impact Ionization Cross-Sections of Cl_2. *New J. Phys.* **2004**, *6*, 118.
51. Sigaud, L.; Wolff, W.; Montenegro, E. C. Strong Isotopic Selectivity on Dication Formation of Benzene. *Phys. Rev. A* **2022**, *105*, 032816.
52. Fantuzzi, F.; de Souza, D. W. O.; Nascimento, M. A. C. Chemical Bonding in the Pentagonal-Pyramidal Benzene Dication and Analogous Isoelectronic Hexa-Coordinate Species. *Comput. Theor. Chem.* **2017**, *1116*, 225.
53. Shiki, S.; Ukibe, M.; Sato, Y.; Tomita, S.; Hayakawa, S.; Ohkubo, M. Kinetic-Energy-Sensitive Mass Spectrometry for Separation of Different Ions With the Same m/z Value. *J. Mass Spectrom.* **2008**, *43*, 1686.
54. Daly, N. R.; Powell, R. E. Electron Collisions in Nitrogen. *Proc. Phys. Soc.* **1966**, *89*, 273.
55. Märk, T. D. Cross Section for Single and Double Ionization of N2 and O2 Molecules by Electron Impact From Threshold up to 170 eV. *J. Chem. Phys.* **1975**, *63*, 3731.
56. Straub, H.; Lindsay, B. G.; Smith, K. A.; Stebbings, R. F. Absolute Partial Cross Sections for electron-Impact Ionization of CO_2 From Threshold to 1000 eV. *J. Chem. Phys.* **1996**, *105*, 4015.
57. Rao, M. V. V. S.; Srivastava, S. K. Total and Partial Ionization Cross Sections of NH_3 by Electron Impact. *J. Phys. B: At., Mol. Opt. Phys.* **1992**, *25*, 2175.
58. Rejoub, R.; Lindsay, B. G.; Stebbings, R. F. Electron-Impact Ionization of NH_3 and ND_3. *J. Chem. Phys.* **2001**, *115*, 5053.
59. Bull, J. N.; Lee, J. W. L.; Vallance, C. Electron Ionization Dynamics of N_2 and O_2 Molecules: Velocity-Map Imaging. *Phys. Rev. A* **2015**, *91*, 022704.
60. Mangan, M. A.; Lindsay, B. G.; Stebbings, R. F. Absolute Partial Cross Sections for Electron-Impact Ionization of CO From Threshold to 1000 eV. *J. Phys. B: At., Mol. Opt. Phys.* **1992**, *25*, 2175.
61. Hille, E.; Märk, T. D. Cross Section for Single and Double Ionization of Carbon Monoxide by Electron Impact From Threshold up to 180 eV. *J. Chem. Phys.* **1978**, *69*, 4600.
62. Sant'Anna, M. M.; Montenegro, E. C.; McGuire, J. H. Inversion Relations for Exclusive and Inclusive Cross Sections Within the Independent Electron Approximation. *Phys. Rev. A* **1998**, *58*, 2148.
63. McGuire, J. H. Double Ionization of Helium by Protons and Electrons at High Velocities. *Phys. Rev. Lett.* **1982**, *49*, 1153.
64. Talukder, M. R.; Bose, S.; Takamura, S. Calculated Electron Impact K-Shell Ionization Cross Sections for Atoms. *Int. J. Mass Spectrom.* **2008**, *269*, 118.
65. Larsen, K. A.; Rescigno, T. N.; Streeter, Z. L.; Iskandar, W.; Heck, S.; Gatton, A.; Champenois, E. G.; Severt, T.; Strom, R.; Jochim, B.; Reedy, D.; Call, D.; Moshammer, R.; Dörner, R.; Landers, A. L.; Williams, J. B.; McCurdy, C. W.; Lucchese, R. R.; Ben-Itzhak, I.; Slaughter, D. S.; Weber, T. Mechanisms and Dynamics of the $NH_2^+ + H^+$ and $NH^+ + H^+ + H$ Fragmentation Channels Upon Single-Photon Double Ionization of NH_3. *J. Phys. B: At., Mol. Opt. Phys.* **2020**, *53*, 244003.
66. Guerra, C.; Kumar, S.; Aguilar-Galindo, F.; Díaz-Tendero, S.; Lozano, A. I.; Mendes, M.; Limão-Vieira, P.; García, G. Unexpected Benzene Oxidation in Collisions With Superoxide Anions. *Sci. Rep.* **2021**, *11*, 23125.

CHAPTER TWELVE

Implicit and explicit solvent models have opposite effects on radiation damage rate constant for thymine

Lea Northcote Sørensen* and Stephan P.A. Sauer
Department of Chemistry, University of Copenhagen, Copenhagen, Denmark
*Corresponding author: e-mail address: l.northcote@chem.ku.dk

Contents

Abstract

Solvation can alter reaction mechanisms through hydrogen bonding, ion–dipole interactions, and van der Waals forces to mention a few. In the study of radiation damage to DNA, solvent effects should be included to model the aqueous biological system of cells adequately. In the present study, we have investigated the effects of different solvent models in calculations of Gibbs free energies and reaction rates for hydrogen abstraction of the methyl group of thymine by the hydroxyl radical at the ωB97X-D/6-311++G (2df,2pd) level of theory with the Eckart tunneling correction. The solvent, water, was included either through the implicit polarizable continuum model (PCM) or through explicit modeling of one or two water molecules at the site of reaction as well as a combination of both. Our investigation shows that the implicit solvent model increases the barrier height and decreases the rate constant for hydrogen abstraction, leading to a value in better agreement with experimental results, whereas solvation by explicit

Advances in Quantum Chemistry, Volume 85
ISSN 0065-3276
https://doi.org/10.1016/bs.aiq.2021.10.002
Copyright © 2022 Elsevier Inc.
All rights reserved.

solvent modeling has the opposite effect. Hence, the PCM seems to provide a better description for radiation damage in thymine, which improves the understanding of the reaction mechanisms behind radiation damage to DNA.

Abbreviations

PCM	polarizable continuum model
IEF	integral equation formalism
COSMO	conductor-like screening model
RC	reactant complex
TST	transition state theory
DFT	density functional theory
PC	product complex
IRC	intrinsic reaction coordinate
TS	transition state

1. Introduction

Ionizing radiation causes DNA damage and subsequent cell death in living organisms.[1] Radiation therapy utilizes the harmful effects of radiation in an effort to kill malignant tumor cells in cancer treatment. One of the most prominent methods of radiation therapy is hadron therapy in which beams of charged particles such as protons or C^{6+} are aimed selectively at cancerous tumors based on the deposition of the hadron energy. The energy of the hadron reaches its maximum just before the particle stops, which minimizes damage to healthy tissue surrounding the tumor if aimed appropriately.[2,3] The energy deposition leads to ionization and formation of reactive radical species.[4,5] The nucleobase radicals are highly reactive and potentially lead to strand breaks and reorganization of the DNA, which can result in cell death or mutation.[1,6] Unfortunately, the precision of the ion beams does not allow for exclusive destruction of cancer cells; healthy cells are also exposed to the radiation. Statistically, the first target of the hadron is water due to the large excess of water in biological cells.[7] The radical ions formed, such as the hydroxyl radical, will then react with other molecules in the cell, including DNA.[6] Detailed knowledge of the chemical reactions that ultimately lead to DNA damage is thus of great importance to improve the selectivity of this form of cancer treatment and thereby minimize the risk of side effects. As only a total rate constant for the radiation damage of each of the DNA nucleobases is currently available from experiments,[8] computational efforts have the potential to elucidate the exact mechanisms behind the reactions.

The radicals are thought to react with DNA nucleobases either by abstraction of a hydrogen or by addition of the radical onto the aromatic rings in the nucleobases.[9] Our group has earlier investigated such reactions by performing calculations of reaction kinetics in the gas phase.[10,11] It was found that both addition and abstraction pathways are important for the overall reaction rate. However, the absence of solvent effects undermines the credibility of the conclusions from the previous studies. Solvation has been shown to both increase the barrier for hydrogen abstraction reactions and make LUMO orbitals of the nucleobases more compact.[12] Hence, the effects of solvation on the reactions with adenine were investigated by our group using the polarizable continuum model (PCM), explicit micro-solvation modeling of a single water molecule, and through a combination of the two methods.[13] These solvation models were not shown to alter the preference of the reaction pathways, but they did have an effect on the rate constants and energetics. In particular, the PCM decreased the rate constant and thereby increased the difference from experimental values, whereas the explicit solvent model increased the rate constant, reducing the variance from experiment. This implied that an explicit model might be the better choice of solvent model for the system.

In the current study, we have chosen to investigate solvent effects on the thermodynamics and kinetics of the reaction between the hydroxyl radical and thymine. Thymine is one of the three pyrimidine bases, which have quite different electrostatic potentials and HOMO/LUMO orbitals compared to the purine bases, adenine and guanine.[10] Therefore, it is of interest to study solvent effects on the reaction with thymine, to see if the effects are similar to the previous adenine results. Hydrogen abstraction from the methyl group of thymine appears to constitute the main reaction mechanism in gas phase as it is the primary contributor to the rate constants (see Fig. 1 for the proposed reaction mechanism).[10] Therefore, we focus on this exact

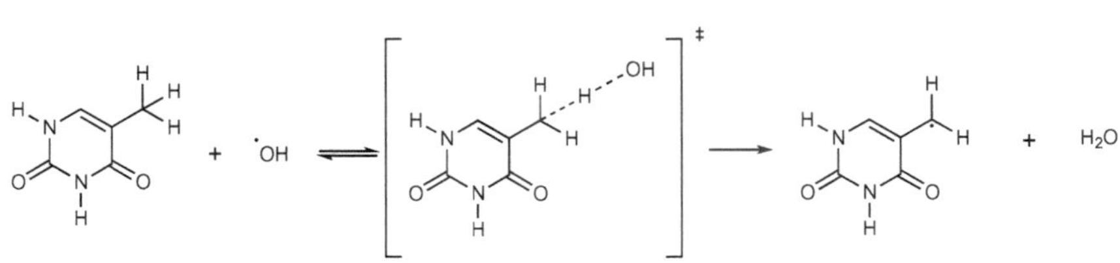

Fig. 1 Abstraction of the methyl hydrogen of thymine by the hydroxyl radical.

reaction in the current work. The effect of solvation is investigated using the PCM, explicit microsolvation modeling of a single and two water molecules, and through a combination of the two.

We believe that inclusion of explicit water molecules in the reaction site will affect the rate constants for hydrogen abstraction because the hydrogen bond restricts accessible reaction angles of the hydroxyl radical and based on our earlier results for the adenine nucleobase.[13] As we previously found an improvement of the rate constants for the reaction with adenine when explicit solvation of a single water molecule was included in the calculations, it is interesting to see the effects of multiple water molecules in the study of thymine.

It should be noted that the incorporation of one of the nucleobases, uracil, in RNA molecules has been shown to affect the rate constants for hydroxyl radical addition reactions. Aranda et al. showed in 2017 using QM/MM MD simulations that incorporation of the nucleobase in RNA gave decreased rate constants for the addition reactions studied.[14] This effect could be explained by reduced accessibility and increased rigidity of the molecule when incorporated in RNA. For thymine, the H_1-hydrogen is actually not present in DNA, as the nucleobase is connected to the sugar-phosphate backbone at this position (see numbering convention in Fig. 2). Furthermore, the H_3-hydrogen participates in hydrogen bonds with the complementary nucleobase when present in double-stranded DNA. The fact that our simplified model of the system cannot take these considerations into account should be acknowledged.

Previous investigations of the effect of solvation for the reaction between the hydroxyl radical and the DNA bases have given contradictory results as shortly reviewed in the following. The PCM model is the most common choice of solvent model used for the reaction. However, the model produces various results based on which nucleobase was under study and when the model was applied in the calculations.

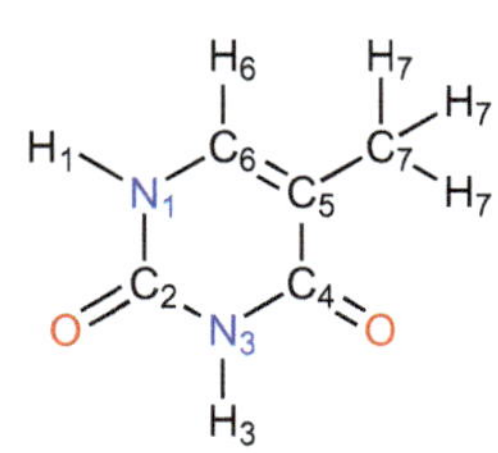

Fig. 2 Atom numbering scheme of thymine.

In 2005, Ji et al. investigated the reactions of the hydroxyl radical with cytosine and included solvation by single-point B3LYP/6-31++G(d,p)// B3LYP/6-31G(d,p) PCM calculations.[15] They found that the barrier for abstraction of hydrogen atoms from the amine group increases when PCM is applied, but the relative order of reaction pathways remains the same. Likewise, in 2011, Sevilla et al. applied density functional theory (DFT) with the B3LYP exchange-correlation functional and the 6-31G* and 6-31++G** basis sets to investigate the reaction between guanine and the hydroxyl radical in gas phase and in solvent with the integral formalism variant of the PCM (IEF-PCM) as single-point energy calculations or in combination with modelling of 12 explicit water molecules.[12] They saw no difference between gas-phase and PCM calculations in the relative order of adduct formation, but for the hydrogen abstraction reaction of the heterocyclic N_1, they did see a stabilizing effect of the PCM. Inclusion of 12 explicit water molecules was found to decrease the importance of the hydrogen abstraction reactions compared to adduct formation. Shukla and co-workers investigated also in 2011 calculations on the hydrogen abstraction reactions from the sugar moiety of 2'-deoxyguanosine by the hydroxyl radical.[16] They completed gas-phase calculations with both B3LYP/6-31+G* and BHandHLYP/6-31+G* level geometry optimizations, and included solvent through PCM single-point calculations at the B3LYP/aug-cc-pVDZ, BHandHLYP/aug-cc-pVDZ and MP2/aug-cc-pVDZ//BHandHLYP/6-31+G* levels. They saw that solvation altered the order of preferred abstraction pathways from the sugar moiety through both energies and rate constants. In 2012, Papiez et al. performed the first QM/MM calculations of the reaction between the hydroxyl radical and guanine-deoxyribose-phosphate in water.[17] They found that the water environment profoundly affects the hydrogen abstraction reaction by the hydroxyl radical because the hydroxyl radical forms hydrogen bonds with surrounding water molecules which restricts the relative orientation of the radicals with respect to the DNA base. Also in 2012, Aravindakumar et al. performed calculations of the reaction between uracil and the hydroxyl radical by using a full geometry optimization in a PCM environment with the B3LYP/6-31+G(d,p) as well as MPW1K/6-31+G(d,p) methods.[18] They further improved the energies by applying CCSD(T)/6-31+G(d,p) single-point corrections in a PCM environment. They saw that the PCM environment gave almost the same energetics and kinetics as gas-phase calculations. They only observed a slight destabilization of the intermediate complexes for addition. No change was found for the abstraction reactions.

Yadav and Mishra investigated the reaction between the hydroxyl radical and the two pyrimidine nucleobases, cytosine and tyrosine, in 2013.[19] They performed gas-phase calculations and included solvent effects by recalculating the energies of the B3LYP/6-31G(d,p) and B3LYP/aug-cc-pVDZ optimized structures in a PCM environment. For cytosine, they found a slight increase in reaction barriers for the abstraction reactions but no change in the order of preferred reactions. No solvation effect was observed for the addition reactions of cytosine.

The fact that different nucleobases are studied makes it difficult to determine whether the discrepancies are due to differences in the computational protocol such as the methods, basis sets, whether PCM was applied in all steps of the calculations, if and how tunnelling corrections were applied, or if they are simply a product of the differences in reactivities of the nucleobases. To evaluate the available thymine studies, they will be compared below.

Only a few studies inspect the effects of solvation. In 2004, Wu et al. showed through B3LYP/6-31G(d,p)-optimized structures that hydrogen abstraction of the methyl group of thymine had a more negative Gibbs free energy of reaction when single-point B3LYP/6-31G(d,p) conductor-like screening model (COSMO) calculations were applied.[20] Also in 2004, Ji et al. investigated the reactions with thymine and included solvation by single-point B3LYP/6-31++G(d,p)//B3LYP/6-31G(d,p) PCM calculations.[21] They found that this solvent model makes addition to C_6 more stable relative to C_5. The results for the hydrogen abstraction reaction are in agreement with a study from 2013 mentioned earlier, in which Yadav and Mishra reported on gas-phase calculations for the reaction of thymine with the hydroxyl radical.[19] All structures were optimized with B3LYP/6-31G (d,p), B3LYP/aug-cc-pVDZ, and BHandHLYP/aug-cc-pVDZ and solvent effects were included as single-point energy PCM calculations at the corresponding level of theory. They found hydrogen abstraction to occur most likely from the methyl group followed by abstraction of the H_1-hydrogen atom (not present in DNA), and that solvation increased the reaction barrier slightly and decreased the Gibbs free energy of reaction. However, for the addition reactions, solvation did not stabilize the C_6 adduct relative to the C_5 in disagreement with the study of Ji et al.[21]. In all previous studies of thymine, solvation is solely applied as single-point PCM calculations on structures optimized without a solvent model. In the current study, the effect of employing three different solvent models in all steps of the calculations is investigated, and an examination of the effects of including more than one explicit water molecule in the microsolvation model is performed.

2. Theory

The large dipole moment of the hydroxyl radical enables it to form strong hydrogen-bonded complexes.[22] Therefore, the reaction mechanism for hydrogen abstraction from thymine is supposed to follow a two-step mechanism, where the first step is the barrierless formation of a reactant complex (RC) followed by formation of the thymine radical and water:

$$R + {}^{\bullet}OH \underset{k_{-1}}{\overset{k_1}{\rightleftharpoons}} [RC]^{\bullet} \xrightarrow{k_2} R^{\bullet} + H_2O \tag{1}$$

where k_1 and k_{-1} are defined as the forward and reverse rate constants for the equilibrium reaction for formation of the RC, and k_2 is the rate constant for formation of the products. The rate constants for the reaction can be modeled using conventional transition state theory (TST).[23] In TST, recrossing effects, which are important for reactions with low barriers in the high temperature region, are not included. Applying a steady-state analysis, the overall rate constant for the hydrogen abstraction can be described by:

$$k = \frac{k_1 k_2}{k_{-1} + k_2} \tag{2}$$

Using the assumption that the rate constant for transforming the RC into reactants is much larger than the rate constant for forming products, such that $k_{-1} \gg k_2$, the overall rate constant for the reaction can be expressed as:

$$k = \frac{k_1 k_2}{k_{-1}} = K_{eq} k_2 \tag{3}$$

This assumption is valid as long as the energy barrier for k_2 is much larger than the energy barrier for k_{-1} or if the entropy change in the reverse direction of the equilibrium is much larger than the formation of products.[24]

Applying basic statistical thermodynamic principles, the equilibrium constant for the first step can be obtained in terms of partition functions (Q):

$$K_{eq} = \frac{Q_{RC}}{Q_R Q_{OH}} \exp\left[\frac{(E_R + E_{OH}) - E_{RC}}{k_B T}\right] \tag{4}$$

where E_{RC} and E_{TS} are the energies of the RC and the transition state (TS), k_B is Boltzmann's constant, and T is the temperature. The rate constant for formation of the products can be evaluated using TST by:

$$k_2 = \kappa(T)\frac{k_B T}{h}\frac{Q^{\ddagger}}{Q_{RC}}\exp\left[\frac{E_{RC} - E_{TS}}{k_B T}\right] \tag{5}$$

where $Q^{\ddagger}$ is the partition function of the activated complex, h is Planck's constant, and $\kappa(T)$ is the tunneling correction factor, which accounts for tunneling effects through the reaction barrier. By combination of Eqs. (3)–(5), the rate constant for the overall reaction is obtained:

$$k = \kappa(T)\frac{k_B T}{h}\frac{Q^{\ddagger}}{Q_R Q_{OH}}\exp\left[-\frac{E_{TS} - (E_R + E_{OH})}{k_B T}\right] \tag{6}$$

It is observed that the RC has no effect on the rate constant in this description. $E_{TS} - (E_R + E_{OH})$ corresponds to the zero-point-corrected energy barrier between the reactants and the TS ($\Delta E^{\ddagger}$). Assuming that the partition functions are uncoupled, each partition function can be factorized into contributions corresponding to the various forms of motion:

$$Q = \frac{Q_{trans}}{V} Q_{elec} Q_{vib} Q_{rot} \tag{7}$$

where the translational partition function is evaluated per unit volume as seen in the equation. Consequently, the ratios of each of the partition functions can advantageously be evaluated individually. As each partition function describes a part of the internal motion in each of the individual molecules, all partition functions must be included when using TST to calculate rate constants. The ratio of the vibrational and rotational partition functions is treated as normal. The translational partition function is obtained using:

$$\left(\frac{(Q^{\ddagger}/V)}{(Q_R/V)(Q_{OH}/V)}\right)_{trans} = \left(\frac{h^2}{2\pi\mu k_B T}\right)^{\frac{3}{2}} \tag{8}$$

where $\mu = m_R m_{OH}/(m_R + m_{OH})$ is the reduced mass of the reactants. Rate constants were calculated using Eq. (3) with the symmetry factor σ being 2 because the TS structure was in all cases nonplanar for this reaction, and a nonplanar structure gives rise to axial chirality. The zero-point-corrected energy of activation, $\Delta E^{\ddagger}$, is the difference in zero-point energy between the TS and the reactants. The electronic partition function is evaluated as:

$$Q_{\text{elec}} = \sum_i g_i e^{-\varepsilon_i/k_B T} \tag{9}$$

where g_i is the degeneracy and ε_i is the electronic energy of the ith level. In nearly all cases, few electronic energy levels other than the ground state need to be considered. An exception is the hydroxyl radical, which has two low-lying electronic states ($^2\Pi_{1/2}$ and $^2\Pi_{3/2}$) separated by 140 cm^{-1}.[25] The electronic partition function for the hydroxyl radical is therefore evaluated as $Q_{\text{elec,OH}} = 2 + 2e^{-140/k_B T} = 3.018$. The final expression for the rate constant is given by:

$$k = \kappa(T)\sigma \frac{k_B T}{h} \frac{Q^\ddagger}{Q_R Q_{OH}} \exp\left[\frac{-\Delta E^\ddagger}{k_B T}\right] \tag{10}$$

where σ is the symmetry factor which denotes the reaction path degeneracy and depends on the symmetry of the reactants and the TS. The temperature-dependent quantum chemical tunneling correction, $\kappa(T)$, which accounts for the tunneling through the reaction barrier, is very important for hydrogen abstraction because it depends exponentially on the mass of the particle. The tunneling factor was calculated by the Eckart approximation in this work, which is an integral over the quantum mechanical transmission probability, $P_{QM}(E)$, for a particle of energy E:[26]

$$\kappa(T) = \frac{e^{-E_{\text{fwd}}/k_B T}}{k_B T} \int_0^\infty P_{QM}(E) e^{-E/k_B T} dE \tag{11}$$

This method considers both the forward and the reverse reaction through incorporation of the forward and reverse barriers, $E_{\text{fwd}} = E_{TS} - E_{RC}$ and $E_{\text{rev}} = E_{TS} - E_{PC}$, in the calculation of $P_{QM}(E)$ where PC is the product complex. It should be noted that these barriers are relative to the reactant and product complexes and not the free species, unlike the barriers used in Eq. (10). In practice, the Eckart correction is often calculated by numerical integration, which is also done in this investigation. A more detailed description of the Eckart tunneling correction can be found in our earlier work.[11]

3. Computational methods

Our group has previously shown that density functional theory (DFT) employing the ωB97X-D exchange-correlation functional,[27] which is optimized for thermochemical calculations, in combination with the Pople triple-zeta 6-311++G(2df,2pd) basis set[28] and the Eckart tunneling correction[26] gives accurate results for the reaction between adenine and the hydroxyl radical.[10] The method gave rate constants within a single order of magnitude of larger coupled-cluster single-point calculations. All calculations in the current work have been carried out using the Gaussian 16 program suite.[29] Solvent models were included in all steps of the calculations. Three different solvent models were employed. The first is the PCM in its integral equation formalism variant (IEF-PCM), which is an implicit solvent model.[30–32] The second is the microsolvation model, where we have included a single explicit water molecule in the vicinity of the reaction site. The water molecule was manually placed in positions that would potentially allow it to participate in hydrogen bonding with the hydroxyl radical as starting structure for optimizations and TS searches. In the third solvent model, the microsolvation model was combined with the PCM. To investigate the effects of having more water molecules explicitly modeled, a microsolvation model with two water molecules was employed as well as a combination of the two water molecule microsolvation and the PCM. Two different conformations of the microsolvation model with two water molecules in the reaction site were obtained and both were included in the study to inspect the importance of different conformers.

In the PCM, the solvent is modelled as a polarizable continuum. Hence, hydrogen-bonding effects of the solvent are not explicitly modelled. To see the effects of hydrogen bonds on the reaction under study, explicit water molecules were included both in gas phase calculations as well as in a PCM environment in a combined solvent model. Combination of explicit solvent molecules with a continuum solvent model could lead to double counting of the electrostatic effects, which should be thought of when interpreting the results. However, we believe that the effect of hydrogen bonding might be more important than this possible double counting.

All reactions were controlled by means of intrinsic reaction coordinate (IRC) calculations from the TS followed by free geometry optimizations of the end-point structures to ensure that the RC and PC were connected through the TS.[33] The stationary points were confirmed by a frequency analysis and visual inspection of the imaginary frequency in the case of TS geometries.

For the microsolvation models, barrier heights, free energies of reaction, and rate constants were calculated by treating the reactants and products as the free hydroxyl radical and the complex between the water molecules and thymine, and the free water molecule and the complex between the water molecules and the thymine radical.

4. Results and discussion

Free energies of activation, $\Delta G^{\ddagger} = \Delta G_{TS} - (\Delta G_{R} + \Delta G_{OH})$, free energies of reaction, $\Delta G_r = (\Delta G_{R^{\cdot}} + \Delta G_{H_2O}) - (\Delta G_R + \Delta G_{OH})$, and rate constants for the hydrogen abstraction of the methyl group of thymine by the hydroxyl radical, which were obtained with our various solvent models, are compared in Tables 1–3 together with the corresponding gas-phase results. In this section, we will discuss the results for the three different solvent models separately followed by a discussion of the effect of having two water molecules rather than a single water molecule in the microsolvation model, and how inclusion of the PCM affects the two water microsolvation results. Finally, we will evaluate the effect of tunneling corrections.

Earlier, experimental investigations by Scholes found the rate constant for the reaction between thymine and the hydroxyl radical to be $0.76 - 1.26 \times 10^{-11} \text{cm}^3 \text{molecules}^{-1} \text{s}^{-1}$.[8] However, as seen in Table 2, the rate constant found in this study is about one order of magnitude larger than the experimental value of Scholes and also the rate constant earlier obtained by our group in 2015.[10] This is due to a different treatment of the partition functions in the current work where the total partition function is calculated

Table 1 Barrier heights ($\Delta G^{\ddagger}$) and reaction free energies (ΔG_r) in parentheses for the abstraction of the methyl group hydrogen of thymine by the hydroxyl radical given in kJ mol^{-1} calculated at the ωB97X-D/6-311++G(2df,2pd) level with different models of solvation.

	Gas phase	Explicit 1 H_2O	Explicit 2 H_2O, TS 1	Explicit 2 H_2O, TS 2
Vacuum	20.48 (−25.20)	12.60 (−122.4)	9.977 (−119.7)	15.23 (−114.5)
PCM	27.83 (−125.8)	19.95 (−128.4)	17.33 (−131.0)	16.48 (−123.9)[a]

[a]Calculated using single-point calculations on the gas phase geometries due to optimization difficulties.
TS 1 is the first transition state located using the given solvation, and TS 2 is a second transition state.

Table 2 Classical rate constants for the abstraction of the methyl group hydrogen of thymine at the ωB97X-D/6-311++G(2df,2pd) level with different models of solvation.

	Gas phase	Explicit 1 H_2O	Explicit 2 H_2O, TS 1	Explicit 2 H_2O, TS 2
Vacuum	1.10×10^{-10}	2.39×10^{-9}	3.70×10^{-9}	7.70×10^{-10}
PCM	4.98×10^{-12}	3.35×10^{-11}	1.89×10^{-10}	3.44×10^{-10}[a]

[a]Calculated using gas-phase partition functions and PCM single-point energies due to optimization difficulties.
TS 1 is the first transition state located using the given solvation, and TS 2 is a second transition state. All values are given in cm^3 molecule^{-1} s^{-1} at ambient conditions ($T = 298$ K and $p = 1$ atm).

Table 3 Eckart tunneling-corrected rate constants for the hydrogen abstraction of the methyl group of thymine at the ωB97X-D/6-311++G(2df,2pd) level with different models of solvation.

	Gas phase	Explicit 1 H_2O	Explicit 2 H_2O, TS 1	Explicit 2 H_2O, TS 2
Vacuum	2.23×10^{-10} (2.02)	3.66×10^{-9} (1.53)	4.16×10^{-9} (1.13)	9.23×10^{-10} (1.20)
PCM	7.35×10^{-12} (1.48)	3.98×10^{-11} (1.12)	1.89×10^{-10a} (1.00)	$3.44 \times 10^{-10a,b}$ (1.00)

[a]No tunneling correction was applied as there was no energy barrier between the reactant complex and the transition state.
[b]Calculated using gas-phase frequencies and partition functions and PCM single-point energies due to optimization difficulties.
TS 1 is the first transition state located using the given solvation, and TS 2 is a second transition state. All values are given in cm^3 molecule^{-1} s^{-1} at ambient conditions ($T = 298$ K and $p = 1$ atm). Tunneling corrections, κ, are given in parentheses.

as the product of the translational, vibrational, rotational, and electronic partition functions as described in Section 2. TST is observed to overestimate rate constants due to the absence of recrossing effects.[34] Therefore, the results of the current study are reasonable.

4.1 Solvent effects of the PCM solvent model

As seen in Fig. 3, the PCM-optimized geometry for the TS of the hydrogen abstraction reaction is clearly related to the gas-phase structure with the hydroxyl radical being located in the same area. From the results in Table 1, which shows barrier heights and reaction free energies, we conclude that the barrier for abstraction of the methyl hydrogen of thymine increases by 7.35 kJ mol^{-1}, as expected with introduction of solvation through PCM. Interestingly, the reaction free energy, shown in parentheses in Table 1, shows an enormous increase in energy gain from abstraction of the methyl hydrogen in a PCM environment compared to the gas-phase result. The energy gain goes from −25.20 kJ mol^{-1} in the gas phase to −125.8 kJ mol^{-1} when PCM is applied. This suggests that the PCM environment stabilizes the thymine radical better than it stabilizes the hydroxyl radical.

From Table 2, it is seen that PCM solvation decreases the rate constant of the abstraction reaction compared with the gas-phase rate constant. The rate constant becomes smaller by two orders of magnitude from 1.10×10^{-10} cm^3 molecule^{-1} s^{-1} to 4.98×10^{-12} cm^3 molecule^{-1} s^{-1}. The PCM rate constant is in better agreement with the experimental results of Scholes[8] than the gas-phase result.

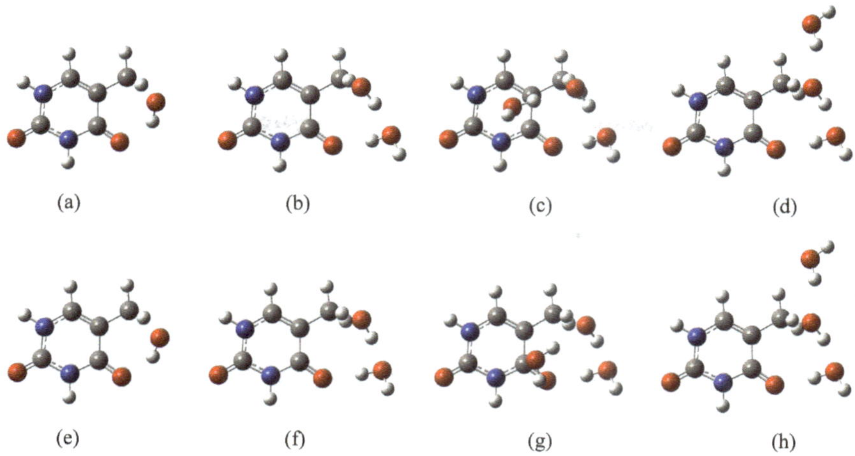

Fig. 3 Structures of the transition states for the hydrogen abstraction reaction of thymine by the hydroxyl radical as found by the ωB97X-D/6-311++G(2df,2pd) protocol with different treatments of the solvent: (A) a gas-phase calculation, (B) a gas-phase calculation with a single water molecule included, (C) is the first transition state of gas-phase calculations with two water molecules included, and (D) a second transition state. (E)–(H) The corresponding calculations with PCM treatment of the solvent included in all steps of the calculations.

4.2 Solvent effects of microsolvation with a single water molecule

To investigate the effects of explicit solvent modeling on the reaction rate constants, we included a single water molecule in otherwise gas-phase calculations, and optimized all intermediate structures with the water molecule residing in the vicinity of the reaction site. Using explicit solvation, it is possible to place solvent molecules at a position where they are able to hydrogen bond to the molecules studied. Any hydrogen bonding between a solvent and the hydroxyl radical will potentially result in more steric strain on the hydrogen bond between the hydroxyl radical and the nucleobase.

From Fig. 3, it can be observed that inclusion of a water molecule alters the geometry of the TS relative to the gas-phase calculation. The TS shows a clear hydrogen bonding interaction between the water molecule and the hydroxyl radical as desired, as do all other found intermediate structures.

In contrast to the PCM calculations, introduction of microsolvation by a molecule decreases the barrier height for abstraction of a methyl hydrogen with 7.88 kJ mol^{-1} from 20.48 to 12.60 kJ mol^{-1}. This effect is interesting, as formation of hydrogen bonds to the solvent is thought to increase the barrier of formation of the TS. Comparing the TS structure from the microsolvation calculation with the geometry of the TS of the gas-phase calculation, as can be seen in Fig. 4, the extra water molecule is observed to improve the conditions for hydrogen bonding by formation of a nine-membered ring TS structure. The water molecule places itself between the hydroxyl radical hydrogen and the oxygen of the carbonyl group neighbor to the methyl-bearing carbon, whereby the distance between the atoms in the ring is shortened to a range more optimal for hydrogen bonding. This could stabilize the TS, decreasing its energy, and thereby the barrier height of the reaction.

The energetic gain is increased from −25.20 to −122.4 kJ mol^{-1} resembling the effects on Gibbs free energy of reaction by the PCM, though slightly smaller. This indicates that having a water molecule interacting with the formed thymine radical almost resembles the effects of applying a dielectric environment.

Inclusion of the extra water molecule is found to increase the rate constant by one order of magnitude to 2.39×10^{-9} cm^3 molecule^{-1} s^{-1}. This means that addition of a water molecule has the opposite effect on the kinetic results compared to treating solvation by the PCM, which is related to the decreased activation barrier. It should be noted that the rate constants were calculated using the partition functions of the complex of thymine and the water molecule, rather than completely free reactants. This results in a model where the nucleobase is already solvated when the incoming hydroxyl radical arrives. This description of the system resulted in large rate constants compared to the experimental value, but if the partition functions for the free reactants were used, an even larger rate constant of 10^{-7} cm^3 molecule^{-1} s^{-1} was obtained, justifying the chosen calculation of the rate

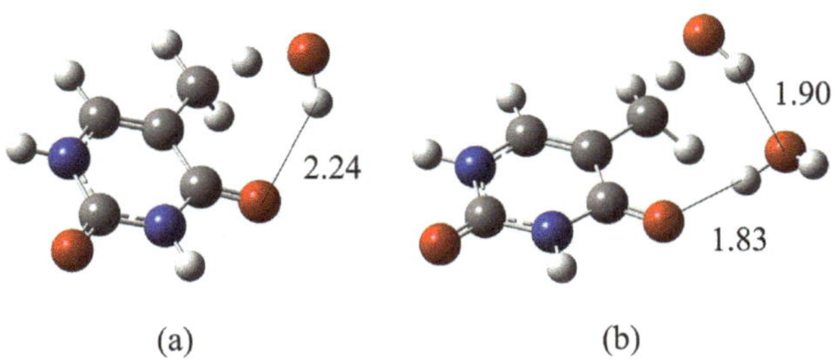

Fig. 4 Selected bond lengths in Å for the transition-state structures of the hydrogen abstraction reaction of thymine by the hydroxyl radical found by the ωB97X-D/6-311++G(2df,2pd) protocol with different treatments of the solvent: (A) the gas-phase structure and (B) the gas-phase calculation with a single water molecule included.

constant. The resulting rate constant is even further from experimental values, which indicates that the explicit model alone does not appear to be the best description of the system under investigation.

4.3 Combination of the PCM and microsolvation with a single water molecule

For our next solvent model, we combined the implicit PCM and the explicit microsolvation model. First, we added a single water molecule to the intermediate structures and then we submerged the system in a PCM environment.

The structure of the TS of the combined solvent model resembles the corresponding microsolvated gas-phase structure with small differences. The barrier height is slightly increased compared to the microsolvation results and the reaction free energy is decreased significantly. The rate constant of the reaction is decreased by less than one order of magnitude compared to the gas-phase result. This method is the most accurate of the methods investigated with the exception of inclusion of more explicit water molecules in combination with the PCM. However, the rate constant for the PCM solvated reaction is in better agreement with the experimental values, $0.76-1.26 \times 10^{-11}$ $cm^3 molecule^{-1} s^{-1}$,[8] compared to the combination of the implicit and explicit solvent.

4.4 Solvent effects of microsolvation with two water molecules

To see the effects of including more than one explicitly modeled water molecule, we performed gas-phase calculations with two water molecules placed close to the reaction site. We found two different conformations of the transition state, TS 1 and TS 2, as seen in Fig. 3.

Inclusion of two explicit water molecules only affects the structure of the TS slightly in both conformers. One of the molecules is placed at about the same position as the single water molecule in the simple microsolvation model, and the other water molecule hydrogen bonds with the hydroxyl radical from two different positions, giving rise to the different TSs. It is seen that the position of this second water molecule affects the barrier height with the first TS having a smaller barrier compared to the second structure by a difference of 5.253 kJ mol^{-1}. The first of the TSs has, among all of the TSs found in this study, the lowest activation barrier of 9.977 kJ mol^{-1}. As a slight difference in the geometry of the two TSs is seen, the first TS must give the optimal conditions for hydrogen bonding, thereby lowering the energy of the TS. The energy gain was larger for both conformers compared to the gas-phase calculations but smaller than the energy gain of the reaction with only a single water molecule included. Comparing the two conformers, the first conformer has a larger energy gain than the second.

The rate constant found for the first of the two TSs with two water molecules explicitly modeled is the largest unmodified rate constant found in this study (see Table 2). It has a value of 3.70×10^{-9} cm^3 molecule^{-1} s^{-1}, which decreases the agreement with the experimental range further compared to the rate constants of the gas phase and explicit solvation with a single water molecule calculation. The rate constant for the second conformer is one order of magnitude smaller than the first conformer. This is in agreement with the difference in barrier heights.

4.5 Combination of the PCM and microsolvation with two water molecules

Combination of the explicit solvent model with two water molecules included in gas-phase calculations and the implicit PCM was the last variation of solvent models investigated in this study. For the second TS, we had trouble locating an optimized geometry for the TS in a PCM environment. For this conformer, it was necessary to carry out the PCM energy calculation as a single-point calculation on the gas-phase optimized geometry of the TS structure and to use the gas-phase frequencies and partition functions in the evaluation of the rate constants.

The geometry of the first TS underwent small changes when a PCM environment was applied. The geometry of the second TS structure is, of course, identical with the vacuum structure because the optimization failed with PCM. Applying a PCM environment raised the barrier height for the first TS and gave a larger energy gain of reaction, like the single water molecule microsolvation (see Table 1). This was to a smaller extend also true for the second TS, probably because the geometry was not optimized in the PCM environment. This again illustrates how the effects of the PCM and the microsolvation counteract each other on the system under examination. Likewise, the rate constants for the combined solvent models, Table 2, were smaller compared to the two water microsolvation gas-phase results. The difference was about one order of magnitude for the first TS compared to two orders of magnitude for the single water molecule model. This indicates that the effect of the PCM environment is subsidized by inclusion of more water molecules in the solvent shell. A plot of the trendlines for the vacuum rate constants and the PCM rate constants is shown in Fig. 5, where it can be seen that the trendlines seem to approach each other. Because the two

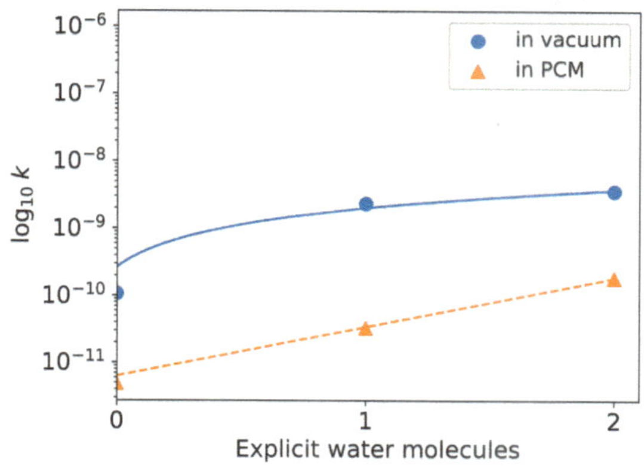

Fig. 5 Classic rate constants given in cm^3 $molecule^{-1}$ s^{-1} in vacuum and in a PCM environment with respect to the number of water molecules explicitly modeled.

solvent models show opposing effects on the rate constant, combination of the two models does not infer double counting effects based on our current results. However, to ensure this conclusion, the full first solvation shell could be included and this system emerged in a PCM environment such that double counting effects cannot occur.

The rate constants for the combination of the two models with two water molecules explicitly modeled were the least altered rate constants compared to the gas-phase results, as they only increased the rate constant slightly. Hence, the resulting rate constants agree less with experiment compared to the gas-phase.

4.6 Tunneling corrections

Tunnelling effects were largest for the gas-phase calculations where they doubled the rate constant for the reaction from 1.10×10^{-10} to 2.23×10^{-10} cm^3 $molecule^{-1}$ s^{-1} as can be seen from Table 3. The resulting rate constant deviates more from the experimental value than the unmodified rate constant. The effect of tunneling was in general smaller for all solvent models where a PCM environment was applied, and explicit solvation decreased the importance of tunneling corrections as well. Hence, for the combination of solvent models, the tunneling corrections had the smallest importance for the rate constants with no tunneling correction applicable for the combined solvent models with two water molecules explicitly modeled, due to the TS and the RC being of identical energies.

The Eckart-corrected rate constant for the PCM-optimized reaction was the rate constant in best agreement with the experimental range of all rate constants calculated. It had a value of 7.35×10^{-12} cm^3 molecule^{-1} s^{-1}, which is close to the experimental range of 0.76–1.26$\times$ 10^{-11} cm^3 molecule^{-1}s^{-1}.[8] Thus, the PCM with Eckart tunneling correction seems to describe the reaction studied the best of the different solvent models investigated in this work.

5. Conclusion

We have investigated the importance of including solvation in calculations of barrier heights, reaction free energies, and rate constants for the methyl hydrogen abstraction reaction of the thymine nucleobase by the hydroxyl radical, thereby fulfilling the aim of this investigation. Solvation was described by means of the PCM, explicit microsolvation of one and two water molecules, and by combination of the two.

The effects of explicit solvation found for thymine in the current study are in contrast to the results obtained for adenine by our group in 2016.[13] For adenine, explicit solvation increased two of the barrier heights for abstraction by 2 and 10 kJ mol^{-1}, two of the barriers were decreased by around 1 kJ mol^{-1}, and the final barrier was decreased remarkably. The significantly decreased barrier stems from a stabilization of this TS by the water molecule. The less affected abstractions are of aromatic hydrogens, while the two abstractions, which experience an increase in reaction barrier, are the hydrogens of the primary amine of adenine. Several factors might explain the opposing effects of explicit solvation on the barrier heights between the methyl group of thymine and the amine group of adenine. First, the radical is formed on a carbon in thymine, whereas a nitrogen radical is formed in adenine. This results in a less stable free radical of adenine due to the more electronegative radical-bearing atom. In either case, resonance stabilization of the free radical is possible, as both of the radical centers are adjacent to a π-bond. However, for adenine, the resonance structure will result in the free radical center on another nitrogen atom, while the resonance structure results in the radical center being on a carbon atom in thymine, again stabilizing the thymine radical relative to the adenine radical. For thymine, the resonance structure will result in a free radical which is adjacent to a nitrogen. The nitrogen can donate electron density from its lone pair orbital to the p-orbital of the radical center which stabilizes the radical further. All of these factors could contribute to the stability of the TS and the resulting radical, and explain the different responses of the systems to solvation.

In the present study, explicitly modeled water molecules decreased the reaction barrier heights for the abstraction of the methyl hydrogen of thymine by the hydroxyl radical. In contrast, application of the PCM increased the barrier heights by 7.35 kJ mol^{-1} in all calculations.

The two solvent models studied have been shown to impact the rate constant of the hydrogen abstraction reaction of the methyl group of thymine in divergent ways. Based on the rate constants, the PCM appears to be the better choice of solvent model for the system, as the PCM rate constant is in agreement with the experimental results.[8] However, as the gas-phase rate constant varies by one order of magnitude from the experimental value, it could be questioned whether the method used in this work is appropriate for the system studied.

The tunneling corrections are found to be of decreasing importance with more elaborate description of the solvent, resulting in no tunneling effect applicable when the microsolvation model with two water molecules included is combined with the PCM. The rate constant which overall shows the best agreement with the experimental results is the Eckart-corrected PCM rate constant. The PCM with Eckart tunneling correction is therefore chosen as the best model for the thymine abstraction reaction, unlike the case of the hydroxyl radical reaction with adenine where the explicit solvent model showed the best results.

Based on the difference in conclusion of the current study with our previous study on adenine, it could be interesting in the future to investigate the effects of solvation on abstraction of one of the aromatic hydrogens of thymine to see if the same trends are present as for the aromatic hydrogens of adenine, even though these hydrogens are not relevant in DNA damage. Furthermore, a better solvent model where the first solvation shell is fully included in the calculations would be interesting to perform to get an even more realistic model of the system.

Acknowledgements

We would like to dedicate this paper to John R. Sabin on the occasion of his retirement as series editor of the Advances in Quantum Chemistry.

SPAS is not only grateful for Jack's many years service to the scientific community as editor of this series, but wants to express his deep gratitude for more than 30 years of scientific collaboration with Jack and for so far 26 joint publications. Jack had introduced SPAS to stopping powers, mean excitation energies, and radiation damage in general so many years ago. So we thought that a paper on the more chemical side of radiation damage to biological material could be an appropriate way of saying "Thank You Jack!"

References

1. Vieira, A. J. S. C.; Steenken, S. Pattern of OH Radical Reaction With N6, N6-Dimethyladenosine. Production of Three Isomeric OH Adducts and Their Dehydration and Ring-Opening Reactions. *J. Am. Chem. Soc.* **1987**, *109*, 7441–7448.
2. Levin, W. P.; Kooy, H.; Loeffler, J. S.; DeLaney, T. F. Proton Beam Therapy. *Br. J. Cancer* **2005**, *93*, 849–854.
3. Fokas, E.; Kraft, G.; An, H.; et al. Ion Beam Radiobiology and Cancer: Time to Update Ourselves. *Biochim. Biophys. Acta Rev. Cancer* **2009**, *1796*, 216–229.
4. Sabin, J. R.; Oddershede, J.; Cabrera-Trujillo, R.; Sauer, S. P. A.; Deumens, E.; Öhrn, Y. Stopping Power of Molecules for Fast Ions. *Mol. Phys.* **2010**, *108*, 2891–2897.
5. Sauer, S. P. A.; Sabin, J. R.; Oddershede, J. Calculation of Mean Excitation Energies. *Adv. Quantum Chem.* **2019**, *80*, 225–245
6. Vieira, A. J. S. C.; Steenken, S. Pattern of Hydroxyl Radical Reaction With 6- and 9-Substituted Purines: Effect of Substituents on the Rates and Activation Parameters of the Unimolecular Transformation Reactions of Two Isomeric Hydroxyl Adducts. *J. Phys. Chem.* **1987**, *91*, 4138–4144.
7. Sabin, J. R.; Oddershede, J.; Sauer, S. P. A. On the Determination of the Mean Excitation Energy of Water. *Adv. Quantum Chem.* **2013**, *65*, 63–77.
8. Scholes, G. Radiolysis of Nucleic Acids and Their Components in Aqueous Solutions. In *Radiation Chemistry of Aqueous Systems*; Stein, G., Ed.; John Wiley and Sons, Inc.: New York, 1968; pp. 259–285.
9. Wala, M.; Bothe, E.; Görner, H.; Schulte-Frohlinde, D. Quantum Yields for the Generation of Hydrated Electrons and Single-Strand Breaks in Poly(C), Poly(A) and Single-Stranded DNA in Aqueous Solution on 20 ns Laser Excitation at 248 nm. *J. Photochem. Photobiol. A Chem.* **1990**, *53*, 87–108.
10. Milhøj, B.; Sauer, S. P. A. Insight Into the Mechanism of the Initial Reaction of an OH Radical With DNA/RNA Nucleobases: A Computational Investigation of Radiation Damage. *Chem. Eur. J.* **2015**, *21*, 17786–17799.
11. Milhøj, B.; Sauer, S. P. A. Kinetics and Thermodynamics of the Reaction Between the ·OH Radical and Adenine–A Theoretical Investigation. *J. Phys. Chem. A* **2015**, *119*, 6516–6527.
12. Kumar, A.; Pottiboyina, V.; Sevilla, M. D. Hydroxyl Radical (OH·) Reaction With Guanine in an Aqueous Environment: A DFT Study. *J. Phys. Chem. B* **2011**, *115*, 15129–15137.
13. Milhøj, B.; Sauer, S. P. A. The Effect of Solvation on the Radiation Damage Rate Constants for Adenine. *ChemPhysChem* **2016**, *17*, 3086–3095.
14. Aranda, J.; Francés-Monerris, A.; Tu nón, I.; Roca-Sanjuán, D. Regioselectivity of the OH Radical Addition to Uracil in Nucleic Acids. A Theoretical Approach Based on QM/MM Simulations. *J. Chem. Theory Comput.* **2017**, *13*, 5089–5096.
15. Ju Ji, Y.; Yuan Xia, Y.; Wen Zhao, M.; et al. Theoretical Study of the OH Reaction with Cytosine. *J. Mol. Struct. - THEOCHEM* **2005**, *723* (1), 123–129.
16. Shukla, P. K.; Kumar, N.; Mishra, P. C. Hydrogen Atom Abstraction Reactions of the Sugar Moiety of 2'-Deoxyguanosine with an OH Radical: A Quantum Chemical Study. *Int. J. Quantum Chem.* **2011**, *111*, 2160–2169.
17. Abolfath, R. M.; Abolfath, P. K.; Rajnarayanam, R.; et al. Multiscale QM/MM Molecular Dynamics Study on the First Steps of Guanine Damage by Free Hydroxyl Radicals in Solution. *J. Phys. Chem. A* **2012**, *116*, 3940–3945.
18. Prasanthkumar, K. P.; Suresh, C. H.; Aravindakumar, C. T. Theoretical Study of the Addition and Abstraction Reactions of Hydroxyl Radical with Uracil. *Radiat. Phys. Chem.* **2012**, *81*, 267–272.

19. Yadav, A.; Mishra, P. C. Reactivities of Hydroxyl and Perhydroxyl Radicals Toward Cytosine and Thymine: A Comparative Study. *Int. J. Quantum Chem.* **2013**, *113*, 56–62.
20. Wu, Y.; Mundy, C. J.; Colvin, M. E.; Car, R. On the Mechanisms of OH Radical Induced DNA-Base Damage: A Comparative Quantum Chemical and Car–Parrinello Molecular Dynamics Study. *J. Phys. Chem. A* **2004**, *108*, 2922–2929.
21. Ji, Y.; Xia, Y.; Zhao, M.; Li, F.; Huang, B. Reactions of With Thymine Studied Using Density Functional Theory. *Int. J. Quantum Chem.* **2005**, *101*, 211–218.
22. Elm, J.; Jørgensen, S.; Bilde, M.; Mikkelsen, K. V. Ambient Reaction Kinetics of Atmospheric Oxygenated Organics With the OH Radical: A Computational Methodology Study. *Phys. Chem. Chem. Phys.* **2013**, *15*, 9636–9645.
23. Eyring, H. The Activated Complex in Chemical Reactions. *J. Chem. Phys.* **1935**, *3*, 107–115.
24. Galano, A.; Alvarez-Idaboy, J.; Bravo-Pérez, G.; Ruiz-Santoyo, M. Mechanism and Rate Coefficients of the Gas Phase OH Hydrogen Abstraction Reaction From Asparagine: A Quantum Mechanical Approach. *J. Mol. Struct. THEOCHEM* **2002**, *617*, 7786.
25. Li, G.-X.; Gao, T.; Chen, D.; Li, Y.-X.; Zhang, Y.-G.; Zhu, Z.-H. The Splitting of Low-Lying States for Hydroxyl Molecule Under Spin-Orbit Coupling. *Chinese Phys.* **2006**, *15*, 998–1003.
26. Eckart, C. The Penetration of a Potential Barrier by Electrons. *Phys. Rev.* **1930**, *35*, 1303–1309.
27. Chai, J. D.; Head-Gordon, M. Long-Range Corrected Hybrid Density Functionals With Damped Atom-Atom Dispersion Corrections. *Phys. Chem. Chem. Phys.* **2008**, *10*, 6615–6620.
28. Krishnan, R.; Binkley, J. S.; Seeger, R.; Pople, J. A. Self-Consistent Molecular Orbital Methods. XX. A Basis Set for Correlated Wave Functions. *J. Chem. Phys.* **1980**, *72*, 650–654.
29. Frisch, M. J.; Trucks, G. W.; Schlegel, H. B.; et al. *Gaussian 16, Revision B.01*; 2016.
30. Cancès, E.; Mennucci, B.; Tomasi, J. A New Integral Equation Formalism for the Polarizable Continuum Model: Theoretical Background and Applications to Isotropic and Anisotropic Dielectrics. *J. Chem. Phys.* **1997**, *107*, 3032–3041.
31. Cancès, E.; Mennucci, B. New Applications of Integral Equations Methods for Solvation Continuum Models: Ionic Solutions and Liquid Crystals. *J. Math. Chem.* **1998**, *23*, 309–326.
32. Tomasi, J.; Mennucci, B.; Cancès, E. The IEF Version of the PCM Solvation Method: An Overview of a New Method Addressed to Study Molecular Solutes at the QM Ab Initio Level. *J. Mol. Struct. - THEOCHEM* **1999**, *464*, 211–262.
33. Fukui, K. The Path of Chemical Reactions – The IRC Approach. *Acc. Chem. Res.* **1981**, *14*, 363–368.
34. Bao, J. L.; Truhlar, D. G. Variational Transition State Theory: Theoretical Framework and Recent Developments. *Chem. Soc. Rev.* **2017**, *46*, 7548–7596.

CHAPTER THIRTEEN

Model dielectric functions for ion stopping: The relation between their shell corrections, plasmon dispersion and Compton profiles

Maarten Vos[a,*] and Pedro L. Grande[b]

[a]Department of Materials Physics, Research School of Physics, Australian National University, Canberra, ACT, Australia

[b]Ion Implantation Laboratory, Institute of Physics, Federal University of Rio Grande do Sul, Porto Alegre, RS, Brazil

*Corresponding author: e-mail address: maarten.vos@anu.edu.au

Contents

Abstract

We describe a set of energy (ω) and momentum (q)-dependent dielectric functions with the same shape of the loss function ($\text{Im}[-1/\epsilon(\omega, q)]$) in the optical limit ($q = 0$) and thus the same mean ionization energy I but different behavior away from $q = 0$. The corresponding proton stopping values differ especially at lower energies.

Advances in Quantum Chemistry, Volume 85
ISSN 0065-3276
https://doi.org/10.1016/bs.aiq.2022.07.001

Copyright © 2022 Elsevier Inc.
All rights reserved.

Within the Bethe formula the stopping only depends on the mean excitation energy *I*. These models display thus different shell corrections, defined as the deviation from the stopping from the Bethe values. Shell-correction contributions originate equally from collisions with low momentum transfer and with high momentum transfer. Intermediate q collisions do not contribute to the shell corrections. At high q the shell corrections are related to the width of the loss function at these q values (which is proportional to the momentum distribution of the electrons ["Compton profiles"]) or, for classical models, from a shift in position. The low-q contribution is related to the plasmon dispersion.

1. Introduction

The interaction of fast, charged particles with matter has been a central topic of research for a long time. Bohr used it to test his emerging understanding of the quantum nature of matter,[1] Bethe derived a first description, fully consistent with modern quantum physics,[2] and further refinements were due to Lindhard,[3] to name only the major players. Besides its fundamental interest the topic is studied for its importance in technological fields like ion beam analysis and ion beam modification of materials. More recently, it has attracted attention within the context of ion-beam-based cancer therapy. It is also used to calculate the excitation spectra after interacting with hypothetical dark matter.[4] The field has been thoroughly reviewed by Sigmund[5] and is described in the context of materials science by Nastasi et al.,[6] and in medical physics by Nikjoo et al.[7] In medical physics the dielectric function is used in track simulation of protons and electrons for charged particle dosimetry see, e.g., Ref. 8,9.

In the first Born approximation (FBA) the interaction between projectile and target is considered weak. This is the case when the charge of the projectile is small and its velocity is large. Then one is in the linear regime and first-order perturbation theory should suffice. The projectile can then be described as a plane wave and cross sections are proportional to the square of its charge. Under these conditions, ion stopping can then be described in terms of the momentum (q) and energy (ω)-dependent dielectric function $\epsilon(\omega, q)$.[5,10] Within this framework obtaining the right stopping values requires knowledge of $\epsilon(\omega, q)$ (or equivalently knowledge of $\mathrm{Im}[-1/\epsilon(\omega, q)]$), but generally, there are no experimental data or ab initio calculations of this quantity, especially away from $q = 0$. Therefore, model dielectric functions are often employed.

An overview of a variety of model dielectric functions used in the literature was given by Nikjoo et al.[10] and within the contest of proton stopping in water by Emfietzoglou et al.[11] The effect of calculating the dielectric function beyond the random phase approximation (RPA) is discussed in Ref. 12. Novel approaches of including the q-dependence to the dielectric function were discussed by Chantler and Bourke[13,14] will not be considered here.

We want to study how ion stopping depends on the extension scheme used for the dielectric function away from $q = 0$ in particular how the stopping of different models differs from the Bethe value. The dielectric functions are either based on classical physics (Drude–Lindhard and extended Drude) or formulated in terms of the RPA and based on either:

- plane waves with occupied states up to k_f (Lindhard,[3] Mermin[15])
- plane waves with a Gaussian dependence of the occupation number (Kaneko[16])
- hydrogenic wave functions (then, in the dilute limit the loss function can be expressed in terms of the generalized oscillator strength (GOS))
- eigenfunctions of the harmonic potential[17,18].

These models can be modified for the presence of a band gap U as demonstrated by Levine and Louie for the Lindhard dielectric function. This modification was not strictly derived but its merits is that the obtained dielectric function resembles that of a semiconductor/insulator and adheres to sum rules and Kramers–Kronig relations. All models converge to the Bethe stopping values at high proton energies, but the rate of convergence differ greatly, i.e., they have different shell corrections.

We aim to keep the discussion as simple as possible, suitable as an introduction to the topic, and will only consider proton stopping, to avoid largely the problems due to different charge states, and assume that the projectile velocity is low enough for relativistic effects to be neglected. Unless otherwise stated we use atomic units ($\hbar = 1$, $m_e = 1$, $\frac{e^2}{4\pi\varepsilon_0} = 1$).

Calculations were done with the Chapidif python/C++ code which is available on request from one of the authors (M. Vos).

1.1 Stopping in the linear regime

The relation between stopping and the dielectric function can be derived using purely classical physics arguments, or based on quantum physics.

Within the classical picture the potential due to the fast projectile causes a change in potential varying in space and time. This can be decomposed in different Fourier components, and then the induced charge redistribution in the target is described by $\epsilon(\omega, q)$. The redistributed charge causes an electrostatic field at the projectile position (the wake field) which causes a decelerating force. The stopping is then just the magnitude of the projectile charge times this induced field (see Abril et al.[19] for details).

Within quantum physics one calculates, based on perturbation theory, the probability that, per unit length traveled the projectile with energy E_0 loses ω energy and changes momentum by q. This is the double differential inverse inelastic mean free path $\frac{d\Lambda}{d\omega dq}$: It is given by[7]:

$$\frac{d\Lambda}{d\omega dq} = \frac{2Z^2}{\pi v^2}\frac{1}{q}\,\mathrm{Im}\left[\frac{-1}{\epsilon(\omega, q)}\right]. \tag{1}$$

where the Z and v are the charge and velocity of the projectile and $\mathrm{Im}\left[\frac{-1}{\epsilon(\omega, q)}\right]$, often called the energy loss function (ELF) is a target property. The fact that $\frac{d\Lambda}{d\omega dq}$ can be written as a product of two factors, one describing the projectile and one describing the target, is a general property of the FBA. Note that the quantum description also describes the deflections due to the projectile–target interactions, which are particularly important for light projectiles (electrons), but were absent in the classical approach sketched before. Also, as we will see, the quantum description makes it possible to calculate other properties such as the inelastic mean free path and straggling that do not easily follow from the classical description. See the book by Nikjoo et al.[7] for a more complete introduction to the dielectric theory within quantum physics.

1.2 Physical observable quantities

In the context of describing the charged particle interaction with matter there are three quantities that are often used to characterize this interaction: the mean-free-path (usually calculated when the charged particle is an electron) and the stopping and straggling (calculated both for electrons and ions). As we will see these quantities are obtained by integrating the loss function ($\mathrm{Im}[-1/\epsilon(\omega, q)]$) suitably weighted over the accessible $q - \omega$ space. However, one can also measure the dielectric function directly, e.g., the dielectric function at $q = 0$ by optical means by measuring absorption and/or reflection as a function of the energy of a photon.[20] Alternatively, using thin samples the loss function can be obtained at $q \simeq 0$ by measuring in the forward direction the energy of a transmitted high-energy electron beam.[21] The experimental knowledge of the dielectric function away from $q = 0$ is much more limited. It can be obtained by moving the detector in a transmission electron microscope away from the forward direction ($\theta_{\mathrm{scat}} \neq 0$) (see, e.g., Ref. 22 for the case of Al) or it can be obtained from inelastic X-ray scattering experiments as was done, e.g., for the case of water by Hayashi et al.[23] In the high-q limit such an experiment is often referred to as Compton spectroscopy[24] and here more data are available.

The mean free path λ for an ion with charge Z is given by:

$$\frac{1}{\lambda} = \int \frac{d\Lambda}{d\omega dq} d\omega dq = \frac{2Z^2}{\pi v^2} \int_0^{E_0} d\omega \int_{q_-}^{q_+} \frac{dq}{q} \operatorname{Im}\left[\frac{-1}{\epsilon(\omega, q)}\right]. \tag{2}$$

If the integration is only over q then the obtained quantity is the differential inverse inelastic mean free path (DIIMFP). For a given energy loss value the limit of integration over momentum ($q_{\pm}$) in Eq. (3) depends on ω, see Fig. 1A.

In order to obtain the stopping one weights the integral over the loss function by ω

$$-\frac{dE}{dx} = \int \omega \frac{d\Lambda}{d\omega dq} d\omega dq = \frac{2Z^2}{\pi v^2} \int_0^{E_0} \omega \, d\omega \int_{q_-}^{q_+} \frac{dq}{q} \operatorname{Im}\left[\frac{-1}{\epsilon(\omega, q)}\right]. \tag{3}$$

Now it is customary to interchange the order of integration. Then the limits of the accessible energy losses will depend on q, as explained in Fig. 1B, and extends from 0 to qv.

Fig. 1 In (A) we show the kinematics at constant energy loss ω. The transferred momentum q is the difference in momentum particle before ($|k_0| = \sqrt{2ME_0}$) and after ($|k_1| = \sqrt{2M(E_0 - \omega)}$) the energy loss event and varies from $q_- = |k_0| - |k_1|$ for $\theta = 0°$ to $q_+ = |k_0| + |k_1|$ for $\theta = 180°$. In (B) we show the kinematics at constant magnitude of $\boldsymbol{q}$. The largest energy loss is now (for $\theta = 180°$) $\omega_{\max} = k_0^2/2M - (k_0 - q)^2/2M \simeq qv_0$. The *dashed part* of the *small circle* corresponds to energy gain, and cannot be accessed.

$$-\frac{dE}{dx} = \frac{2Z^2}{\pi v^2}\int_0^\infty \frac{dq}{q}\int_0^{qv} \omega\, d\omega\, \mathrm{Im}\left[\frac{-1}{\epsilon(\omega, q)}\right] \tag{4}$$

This is the usual equation for ion stopping in terms of the dielectric function. Thus for a given q the maximum value of ω contributing to the stopping is qv.

For the straggling the weighting factor in the DIIMFP is ω^2:

$$\begin{aligned}\frac{d\Omega^2}{dx} &= \int \omega^2 \frac{d\Lambda}{d\omega dq} d\omega dq \\ &= \frac{2Z^2}{\pi v^2}\int_0^{E_0} \omega^2\, d\omega \int_{q_-}^{q_+} \frac{dq}{q}\, \mathrm{Im}\left[\frac{-1}{\epsilon(\omega, q)}\right].\end{aligned} \tag{5}$$

Weighting the integral over the double differential inverse inelastic mean free path by ω and even more so by ω^2 changes the range of ω values that has to be considered for the integration. A reasonable estimate of λ can usually be obtained from an integration up to 100 eV. For stopping values contributions up to $\omega = 1$ keV are significant, whereas for straggling contributions up to a few times the binding energy of the inner shell have to be considered.

1.3 Sum rules, Kramers–Kronig relations and mean excitation energy

There are sum rules that should apply to any dielectric function. These are nicely summarized in Ref. 25. For example, there is the f-sum rule:

$$\frac{1}{2\pi^2}\int_0^\infty \omega' \mathrm{Im}[\epsilon(\omega', q)] d\omega' = n, \tag{6}$$

which should be fulfilled at any q. Here n is the electron density of the system described by the dielectric function. Similarly the Bethe sum rule:

$$\frac{1}{2\pi^2}\int_0^\infty \omega' \mathrm{Im}\left[\frac{-1}{\epsilon(\omega', q)}\right] d\omega' = n, \tag{7}$$

Causality implies that the real and imaginary part of $\epsilon(\omega, q)$ are linked via Kramers–Kronig relations:

$$\mathrm{Re}[\epsilon(\omega, q)] - 1 = \frac{2}{\pi}\mathcal{P}\int_0^\infty \frac{\omega' \mathrm{Im}[\epsilon(\omega', q)]}{(\omega')^2 - \omega^2} d\omega' \tag{8}$$

with $\mathcal{P}$ the Cauchy principal value (this relation should again apply for any q). and similarly:

$$\mathrm{Re}\left[\frac{1}{\epsilon(\omega, q)}\right] - 1 = \frac{2}{\pi}\mathcal{P}\int_0^\infty \frac{\omega' \mathrm{Im}\left[\frac{1}{\epsilon(\omega', q)}\right]}{(\omega')^2 - \omega^2} d\omega' \tag{9}$$

The real and imaginary parts of $\mathrm{Im}[1/\epsilon(\omega, q)]$ are thus Kramers–Kronig pairs as well. All dielectric functions used in this work fulfill all these sum rules for any q with the electron density n being the actual electron density of the system (either the valence electron density if only low losses are considered or the total electron density if the integration of Eqs. (7) and (6) extends over all energies.)

For very large momentum transfer it is generally accepted that one has binary collisions with a single electron. This implies that the energy transfer should be centered around $q^2/2$. All dielectric functions considered here will have for their loss functions centered near $q^2/2$ for large q values.

For each dielectric function one can calculate the mean excitation energy I, as introduced by Bethe. It is obtained from the ELF in the optical limit ($q = 0$) and given by:

$$I = \exp\left[\frac{\int_0^\infty \omega' \ln\omega' \,\mathrm{Im}\left[\frac{-1}{\epsilon(\omega', 0)}\right] d\omega'}{\int_0^\infty \omega' \,\mathrm{Im}\left[\frac{-1}{\epsilon(\omega', 0)}\right] d\omega'}\right] \tag{10}$$

This quantity is sufficient to describe the stopping at very high projectile energies.

2. Model dielectric functions (DF)

2.1 Ad hoc DF

Before we describe the more traditional dielectric functions let us first define a rather ad hoc (AH) dielectric function

$$\mathrm{Im}\left[\frac{-1}{\epsilon(\omega, q)}\right] = \frac{A_{\mathrm{AH}}\delta(\omega - \omega_p(q))}{\omega_p(q)} \tag{11}$$

with $\omega_p(q) = \max(\omega_{\mathrm{AH}}, \frac{q^2}{2})$. We refer to this q dependence as "Bethe dispersion."

Collisions with q values such that $q^2/2 < \omega_{AH}$ are considered "soft" collisions, otherwise they are considered "hard" (billiard ball-type) collisions. From Eq. (7) it is clear that this dielectric function has indeed for any q the same Bethe sum and the electron density of this system is given by $n = A_{AH}/(2\pi^2)$. Using Kramers–Kronig (Eq. 9) one can show that $\mathrm{Re}\left[\frac{1}{\epsilon(0,0)}\right] = 0$, as required for a metal, if $\omega_{AH}^2 = 4\pi n$ and $\mathrm{Re}\left[\frac{1}{\epsilon(0,0)}\right] > 0$ for $\omega_{AH}^2 > 4\pi n$ values. Densities for which $\omega_{AH}^2 < 4\pi n$ would result in negative values of $\mathrm{Re}\left[\frac{1}{\epsilon(0,0)}\right]$ and these densities are considered nonphysical.

The AH dielectric function obeys the sum rules and corresponds at large q values to scattering from stationary electrons. Moreover, it can be shown, see Appendix B, that calculating the stopping by evaluating Eq. (3) one obtains the famous Bethe stopping equation:

$$-\frac{dE}{dx} = \frac{4\pi n Z^2}{v^2} \ln\left(\frac{2v^2}{\omega_p(0)}\right) = \frac{4\pi n Z^2}{v^2} \ln\left(\frac{2v^2}{I}\right). \tag{12}$$

Thus, by comparing a model loss functions with this AH loss function one can understand why the stopping at a certain ion energy for this model loss function is larger or smaller than the Bethe value. Sigmund defines the shell effects as the difference of the actual stopping from the Bethe value (Eq. 6.100 in Ref. 5). We will use this AH dielectric function as a reference, when we introduce other dielectric function.

2.2 Examples of classical dielectric functions

Classical dielectric functions are quite simple and convenient since all other, more advanced, DF convergence to them in giving limits, generally high frequencies ω or small moment transfer q (optical limit). They were derived in different contexts in classical physics for long wavelengths and can be written as

$$\varepsilon(\omega) = 1 - \frac{\omega_p^2}{(\omega(\omega + i\Gamma) - \omega_0^2)} \tag{13}$$

where ω_p is the plasmon frequency (also Langmuir frequency) for free electron oscillations determined from the electron density n ($\omega_p^2 = 4\pi n$) and ω_0 is the harmonic oscillator frequency representing bound electron oscillations. $\mathrm{Re}[\epsilon(\omega)] = 0$ at $\omega = \sqrt{\omega_p^2 + \omega_0^2}$ for vanishing damping constant Γ. This causes a maximum in the corresponding ELF.

2.2.1 Extended Drude

Here we present briefly the model dielectric functions we will be using. Most of them have been described extensively in the literature before, e.g., see Refs. 10, 21,26–28 and the first two (extended Drude and Drude–Lindhard) can be seen as simple extensions of Eq. (13) to finite q values.

The extended Drude dielectric function $\epsilon^{\text{ED}}(\omega, q)$:

$$\begin{aligned} \text{Re}\left[\epsilon^{\text{ED}}(\omega, q)\right] &= 1 - \frac{A_1(\omega^2 - \omega_1(q)^2)}{\left(\omega^2 - \omega_1(q)^2\right)^2 + \Gamma_1^2\omega^2} \\ \text{Im}\left[\epsilon^{\text{ED}}(\omega, q)\right] &= \frac{A_1\Gamma_1\omega}{\left(\omega^2 - \omega_1(q)^2\right)^2 + \Gamma_1^2\omega^2} \end{aligned} \tag{14}$$

is obtained from Eq. (13) after associating $A_1 = \omega_p^2$ and $\omega_1(0) = \omega_0$. Therefore, A_1 (in units of (energy)^2) relates to the density of electrons and $\omega_1(q)$ is the excitation energy of this oscillator at momentum q. Γ_1 determines the width of the excitation. In this work we will use for the dispersion: $\omega_1(q) = \omega_1(0) + q^2/2$. For a free electron metal $\omega_1(0) = 0$ and then the original Drude model is retrieved.

Note that the refractive index n is given by:

$$n^2 = \text{Re}\left[\epsilon(0, 0)\right] = 1 + \frac{A_1}{\omega_1(0)^2}, \tag{15}$$

and the maximum of the loss function at $q = 0$ is at $\sqrt{A_1 + \omega_1^2}$. For $\omega_1(0) = 0$ (i.e., the material consist of free electron) $\text{Re}[\epsilon(0, 0)]$ is indeed infinite, as a DC field is completely screened in a metal.

The energy of oscillator 1 can depend on q (dispersion). This dependence will be assumed here to have a simple form (using atomic units):

$$\omega_1(q) = \omega_1 + \alpha_1 \frac{q^2}{2} \tag{16}$$

but more complex dependencies (e.g., full dispersion[29]) could be used as well. The maximum in the loss function will then be at $\sqrt{\omega_p^2 + (\omega_1^2 + \alpha_1 q^2/2)^2}$. Only for $\alpha_1 = 1$ does the peak position of the loss function approaches the free electron value ($q^2/2$) at large q, and hence *only this value is used in this work.*

An example of the extended Drude dielectric function is compared with the AH dielectric function in Fig. 2. A_1 was chosen to be 225 eV^2 (corresponding to a plasmon energy of 15 eV), $\omega_1 = 0$ eV and $\alpha_1 = 1$. These values are chosen to mimic the behavior of the valence band of Al.

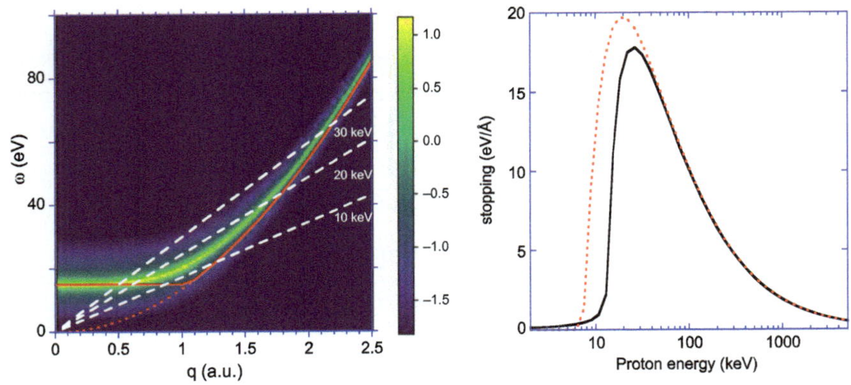

Fig. 2 In the *left panel* we show the log of the loss function $\mathrm{Im}[-1/\epsilon(\omega, q)]$ for the extended Drude dielectric function with $\omega_p = 15$ eV and as a *solid red line* the Bethe dispersion. The *dotted line* is the energy transfer to a stationary electron ($q^2/2$) and the *white dashed lines* are the qv lines at the energy as indicated. Only the loss function below the qv line contributes to the stopping at that energy. The Drude dispersion follows the Bethe dispersion closely. Hence the stopping (*right panel*) corresponding to this Drude dielectric function (*solid black line*) is quite similar to the Bethe one (*dotted red line*).

The dispersion is very close to the Bethe dispersion with $\omega_{\mathrm{AH}} = 15$ eV and an electron density n such that $\omega_{\mathrm{AH}}^2 = 4\pi n$, and the proton stopping follows the Bethe stopping below closely, with differences appearing at relative low energy (40 keV).

The dielectric function described above corresponds to a metal. One can use the extended Drude model as well to describe an insulator by choosing $\omega_1 \neq 0$. For example, by taking $A_1 = 112.5$ eV^2 and $\omega_1 = 10.6$ eV one obtains again a loss function with a peak at 15 eV, but now only half its height. The maximum of the loss function would then follow $\sqrt{A_1 + (\omega_1 + q^2/2)^2}$

2.2.2 Drude–Lindhard

A second, but closely related, model is often referred to as the Drude–Lindhard (DL) model.[30] Here ϵ^{DL} is defined in terms of $1/\epsilon(\omega, q)$ rather than $\epsilon(\omega, q)$ itself:

$$\mathrm{Im}\left[\frac{-1}{\epsilon^{\mathrm{DL}}(\omega, q)}\right] = C_1 \frac{\omega \Gamma_i \omega_1(0)^2}{\left(\omega^2 - \omega_1(q)^2\right)^2 + \omega^2 \Gamma_1^2} \tag{17}$$

and for the real part:

$$\mathrm{Re}\left[\frac{1}{\epsilon^{\mathrm{DL}}(\omega, q)}\right] = 1 + C_1 \frac{(\omega^2 - \omega_1(q)^2)\omega_1(0)^2}{(\omega^2 - \omega_1(q)^2)^2 + \omega^2\Gamma_1^2} \tag{18}$$

For a free electron metal the DL dielectric function is at $q = 0$ equivalent to the Drude dielectric function if $C_1 = 1$ and $\omega_1^{\mathrm{DL}} = \sqrt{A_1}$, but even then, away from $q = 0$, when using again Eq. (16) their effective dispersion will differ. The maximum of the loss function is now simply at $\omega_1 + \alpha_1 q^2/2$.

For $0 < C_1 < 1$, one obtains $1/n^2 = \mathrm{Re}[1/\epsilon(0, 0)] = 1 - C_1$ and the model describes an insulator. C_1 values > 1 result in negative values for $\mathrm{Re}[\epsilon(0, 0)]$ and are again considered nonphysical. At $q = 0$ one gets thus the equivalent extended Drude ELF if one uses in addition of $\omega_1^{\mathrm{DL}} = \sqrt{A_1 + \omega_1^{\mathrm{ED}2}}$ also $1/(1 - C_1) = n^2 = 1 + A_1/\omega_1^{\mathrm{ED}2}$. For example, extended Drude with $A_1 = 9\ \mathrm{eV}^2$, $\omega^{\mathrm{ED}} = 3$ eV and $\Gamma_1 = 1$ eV is at $q = 0$ equivalent to the Drude–Lindhard model with parameters: $C_1 = 0.5$, $\omega_1^{\mathrm{DL}} = \sqrt{18}$ eV and $\Gamma_1 = 1$ eV. This is illustrated in Fig. 3

In Fig. 4 we compare again the loss function and associated proton stopping with the AH dielectric function with the same n and I values. Now the dispersion at small q values is stronger than in the extended

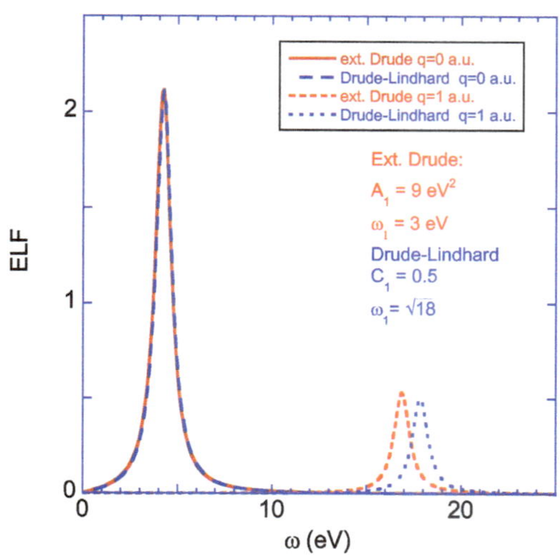

Fig. 3 The loss function at $q = 0$ a.u. and $q = 1$ a.u. for the extended Drude and Drude–Lindhard models for the parameters as indicated. In both cases $\alpha_1 = 1$ and $\Gamma = 1$ eV, the ELF at $q = 0$ a.u. are identical, but not at $q = 1$ a.u.

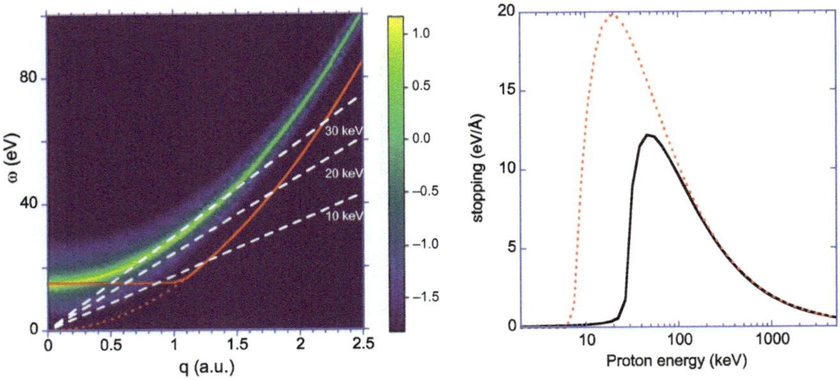

Fig. 4 Same as Fig. 2 but now for the Drude–Lindhard dielectric function. The electron density (and hence the plasmon energy at $q = 0$) is the same. The dispersion is now faster, and hence the stopping is more severely reduced, compared to the Bethe stopping

Drude case. As a consequence, the stopping drops off more quickly at low projectile energies. One could change this by adopting another dispersion relation for Drude–Lindhard:

$$\omega_1(q) = \sqrt{\omega_1(0)^2 + \left(\alpha_1 \frac{q^2}{2}\right)^2}. \tag{19}$$

Then both approaches would give again the same result.

2.3 Examples of quantum-physics-based dielectric functions

For an isotropic medium made of independent atoms (or molecules) with electron density n the longitudinal dielectric function can be defined in the framework of the RPA. For simplicity we restrict ourselves to the RPA here, for fully-quantitative description extensions beyond the RPA are required see, e.g., Refs. 31, 32. In the nonrelativistic approximation the RPA is given by[5]:

$$\epsilon(\omega, q) = 1 + \omega_p^2/q^2 \sum_j |F_{j0}(q)|^2 \left(\frac{1}{\omega_{j0} - \omega - i\Gamma} + \frac{1}{\omega_{j0} + \omega + i\Gamma} \right), \tag{20}$$

where $\omega_p^2 = 4\pi n$, ω_{j0} is the transition energy from the ground state ($j = 0$) to an excited state j.

$F_{j0}(q)$ is the form factor, it should be evaluated for the appropriate basis set. This has been done for plane waves (leading to the Lindhard dielectric function), harmonic oscillator eigenstates (leading to the Belkacem–Sigmund dielectric function) and here we do it as well based on hydrogenic wavefunctions. We will discuss theses cases briefly.

2.3.1 Lindhard and Mermin dielectric function

The loss functions described so far have their roots in classical physics. Lindhard derived, based on quantum physics, a dielectric function for a free electron gas $\epsilon^{\mathrm{L}}(\omega, q)$[3] (see Appendix A for details). Here the loss function consists of a delta function (describing collective excitations, or "plasmons") and a continuous part (describing single-particle excitations). Mermin added relaxation time to the Lindhard dielectric function which transforms the delta function to a peak with finite width[15]:

$$\epsilon^{\mathrm{M}}(\omega, q) = 1 + \frac{(1 + i\Gamma/\omega)(\epsilon^{\mathrm{L}}(\omega + i\Gamma, q) - 1)}{1 + i\Gamma/\omega[\epsilon^{\mathrm{L}}(\omega + i\Gamma, q) - 1]/[\epsilon^{\mathrm{L}}(0, q) - 1]} \tag{21}$$

It is instructive to see the effect of the Mermin correction on the Lindhard dielectric function with finite Γ in some detail. For this we take parameters as are suitable for Al, but an exaggerated width of $\Gamma = 6$ eV. $\mathrm{Re}[\epsilon(0, 0)]$ of the Lindhard dielectric function does not go to infinity as expected for a metal, after the Mermin correction it does. The ELF at $q = 0$ is wider than 6 eV in the Lindhard case, but the 6 eV width is recovered after the Mermin correction. Both dielectric functions fulfill the F and Bethe sum rules (Eqs. 6 and 7) with the same number of electrons per unit volume. The Lindhard dielectric function has a larger I value (as it is broader) and has thus somewhat smaller stopping compared to that obtained after the Mermin correction. This is all illustrated in Fig. 5. Clearly the Mermin dielectric function is to be preferred over the Lindhard one, for finite Γ values, since the latter does not yield the correct dc conductivity and Thomas-Fermi static screening.[33] The Lindhard dielectric function with finite Γ thus fails to describe a free electron gas but is still a "valid" dielectric function in the sense that it adheres to sum rules and the real and imaginary part are related by Kramers–Kronig relations.

The Mermin loss function, again compared to our "AH" loss function, is shown in Fig. 6. Compared to Drude–Lindhard there is now a broad low-intensity structure visible covering at $\omega = 0$ the region between $k = 0$ and $2k_f$. In the stopping curve this has the effect of significant stopping, even at 1 keV, as the qv line always intersect with the electron–hole pair excitation region.

2.3.2 Levine–Louie dielectric function for insulators

The derivation of the Lindhard loss function (and hence the Mermin one) assumes a free electron gas, and its use to describe an insulator is thus highly questionable. Levine and Louie derived a quantum-physics based loss function for an insulator (ϵ^{LL}) by transforming the energy scale according to $\omega \rightarrow \sqrt{(\omega^2 + U^2)}$ with U a parameter that is related to the band

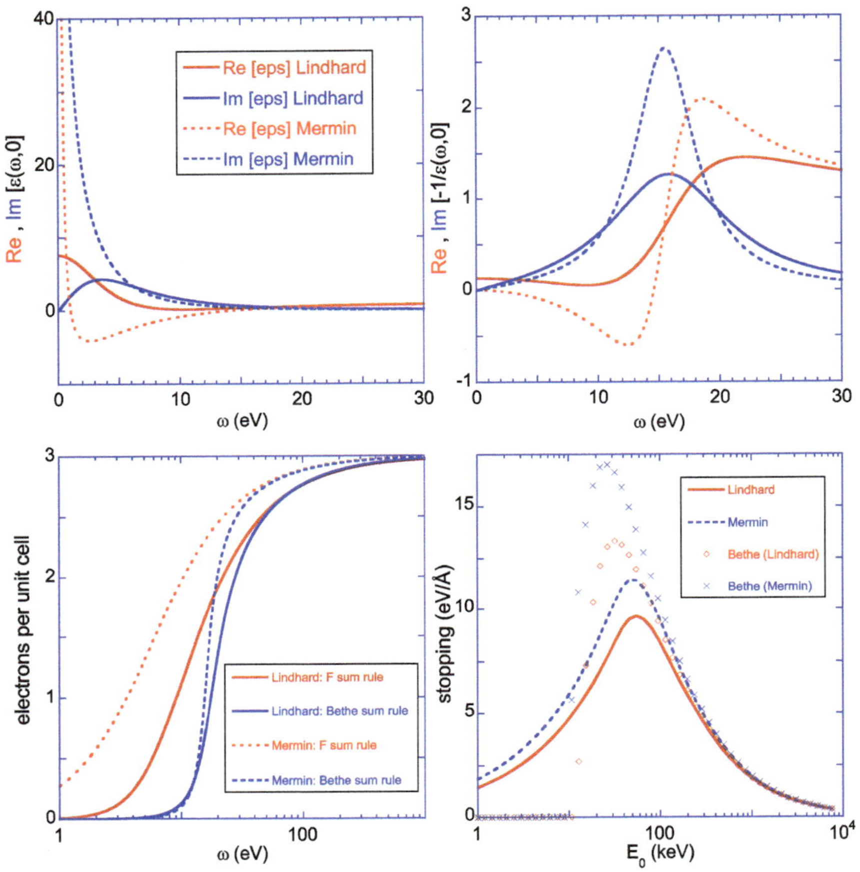

Fig. 5 A comparison of Re[$\epsilon(\omega, 0)$] and Im[$\epsilon(\omega, 0)$] of the Lindhard dielectric function with a large Γ value (6 eV) with and without the Mermin correction. All other parameters are taken as applied to the aluminum valence band (*top left panel*). The same comparison but for Re[$1/\epsilon(\omega, 0)$] and Im[$-1/\epsilon(\omega, 0)$] is given in the *top right panel*. The *lower left panel* shows the sum rule integrals (Eqs. 7 and 6) as a function of ω as the upper limit of the integral. The *lower right panel* shows their stopping values as well as the Bethe stopping for the corresponding *I* values. The differences are discussed in the main text.

gap[34] (see Appendix B for full expressions). It reverts to the Lindhard function for $U \rightarrow 0$. Archubi and Arista studied the effect of the U parameter on the stopping, inelastic mean free path and straggling for electrons, positrons and protons using the ϵ^{L} (or equivalently, ϵ^{LL} with $U = 0$) and ϵ^{LL} dielectric function with $U \neq 0$.[35] Here we add a relaxation time to this function by replacing ϵ^{L} in Eq. (21) by ϵ^{LL} and refer to the resulting dielectric function as ϵ^{MLL} (Mermin–Levine–Louie). Using this approach one has $\sum_i C_i^{MLL} = 1$

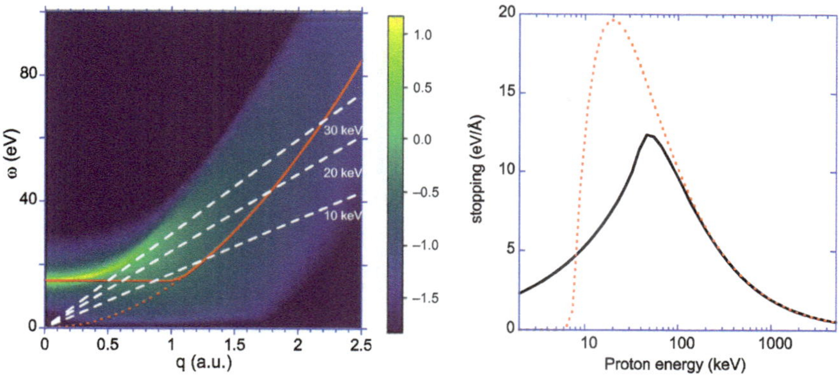

Fig. 6 Same as Fig. 2 but now for the Mermin dielectric function. The electron density is again the same. The dispersion of the plasmon is similar to Drude–Lindhard but now there is a second broad low-intensity branch due to electron–hole excitations that extends below the *qv* line, even at low energies. As a consequence the stopping is larger at low energies.

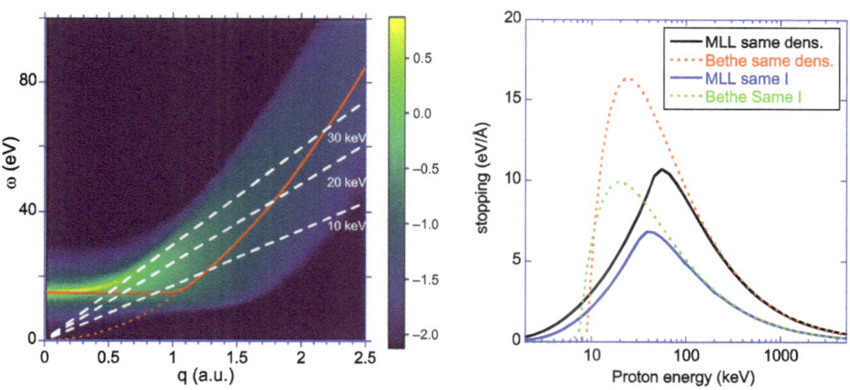

Fig. 7 Same as Fig. 6 but now for a MLL dielectric function corresponding to a material with half the electron density (i.e., $\omega_{\mathrm{pl}} = 15/\sqrt{2}$) but with a band gap *U* of 10.6 eV. In this way the peak in the ELF is again at 15 eV and the *I* value is not changed. The structure of the loss function shows again a plasmon branch and a branch due to electron–hole excitation, but the intensity of the latter is suppressed in the gap. For large proton energies the stopping is now half that of the Mermin case (due to lower electron density) but at low energies the stopping is reduced by more than a factor of 2. Also shown is the stopping when the density *n* is left unchanged, but the same gap is introduced. This shifts the maximum of the ELF to $\sqrt{15^2 + 10.6^2}$ and changes thus the *I* value.

as a boundary condition, but by choosing the appropriate ω_i and U values, one can, in the optical limit, obtain again the same loss function as the Drude–Lindhard model with $C_1 < 1$. For details see Ref. 27. An example of a MLL loss function and the corresponding stopping is given in Fig. 7.

2.3.3 Dielectric function using Gaussian occupation

In the Lindhard (free electron) approach all states are fully occupied for $|q| < k_f$ with k_f the Fermi wave vector and empty for $|q| > k_f$. Deeper levels have often an occupation that decreases more gradually, and for s levels the momentum occupation resembles roughly a Gaussian shape. This led Kaneko to derive a dielectric function based on a Gaussian occupation proportional to: $\exp(-q^2/Q^2)$ with Q the characteristic momentum for that level. In the most simple case (one electron in the shell) this gives for the plasmon energy: $\omega_p^2 = Q^3/\sqrt{\pi}$. For $Q = 0.64$ we obtain thus again a plasmon energy of 15 eV, just as in our earlier examples. The corresponding loss function and stopping are presented in Fig. 8.

2.3.4 GOS-based dielectric functions

For the description of deeper levels a plane-wave-derived description (as is the Lindhard dielectric function and its derivatives) may not be the most appropriate starting point. Hydrogenic wavefunctions would appear more appropriate. Then the dielectric function is closely related to the GOS (Generalized Oscillator Strengths)

Considering the GOS[21]:

$$f_{j0}(q) = 2\omega_{j0}\frac{|F_{j0}(q)|^2}{q^2} \tag{22}$$

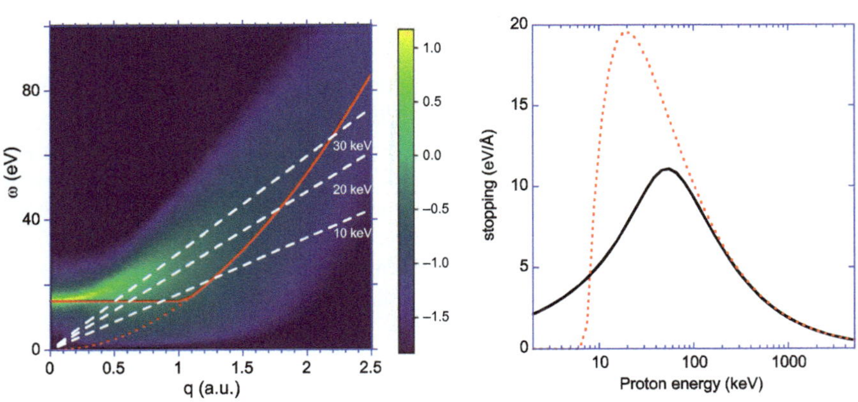

Fig. 8 Same as Fig. 2 but now for the Kaneko dielectric function using $Q = 0.64$ a.u. The $\mathrm{Im}[-1/\epsilon(\omega, q)]$ distribution resembles the Mermin one, except that the edges of the electron–hole excitation branch is now not sharp, due to the more gradual decrease of the momentum density. The stopping curve is again similar to the Mermin one except for the absence of the little kink near the maximum.

and continuum states we can rewrite Eq. (20) as

$$\epsilon(\omega, q) = 1 + \omega_p^2 \int_0^\infty \frac{d\omega'}{2\omega'} \frac{df}{d\omega'}(\omega', q) \left(\frac{1}{\omega' - \omega - i\Gamma} + \frac{1}{\omega' + \omega + i\Gamma} \right) \quad (23)$$

where ω_{j0} and $f_{j0}(q)$ are replaced by ω' and $\frac{df}{d\omega'}(\omega', q)$, respectively. The real and imaginary parts of Eq. (23) can be evaluated using the mathematical relation

$$\lim_{\Gamma \to 0^+} \frac{1}{x - a + i\Gamma} = \mathcal{P}\frac{1}{x - a} - i\pi\delta(x - a) \quad (24)$$

where, $\mathcal{P}$ indicates the Cauchy principal value of the integral, and read

$$\mathrm{Re}[\epsilon(\omega, q)] = 1 - \omega_p^2 \mathcal{P}\left[\int_0^\infty d\omega' \frac{df}{d\omega'}(\omega', q) \frac{1}{\omega^2 - \omega'^2} \right] \quad (25)$$

and

$$\mathrm{Im}[\epsilon(\omega, q)] = \frac{\omega_p^2 \pi}{2\omega} \frac{df}{d\omega}(\omega, q) \quad (26)$$

In the dilute limit ($\omega_p \to 0$) $\mathrm{Im}[\epsilon(\omega, q)]$ is small and $\mathrm{Re}[\epsilon(\omega, q)] \approx 1$. Then Im $[-1/\epsilon(\omega, q)] \approx \mathrm{Im}[\epsilon(\omega, q)]$ and ω-times the loss function is proportional to the GOS. The loss function and the corresponding stopping is shown in Fig. 9.

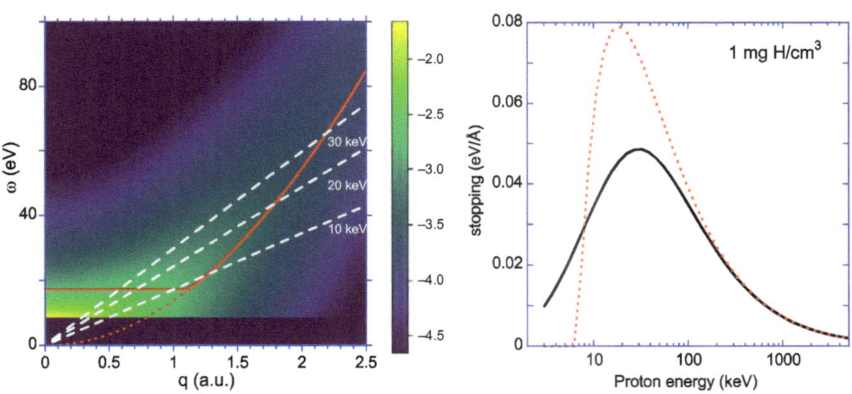

Fig. 9 Same as Fig. 7 but now for the GOS of H 1s atoms. The energy of the edge was taken at 8.4 eV, rather than the H 1s binding energy (13.6 eV) or minimum excitation energy (10.4 eV), to approximately take into account excitations to bound states. With this value the sum rule gives 1 electron per hydrogen atom and the mean excitation energy of $I = 14.9$ eV is obtained in close agreement with literature value of $I = 15.1$ eV. The *right panel* shows the stopping for atomic H with a density of 1 mg/cm^3 (*solid line*) together with the Bethe stopping for this I value and density (*dashed line*).

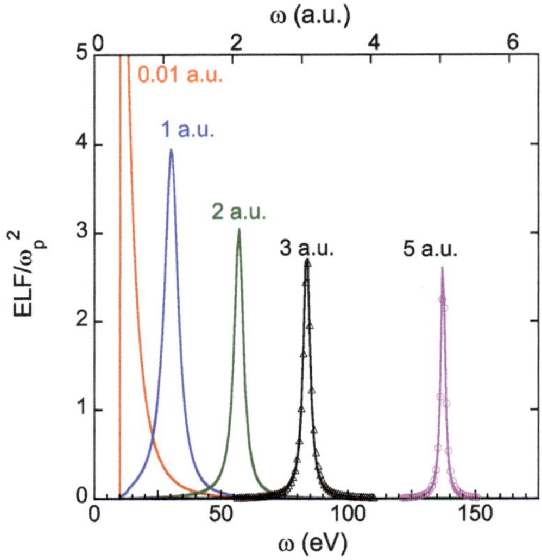

Fig. 10 The influence of the density on the loss function of H atoms. The loss function is divided by $\omega_p^2 = 4\pi n$, i.e., is proportional to the loss per atom. The peak in the loss function shifts to larger values and is ω_p at the larger densities. Then it resembles the Mermin loss function shown here by *triangles* and *circles* for the $\omega_p = 3$ and $\omega_p = 5$ cases. ($\omega_p = 5$ a.u. corresponds to a H density of 21 g/cm^3).

If one increases the H density than the situation changes as shown in Fig. 10. With increasing density the loss function shifts to larger ω values and the sharp onset of the loss function disappears: the loss function becomes more symmetrical. At the largest density considered, the peak position is very close to the plasmon energy of that density.

2.3.5 Dielectric function based on Harmonic oscillator

An RPA-level dielectric function was derived by Belkacem and Sigmund assuming the electrons are bound in a harmonic potential with resonance frequency ω_0.[17] Here the final state always remains bound. In the dilute limit the ELF has contributions at energy losses $n\omega_0$ with n an integer. If significant broadening Γ is applied then Mermin-type corrections have to be considered.[18] For $q = 0$ all intensity is in the $n = 1$ branch (dipole limit). The peak positions change at significant densities. Then there is a q-dependence of the energy loss, especially at small n values. This q-dependence increases the position of the $n = 1$ peak from ω_0 to $\sqrt{\omega_o^2 + \omega_p^2}$ with ω_p again the plasmon energy corresponding to the density of the system.

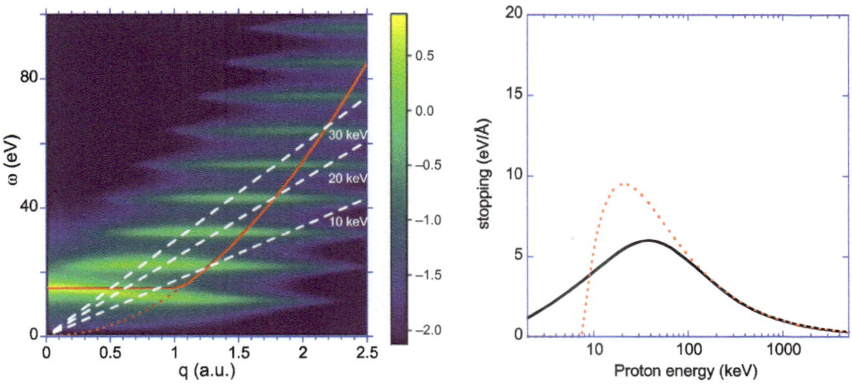

Fig. 11 Same as Fig. 7 but now for the Belkacem–Sigmund loss dielectric function. The density is the same as the reduced density of the MLL case, and the oscillator energy (gap) is again 10.4 eV.

The loss function and stopping for parameters such that at $q = 0$ the ELF of this model is the same as that used for MLL is shown in Fig. 11. The discrete nature of the final state is clearly visible in the ELF distribution. The stopping is slightly less than the MLL case with the same density and $U = \omega_0$.

3. Compton profiles

In the preceding we presented many dielectric functions that have, in the optical limit, a maximum at 15 eV loss. To stress their differences it is also worthwhile to show their energy distribution in the high momentum transfer case. This is done in Fig. 12. There is a clear difference between the Compton profiles derived from classical and quantum-physics-based dielectric functions. The classical dielectric function have loss functions at large q with the same width as the loss function at $q = 0$. The quantum-based ones have a much increased width. Interpreting the loss function at high-q in terms of binary collisions of the projectile with a target electron with momentum $\boldsymbol{p}$ one gets for the energy loss:

$$\omega = q^2/2 + \boldsymbol{q} \cdot \boldsymbol{p} \tag{27}$$

The Mermin Compton distribution stops at $p = k_f$ as it is the projection of the density of states within the Fermi sphere on the $\boldsymbol{q}$ direction. Kaneko's Compton profile is a Gaussian with width Q, reflecting the assumed momentum occupation for the initial state. Our MLL dielectric function corresponds to half the electron density of the Mermin one. Hence the

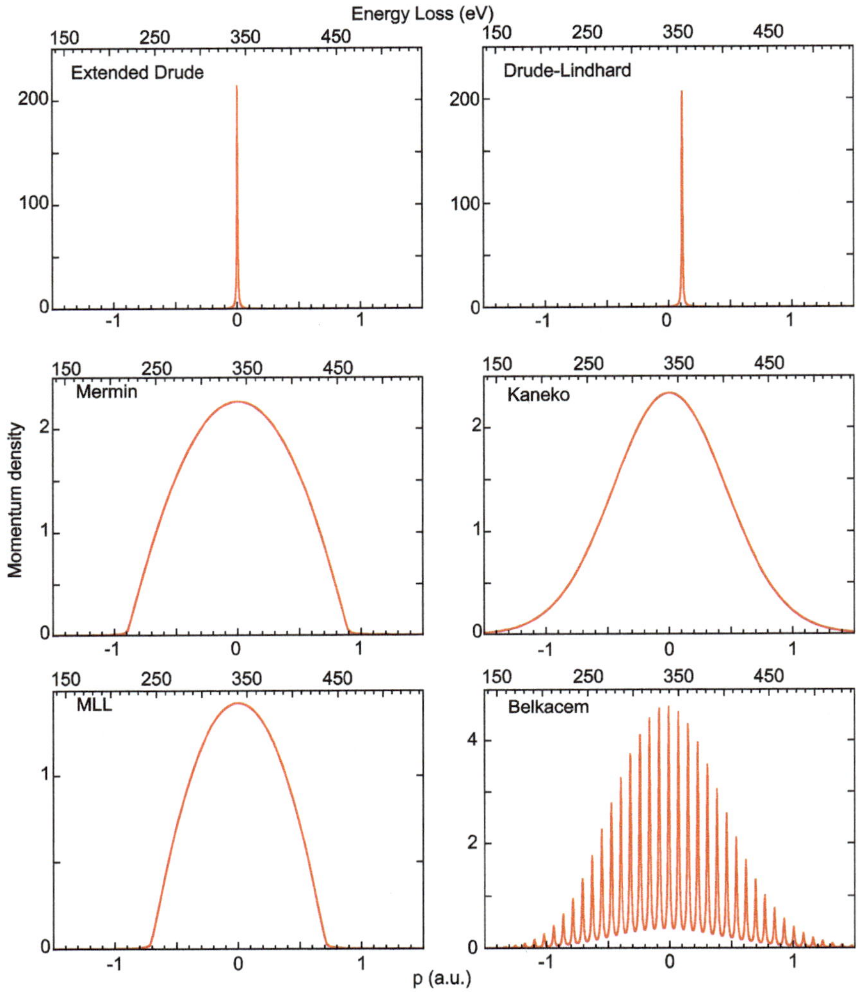

Fig. 12 the loss function at $q = 5$ a.u. for the various dielectric functions as indicated. The scale at the top axis refers to the energy loss ω. At $q = 5$ a.u. we can interpret these loss function as Compton profiles. The corresponding target electron momentum is given by the lower axis.

diameter of its Fermi sphere is reduced by a factor $2^{1/3}$ and thus the Compton profile cuts off at smaller $|p|$ values. The Belkacem Compton profile consist of a set of sharp peaks, separated by the oscillator energy (10.6 eV). Here the final state is always one of the eigenfunctions of the harmonic oscillator and the excited electron always remains bound in the potential. The envelope of this set of peaks (plotted as a function of p) has the same shape as the density of the (momentum-space) ground state of this oscillator.

4. Multiple oscillators

It is generally not possible to model the observed optical ELF with a single oscillator. Then the dielectric function is often described by a sum of oscillators each with different values for the A_i or C_i, ω_i and Γ_i parameter. The question than raises how to add their contributions to $\epsilon(\omega, q) = 1 + \chi(\omega, q)$. Here there are two options:

- add their χ values. This is most appropriate if one considers the contribution due to more than one type of electron inside the same volume. The polarization of one wavefunction affects the electric field seen by the other electron. The induced dipole moment will add up, i.e., their χ contributions will add up. In this mode it appears that the contributions to the ELF of the different components will "repel" each other and, e.g., for the extended Drude model the peaks positions in the ELF deviate from $\sqrt{A_i + \omega_i(q)^2}$.[26] This approach is often used when one aims to only describe the valence band part of the loss spectra. For example, it was used by Tung et al.[36] and Werner et al.[37] to describe the differential inverse inelastic mean free path DIIMFP of several metals as measured in reflection electron energy loss spectroscopy (REELS). In the case of metals one of the parameters ω_i in the extended Drude model is taken to be 0. In the description of REELS experiments some of the α_i parameters are sometimes taken less than one, when one expects that dispersion for low q values is limited. This works for REELS, as the DIIMFP is dominated by contributions at low ω and low q. However, then the dielectric function does not center at $q^2/2$ at high-q values an essential property for getting the right stopping and straggling values.
- add their loss functions. This comes natural if one assumes the different components occupy different parts of the unit cell. For example, it is often assumed that the volume of the unit cell with density n_x behaves as a free electron gas with that density. In that case it is natural to assume that each fraction of the unit cell has its own loss function and one adds the ELF of all components. Within this framework C_i is the fraction of the unit cell that is filled by component i. Clearly $\sum_i C_i <= 1$ (one can leave part of the unit cell empty). In the case of the extended Drude model the fraction of the unit cell occupied by an oscillator is determined by ω_i. If one takes $\omega_i = 0$ then that component will occupy the whole unit cell.

This approach, adding the ELFs from Drude–Lindhard dielectric function, is used, e.g., in the QUASUS program, popular for the analysis of REELS spectra[38] for the determination if the electron inelastic mean free path.

Abril et al. used a sum of Mermin loss functions to fit a optical data[19] or REELS data (e.g., Ref. 39) to describe the proton stopping. Denton et al. used the Mermin dielectric function to calculate the electron inelastic mean free path.[40] Da et al. used a large number of positive and negative Mermin oscillator to fit the ELF of Cu and calculate its inelastic mean free path.[41]

In either way (add χ or add ELF) one obtains a sum dielectric function that adheres to the sum rules and Kramers–Kronig relations but they can be very different! An example of this is given in Fig. 13. Note that adding the ELF in this case results in a metal ($\mathrm{Re}[1/\varepsilon(0,0)] = 0$) whereas adding the χ values results in an insulator ($\mathrm{Re}[1/\varepsilon(0,0)] > 0$). For larger energy losses (beyond the valence band) where ($\mathrm{Re}\ [\epsilon(\omega, q)] \simeq 1$) these considerations are not important and either method gives very similar results.

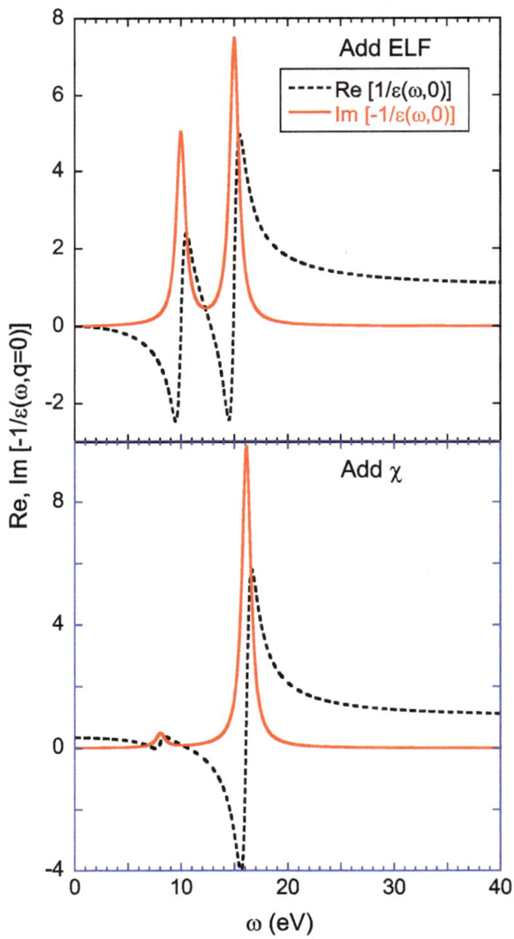

Fig. 13 A Drude–Lindhard dielectric function comprised out of 2 oscillators ($A_1 = 0.5$, $\omega_1 = 10\mathrm{eV}$, $\Gamma_1 = 1$ eV, $A_2 = 0.5$, $\omega_2 = 15\mathrm{eV}$, $\Gamma_2 = 1$ eV) obtained by either adding the ELF (*top*) or their $\chi(q, \omega)$ values.

Modeling the ELF of an actual solid by choosing the most appropriate model and obtaining the best parameters to fit the available experimental data (optical absorption, refractive index, Compton profiles, etc.) is a complex problem, and the procedure one chooses depends, in part on the physics question one wants to answer. We will not attempt it here, but focus instead on the shell corrections in one of the simplest cases: an electron gas that resembles aluminum.

5. Shell corrections in stopping

Shell effects have been ascribed as the difference of stopping as calculated from Eq. (3) from the Bethe value[5,42] and are attributed to velocity distribution of electrons in the medium. The stopping is then usually written as a product of two factors:($-\frac{dE}{dx} = \frac{4\pi n Z^2}{v^2} L$). L can thus be defined in terms of the dielectric function:

$$L = \frac{1}{2\pi^2} \int_0^\infty \frac{dq}{q} \int_0^{qv} \omega \, d\omega \, \mathrm{Im}\left[\frac{-1}{\epsilon(\omega, q)}\right] \tag{28}$$

and, by convention, the shell effect is then given by:

$$\Delta L = L_{\mathrm{Bethe}} - L, \tag{29}$$

with $L_{\mathrm{Bethe}} = \ln(2v^2/I)$. Full expressions for the shell effect in terms of the velocity-distribution momenta are given in Ref. 5 and values for some atoms and nucleobases in Refs. 42–44. These are usually studied at high projectile velocities. Then the qv line is much steeper, and also for the quantum-based dielectric function the cut-off at low-q is due to the intersection of the qv line with the plasmon branch. It turns out that details in the plasmon dispersion near $q = 0$ are also important. Hence we show an enlargement of the loss distribution for four dielectric functions in Fig. 14. For clarity the width Γ was now taken to be only 0.2 eV but further the models are the same as in the previous section. Clearly the dispersion is different for the models. The dispersion is largest for Mermin and Drude–Lindhard (peak position given by $\omega_p + q^2/2$), intermediate for MLL (peak dispersion given by $\sqrt{U^2 + (\omega_p^{\mathrm{MLL}} + q^2/2)^2}$) and smallest for extended Drude (peak dispersion given by $\sqrt{\omega_p^{\mathrm{MLL}\,2} + q^4/4}$). Note however, that the similarity of the Mermin and Drude–Lindhard dispersion is *not* a general property of that dielectric function but a coincidence for the electron density of the valence band of Al. The Mermin dispersion increases with increasing electron density.

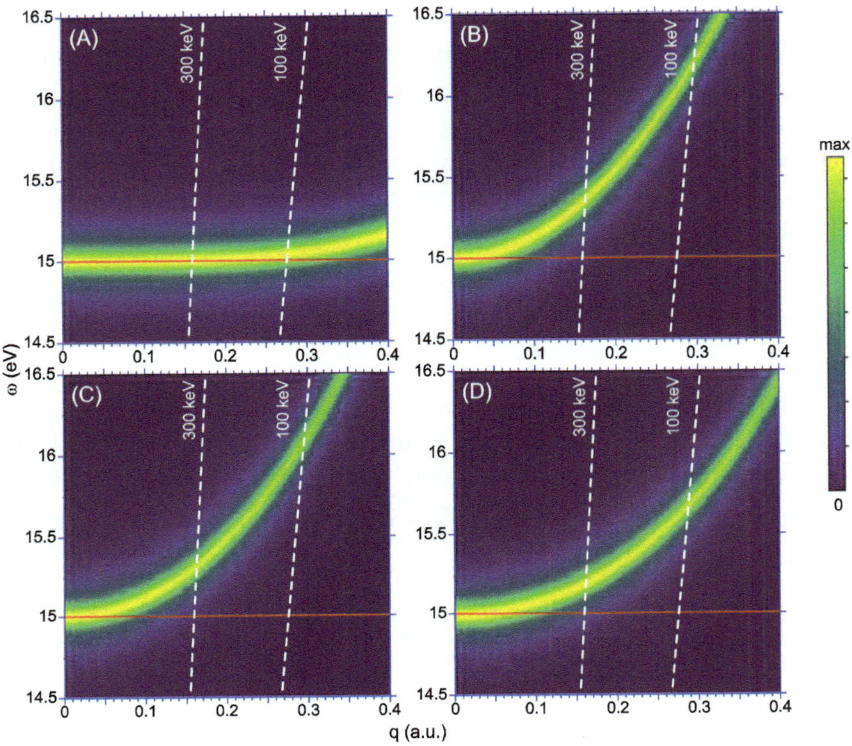

Fig. 14 The ELF distribution of the extended Drude (A), Drude–Lindhard (B), Mermin (C), and Mermin–Levine–Louie (D) dielectric function for small momentum values and energies close to the plasmon energy. The *red line* is again the dispersion of our "ad hoc" dielectric function. The *dahed lines* are the *qv* lines at 100 keV and 300 keV.

The dispersion affects the q value for which the qv line intersects the plasmon branch. For these low q values almost all intensity is in the plasmon branch, and then the Bethe sum rule determines the contribution of that q interval to the stopping.

Besides these differences at low-q there is also a contribution at the high-q crossing of the qv line and the dispersing ELF feature. For the Mermin case, this is illustrated in Fig. 15. The loss function near $q = 10$ a.u. is wide (Compton profile) hence the contribution of the ELF to the stopping decreases gradually when the qv-line crosses the Compton profile. The crossing of the ELF of the "AH" dielectric function is however abrupt, at $q = 10$ a.u. As, when calculating L the ELF contribution is weighted by $1/q$ there remains a net effect when integration is done over all q. It is predicted by the "equipartition principle"[45] the contribution at low and high-q to the shell corrections are equal, and this turns out to be indeed the case.

Shell corrections for Mermin and Drude–Lindhard are very similar. However the latter has no "Compton broadening." The Drude–Lindhard

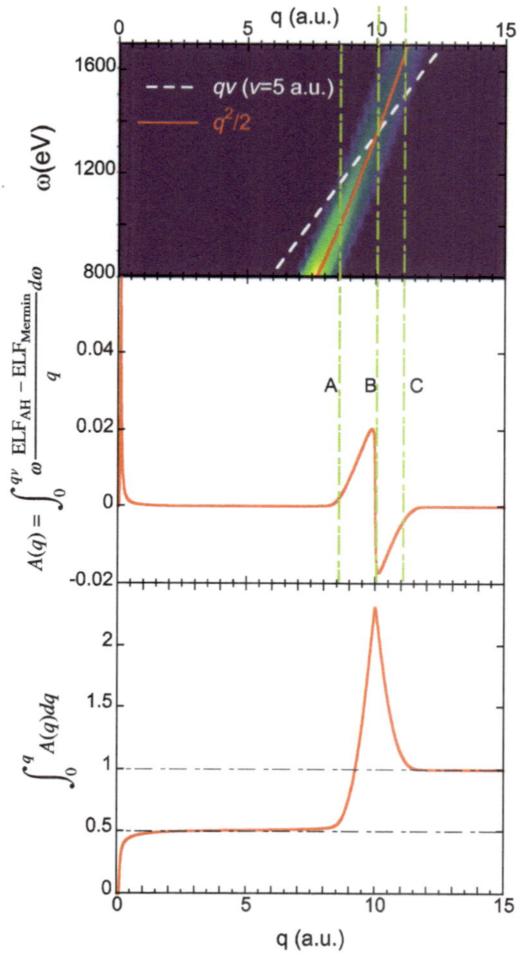

Fig. 15 Analysis of the shell corrections for the Mermin case for protons with $v = 5$ a.u. The *top panel* shows the ELF for the region of the high-q crossing of the qv line and the ELF. The ad hoc dielectric function is the *red line*, the *dashed line* is the qv line for $v = 5$ a.u. (625 keV). The *central panel* shows the shell correction contribution as a function of q. For q values between A and C the contribution of the Mermin loss function to the stopping integral decreases. The Bethe contribution (at $q^2/2$) drops abruptly to zero at B. *Lower panel*—fractional shell correction: Integrating over all q the residual shell correction is equally due to low-q and high-q contribution.

ELF is shifted by 15 eV relative to the Mermin ELF, at large q values, see Fig. 12. In Appendix B it is shown that this shift is responsible for the shell correction in the Drude–Lindhard case. For Drude–Lindhard the equipartition principle is adhered to as well.

In Fig. 16 we show the obtained effects. There is good agreement between the analytical calculation of the shell effect using the formulae

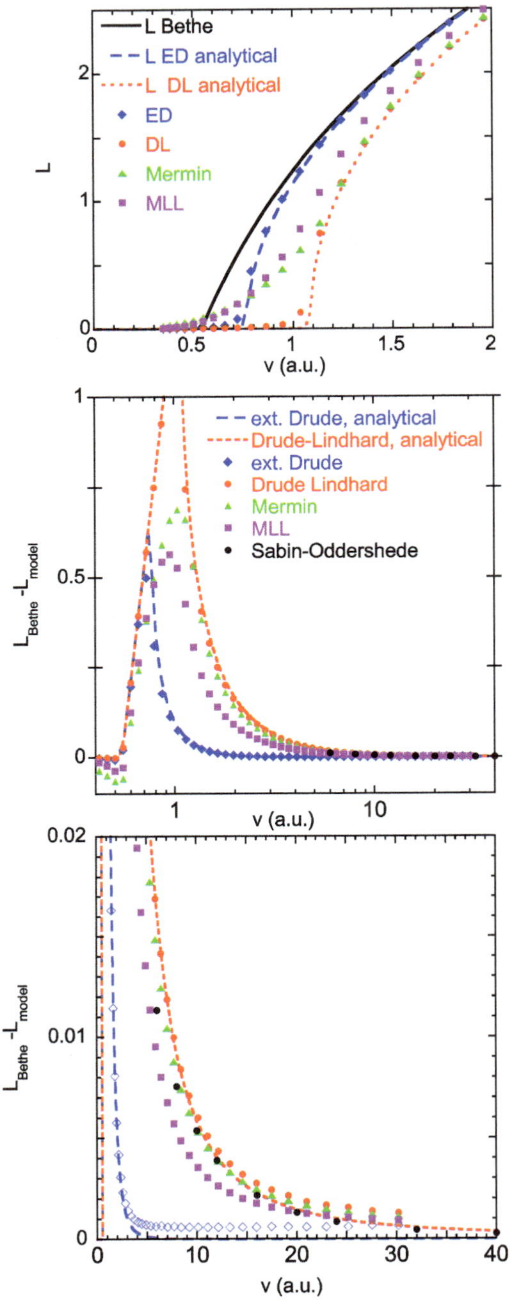

Fig. 16 Top panel shows the *L* values for proton as a function of its velocity *v* for the models as indicated. The lines are *L* values corresponding to the analytical expressions (B.4), (B.6), (B.8), the symbols are obtained by evaluating the stopping integral for the model dielectric functions. The central panel shows the shell effect: $L_{\mathrm{Bethe}} - L_{\mathrm{model}}$, the *lower panel* has an expanded vertical scale. The *black dots* were taken from the calculation by Sabin et al. for the 3s and 3p electrons of *atomic* Al.[42]

derived in Appendix B for extended Drude and Drude–Lindhard and the values obtained by numerically integrating the stopping integral for the corresponding dielectric functions. As expected the extended Drude model, with a dispersion that is closest to our AH dielectric function has the smallest shell effect, virtually absent for $v > 3$ a.u. The shell effect for the Drude–Lindhard decreases almost 10 times slower with increasing v. The shell effect for Mermin is negative at very low v values, as it has significant stopping even at very low v values, but for $v > 2$ a.u. it appears to be about 10% less than the shell effect for Drude–Lindhard. The shell effect for the MLL model is in between extended Drude and Drude–Lindhard, in agreement with the reduced dispersion of its plasmon branch.

Similar considerations apply to straggling, as is sketched in Appendix C.

6. Conclusion

In this chapter we showed the loss function of a large number of model dielectric functions, both classical models and based on quantum physics. All models adhere to sum rules and Kramers–Kronig relations at any q. Their stopping was compared to the Bethe stopping. All models converge to the Bethe stopping, but the rate at which they do differs greatly. In other words, the models predict different shell corrections. It was demonstrated for selected cases that shell corrections have a contribution at low q and at high q, i.e., at the q values the qv line crosses the maxima of the ELF.

The models used were chosen so the ELF at $q = 0$ has a peak at the same energy loss as aluminum. Most materials have a much more complicated loss function at $q = 0$ and require more that one oscillator to be described properly. Considerations how to combine different oscillators were discussed. The modeling of more complex solids over the whole q–ω plane remains a challenging topic. The description given here of the different model dielectric functions may aid the reader in devising schemes that accomplish such a description. That will make shell corrections available for a larger group of materials, extending the work initiated by Sabin and Oddershede over four decades ago.

Acknowledgments

This work was made possible by funding of the Australian Research Council and in part by the Coordenação de Aperfeiçoamento de Pessoal de Nível Superior—Brazil (CAPES)—Finance Code 001, by FINEP, by Conselho Nacional de Desenvolvimento Científico e Tecnológico (CNPq) process number 404301/2016-9 and PRONEX-FAPERGS.

Appendix A. Lindhard dielectric function

For completeness we reproduce here the formula for the Lindhard[3] and Levine–Louie (LL)[34] dielectric function.

The Lindhard dielectric function for an electron gas with density n can be expressed in terms of the Fermi velocity v_f (or Fermi energy $E_f = \frac{1}{2}v_f^2$) and the dimensionless variables $u = \omega/(qv_f)$, $z = q/(2v_f)$ and $\chi^2 = 1/(\pi v_f)$ according to

$$\epsilon^L(\omega, q) = 1 + \frac{\chi^2}{z^2}(f_1(u, z) + if_2(u, z)) \tag{A.1}$$

where

$$f_1(u, z) = \frac{1}{2} + \frac{1}{8z}(g(z-u) + g(z+u)) \tag{A.2}$$

with

$$g(x) = (1-x^2)\ln\left|\frac{1+x}{1-x}\right| \tag{A.3}$$

and

$$\begin{aligned} f_2(u,z) = &\frac{\pi}{2}u\Theta(1-z-u) + \frac{\pi}{8z}(1-(z-u)^2) \\ &\Theta(1-|z-u|)\Theta(z+u-1) \end{aligned} \tag{A.4}$$

where $\Theta(x)$ is the usual step function.

The LL dielectric function is obtained by introducing a gap energy U in the imaginary part of the Lindhard dielectric function according to

$$\epsilon_2^{LL}(\omega, q) = \epsilon_2^L(\omega_-, q)\Theta(|\omega| - U) \tag{A.5}$$

with $\omega_- = \sqrt{\omega^2 - U^2}$. The real part of the LL dielectric function can be obtained straightforwardly from Kramers–Krönig relation and reads

$$\begin{aligned} \epsilon_1^{LL}(\omega,q) = &\epsilon_1^L(\omega_-,q)\Theta(|\omega| - U) \\ &+ \left(1 + \frac{2}{\pi v_f}F(Q, \Delta)\right)\Theta(U - |\omega|) \end{aligned} \tag{A.6}$$

and

$$F(Q,\Delta) = \frac{1}{Q^2} - \frac{\Delta}{2Q^3}\left(\arctan\frac{2Q+Q^2}{\Delta} + \arctan\frac{2Q-Q^2}{\Delta}\right) + \left(\frac{\Delta^2}{8Q^5} + \frac{1}{2Q^3} - \frac{1}{8Q}\right)\ln\left(\frac{\Delta^2+(2Q+Q^2)^2}{\Delta^2+(2Q-Q^2)^2}\right) \tag{A.7}$$

with $Q = q/v_f$ and $\Delta^2 = (U^2 - \omega^2)/E_f^2$.

Appendix B. Stopping formulae and shell corrections for simple dispersion relations

For completeness we show here explicitly that the dispersion relation assumed in Eq. (11) indeed leads to the Bethe-like stopping (Eq. 12). The stopping power is given in terms of the ELF as

$$-\frac{dE}{dx} = \frac{2Z^2}{\pi v^2}\int_0^\infty \frac{dq}{q}\int_0^{qv} \omega\, d\omega\, \mathrm{Im}\left[\frac{-1}{\epsilon(q,\omega)}\right], \tag{B.1}$$

where Z is the projectile charge. Assuming the ELF given by the Drude model with a narrow peak ($\Gamma \to 0$) at $\omega_p(q)$ the inner integral of the above equation will be given by the Bethe sum rule (Eq. 7) for $\omega_p(q) < qv$ and zero otherwise. Therefore,

$$-\frac{dE}{dx} = \frac{4\pi n Z^2}{v^2}\int_0^\infty \frac{dq}{q}\Theta(qv - \omega(q)), \tag{B.2}$$

where $\Theta(x)$ is the Heavy-side function. The condition $\omega_p(q) < qv$ is realized by $q_{\mathrm{min}} < q < q_{\mathrm{max}}$ see Fig. 2. For this case the stopping power reads

$$-\frac{dE}{dx} = \frac{4\pi n Z^2}{v^2}\ln\left(\frac{q_{max}}{q_{min}}\right). \tag{B.3}$$

For the dispersion relation from Eq. (11) we have $q_{min}^{\mathrm{Bethe}} = \omega_p/v$ and $q_{max}^{\mathrm{Bethe}} = 2v$ and therefore we get the Bethe-like formula:

$$-\frac{dE}{dx} = \frac{4\pi n Z^2}{v^2}\ln\left(\frac{2v^2}{\omega_p}\right). \tag{B.4}$$

The parabola and the straight line in the "Bethe dispersion" meet at $q_c = \sqrt{2\omega_p}$. The integral can be divided into two parts $q_{min} < q < q_c$ and $q_c < q < q_{max}$.

$$-\frac{dE}{dx} = \frac{4\pi n Z^2}{v^2}\left(\ln\left(\frac{q_{max}}{q_c}\right) + \ln\left(\frac{q_c}{q_{min}}\right)\right). \tag{B.5}$$

Inspection shows that $q_{max}/q_c = 2v/\sqrt{2\omega_p}$ and $q_c/q_{min} = \sqrt{2\omega_p}/(\omega_p/v)$ are identical. This is an example of the "equipartition principle"[45] which state that the plasmon and electron–hole excitations contribute equally to the stopping.

One can use exactly the same approach to derive an analytical expression for the stopping in the case of simple quadratic dispersion (in the limit of $\Gamma \rightarrow 0$). Then one has to replace simply $q_{\max}$ and $q_{\min}$ by the solutions of $qv = q^2/2 + \omega_{\rm p}$. ($q^{\rm DL}_{\rm min,max} = v \mp \sqrt{v^2 - 2\omega_p}$) One obtains:

$$-\frac{dE}{dx} = \frac{4\pi n Z^2}{v^2}\ln\left(\frac{v + \sqrt{v^2 - 2\omega_{\rm p}}}{v - \sqrt{v^2 - 2\omega_{\rm p}}}\right), \tag{B.6}$$

which should not overestimate the stopping at the stopping maximum as much as the Bethe-like equation does.

$q^{\rm Bethe}_{\rm max} > q^{\rm DL}_{\rm max}$ and similarly $q^{\rm Bethe}_{\rm min} < q^{\rm DL}_{\rm min}$. The contribution to the stopping integral between $q^{\rm DL}_{\rm min}$ and $q^{\rm DL}_{\rm max}$ is the same for the Bethe and DL dielectric function. The shell effect $L_{\rm Bethe} - L_{\rm DL}$ is thus equal to

$$L_{\rm Bethe} - L_{\rm DL} = \ln\left(\frac{q^{\rm DL}_{\rm min}}{q^{\rm Bethe}_{\rm min}}\right) + \ln\left(\frac{q^{\rm Bethe}_{\rm max}}{q^{\rm DL}_{\rm max}}\right) \tag{B.7}$$

The contribution to the shell corrections of each part of Eq. (B.7), namely $\frac{\omega_p}{2v^2}$, are identical at high projectile velocities and reflects the equipartition principle, which is also fulfilled for DL : $q^{\rm DL}_{\rm max}/q_c = q_c/q^{\rm DL}_{\rm min}$ as can be easily verified.

In the extended Drude case for a free electron gas the limits of integration are obtained from $(qv)^2 = A_1^2 + q^4/4$ $\left(q^{\rm ED}_{\rm min,max} = \sqrt{v^2 \mp \sqrt{v^4 - \omega_p^2}}\right)$. Then the stopping becomes:

$$-\frac{dE}{dx} = \frac{4\pi n Z^2}{v^2}\ln\left(\frac{q^{\rm ED}_{\rm max}}{q^{\rm ED}_{\rm min}}\right) \tag{B.8}$$

The extended Drude dielectric function corresponds to the static electron gas, as described by Sigmund and this formula was also derived there (Ref. 5, Eq. 5.77). Of course its shell corrections are given by:

$$L_{\text{Bethe}} - L_{\text{ED}} = \ln\left(\frac{q_{\min}^{\text{ED}}}{q_{\min}^{\text{Bethe}}}\right) + \ln\left(\frac{q_{\max}^{\text{Bethe}}}{q_{\max}^{\text{ED}}}\right) \tag{B.9}$$

Now both contributions are similar, but not exactly the same, as the q value separating hard and soft collisions is different for ED compared to Bethe.

Similarly, when using full dispersion $(\omega_p(q))^2 = \omega(0)^2 + \beta^2 q^2 + q^4/2$) the stopping becomes:

$$-\frac{dE}{dx} = \frac{4\pi n Z^2}{v^2} \ln\left(\frac{\sqrt{-\beta^2 + v^2 + \sqrt{(\beta^2 - v^2)^2 - \omega_p^2}}}{\sqrt{-\beta^2 + v^2 - \sqrt{(\beta^2 - v^2)^2 - \omega_p^2}}}\right), \tag{B.10}$$

with $\beta^2 = 2/3E_f$[46] or $\beta^2 = 6/5E_f$.[47]

Appendix C. Straggling formula for AH and Drude–Lindhard DF

The AH loss function Eq. (11) also leads to the Bohr-like straggling for high projectile velocities according to

$$\begin{aligned}\frac{d\Omega^2}{dx} &= \frac{2Z^2}{\pi v^2}\int_0^\infty \frac{dq}{q}\int_0^{qv} \omega^2\, d\omega\, \mathrm{Im}\left[\frac{-1}{\epsilon(q,\omega)}\right] \\ &= \frac{2Z^2}{\pi v^2}\int_0^\infty \frac{dq}{q}\int_0^{qv} \omega_p(q)\, d\omega\, A_{AH}\delta(\omega - \omega_p(q)),\end{aligned} \tag{C.1}$$

with $\omega_p(q) = \max(\omega_p, q^2/2)$. As in Appendix B, assuming $q_{min} < q < q_{max}$ and $q_c = \sqrt{2\omega_p}$, with $q_{min} < q_c < q_{max}$ we have

$$\begin{aligned}\frac{d\Omega^2}{dx} &= \frac{4\pi n Z^2}{v^2}\left(\omega_p \ln\left(\frac{q_c}{q_{\min}}\right) + \frac{1}{4}\left(q_{\max}^2 - q_c^2\right)\right) \\ &= \frac{4\pi n Z^2}{v^2}\left(\frac{\omega_p}{2}\ln\left(\frac{2v^2}{\omega_p}\right) + \left(v^2 - \frac{\omega_p}{2}\right)\right),\end{aligned} \tag{C.2}$$

which gives the Bohr straggling $\frac{d\Omega_B^2}{dx} = 4\pi n Z^2$ at high projectile velocities and is valid for $2v^2 > \omega_p$. It is similar to one derived for atomic systems[5] presenting a Bethe–Livingston-like shoulder.

Using the same arguments we get for the Drude–Lindhard dielectric function:

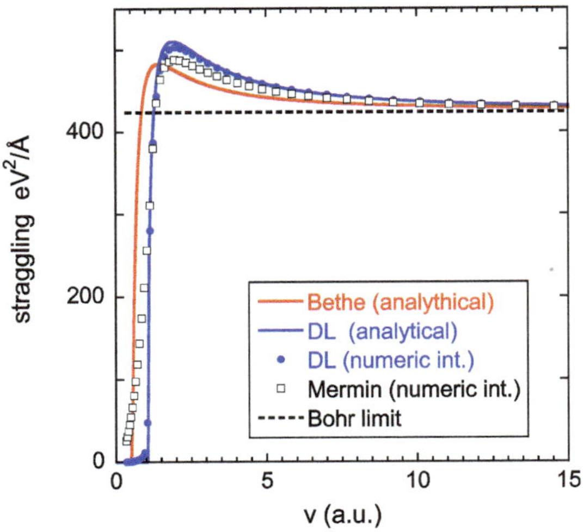

Fig. C.1 The straggling for protons in a plasma with $\omega_p = 15$ eV as calculated analytically for our "ad hoc" (Bethe) dielectric function (Eq. C.2), Drude–Lindhard dielectric function (Eq. C.3) as well as the numerical integration of the same Drude–Lindhard dielectric function and the Mermin dielectric function.

$$\left(\frac{d\Omega^2}{dx}\right)_{\mathrm{DL}} = \frac{4\pi n Z^2}{v^2}\left(\omega_p \ln\left(\frac{q_{\max}}{q_{\min}}\right) + \frac{1}{4}\left(q_{\max}^2 - q_{\min}^2\right)\right) \tag{C.3}$$

with now $q_{\min,\max} = v \mp \sqrt{v^2 - 2\omega_p}$. Results are compared in Fig. C.1.

References

1. Bohr, N. On the Theory of the Decrease of Velocity of Moving Electrified Particles on Passing Through Matter. *Philos. Mag.* **1913**, *25*, 10. https://doi.org/10.1080/14786440108634305.
2. Bethe, H. Zur Theorie des Durchgangs schneller Korpuskularstrahlen durch Materie. *Ann. Phys.* **1930**, *397* (3), 325–400. https://doi.org/10.1002/andp.19303970303.
3. Lindhard, J. On the Properties of a Gas of Charged Particles. *K. Dan. Vidensk. Selsk. Mat.-Fys. Medd.* **1954**, *28* (8), 1–57. http://gymarkiv.sdu.dk/MFM/kdvs/mfm%2020- -29/mfm-28-8.pdf.
4. Knapen, S.; Kozaczuk, J.; Lin, T. Python Package for Dark Matter Scattering in Dielectric Targets. *Phys. Rev. D* **2022**, *105* (1), 015014. https://doi.org/10.1103/physrevd.105.015014.
5. Sigmund, P. Particle Penetration and Radiation Effects. *Springer Series in Solid-State Sciences*, Vol. 151; Springer: Berlin, 2006.
6. Nastasi, M.; Mayer, J.; Hirvonen, J. K. *Ion-Solid Interactions*; Cambridge University Press, 1996. https://doi.org/10.1017/cbo9780511565007.
7. Nikjoo, H.; Uehara, S.; Emfietzoglou, D. *Interaction of Radiation With Matter*; CRC Press, 2012.

8. Garcia-Molina, R.; Abril, I.; Heredia-Avalos, S.; Kyriakou, I.; Emfietzoglou, D. S. A Combined Molecular Dynamics and Monte Carlo Simulation of the Spatial Distribution of Energy Deposition by Proton Beams in Liquid Water. *Phys. Med. Biol.* **2011**, *56*, 6475. https://doi.org/10.1088/0031-9155/56/19/019.
9. Garcia-Molina, R.; Abril, I.; Kyriakou, I.; Emfietzoglou, D. Energy Loss of Swift Protons in Liquid Water: Role of Optical Data Input and Extension Algorithms. In Gomez-Tejedor, G. G., Fuss, M. C., Eds.; Radiation Damage in Biomolecular Systems; Springer, 2012.
10. Nikjoo, H.; Emfietzoglou, D.; Liamsuwan, T.; Taleei, R.; Liljequist, D.; Uehara, S. Radiation Track, DNA Damage and Response—A Review. *Rep. Prog. Phys.* **2016**, *79*, 116601. https://doi.org/10.1088/0034-4885/79/11/116601.
11. Emfietzoglou, D.; Garcia-Molina, R.; Kyriakou, I.; Abril, I.; Nikjoo, H. A Dielectric Response Study of the Electronic Stopping Power of Liquid Water for Energetic Protons and a New I-Value for Water. *Phys. Med. Biol.* **2009**, *54*, 3451. https://doi.org/10.1088/0031-9155/54/11/012.
12. Emfietzoglou, D.; Kyriakou, I.; Garcia-Molina, R.; Abril, I. Inelastic Mean Free Path of Low-Energy Electrons in Condensed Media: Beyond the Standard Models. *Surf. Interface Anal.* **2017**, *49*, 4. https://doi.org/10.1002/sia.5878.
13. Bourke, J. D.; Chantler, C. T. Momentum-Dependent Lifetime Broadening of Electron Energy Loss Spectra: A Self-Consistent Coupled-Plasmon Model. *J. Phys. Chem. Lett.* **2015**, *6*, 314. https://doi.org/10.1021/jz5023812.
14. Chantler, C. T.; Bourke, J. D. New Constraints for Low-Momentum Electronic Excitations in Condensed Matter: Fundamental Consequences From Classical and Quantum Dielectric Theory. *J. Phys. Condens. Matter* **2015**, *27* (45), 455901. https://doi.org/10.1088/0953-8984/27/45/455901.
15. Mermin, N. Lindhard Dielectric Function in the Relaxation-Time Approximation. *Phys. Rev. B* **1970**, *1*, 2362–2363. https://doi.org/10.1103/physrevb.1.2362.
16. Kaneko, T. Wave-Packet Theory of Stopping of Bound Electrons. *Phys. Rev. A* **1989**, *40*, 2188. https://doi.org/10.1103/physreva.40.2188.
17. Belkacem, A.; Sigmund, P. Stopping Power and Energy Loss Spectrum of a Dense Medium of Bound Electrons. *Nucl. Instrum. Methods Phys. Res. Sect. B* **1990**, *48* (1–4), 29–33. https://doi.org/10.1016/0168-583x(90)90066-4.
18. Al-Bahnam, N. J. The Description of Quantum Dielectric Function for Insulators Over Bethe Surface. *Pramana* **2019**, *92* (2), 27. https://doi.org/10.1007/s12043-018-1675-1.
19. Abril, I.; Garcia-Molina, R.; Denton, C.; Pérez-Pérez, F. J.; Arista, N. Dielectric Description of Wakes and Stopping Powers in Solids. *Phys. Rev. A* **1998**, *58*, 357. https://doi.org/10.1103/physreva.58.357.
20. Palik, E. *Handbook of Optical Constants of Solids II*; Academic Press: New York, 1991.
21. Egerton, R. F. *Electron Energy-Loss Spectroscopy in the Electron Microscope*; Plenum Press: New York, 1996.
22. Batson, P. E.; Silcox, J. Experimental Energy-Loss Function, $Im[-\frac{1}{\epsilon}(\omega,q)]$, for Aluminum. *Phys. Rev. B* **1983**, *27*, 5224–5239. https://doi.org/10.1103/PhysRevB.27.5224.
23. Hayashi, H.; Hiraoka, N. Accurate Measurements of Dielectric and Optical Functions of Liquid Water and Liquid Benzene in the VUV Region (1–100 eV) Using Small-Angle Inelastic X-Ray Scattering. *J. Phys. Chem. B* **2015**, *119*, 5609. https://doi.org/10.1021/acs.jpcb.5b01567.
24. Cooper, M. J.; Mijnarends, P. E.; Shiotani, N.; Sakai, N.; Bansil, A. *X-Ray Compton Scattering*; Oxford University Press, 2004.
25. Wooten, F. *Optical Properties of Solids*; Academic Press, 1972.
26. Raether, H. *Excitation of Plasmons and Interband Transitions by Electrons*; Springer Verlag: Berlin, 1979.

27. Vos, M.; Grande, P. Simple Model Dielectric Functions for Insulators. *J. Phys. Chem. Solids* **2017**, *104*, 192. https://doi.org/10.1016/j.jpcs.2016.12.015.
28. Vos, M.; Grande, P. L. Extracting Dielectric Function From High-Energy REELS Measurements. *Surf. Interface Anal.* **2017**, *49*, 809–821. https://doi.org/10.1002/sia.6227.
29. Ferrell, R. A. Characteristic Energy Loss of Electrons Passing Through Metal Foils. II. Dispersion Relation and Short Wavelength Cutoff for Plasma Oscillations. *Phys. Rev.* **1957**, *107*, 450. https://doi.org/10.1103/physrev.107.450.
30. Ritchie, R. H.; Howie, A. Electron Excitation and the Optical Potential in Electron Microscopy. *Philos. Mag.* **1977**, *36*, 463. https://doi.org/10.1080/14786437708244948.
31. Emfietzoglou, D.; Kyriakou, I.; Garcia-Molina, R.; Abril, I.; Nikjoo, H. Inelastic Cross Sections for Low-Energy Electrons in Liquid Water: Exchange and Correlation Effects. *Radiat. Res.* **2013**, *180*, 499. https://doi.org/10.1667/rr13362.1.
32. Emfietzoglou, D.; Kyriakou, I.; Garcia-Molina, R.; Abril, I. The Effect of Static Many-Body Local-Field Corrections to Inelastic Electron Scattering in Condensed Media. *J. Appl. Phys.* **2013**, *114*, 144907. https://doi.org/10.1063/1.4824541.
33. Kliewer, K. L.; Fuchs, R. Lindhard Dielectric Functions with a Finite Electron Lifetime. *Phys. Rev.* **1969**, *181* (2), 552–558. https://doi.org/10.1103/physrev.181.552.
34. Levine, Z. H.; Louie, S. G. New Model Dielectric Function and Exchange-Correlation Potential for Semiconductors and Insulators. *Phys. Rev. B* **1982**, *25*, 6310. https://doi.org/10.1103/physrevb.25.6310.
35. Archubi, C. D.; Arista, N. R. A Comparative Study of Threshold Effects in the Energy Loss Moments of Protons, Electrons and Positrons Using Dielectric Models for Band-Gap Materials. *Eur. Phys. J. B* **2017**, *90*, 18. https://doi.org/10.1140/epjb/e2016-70637-9.
36. Tung, C.; Chen, Y.; Kwei, C.; Chou, T. Differential Cross Sections for Plasmon Excitations and Reflected Electron-Energy-Loss Spectra. *Phys. Rev. B* **1994**, *49*, 16684–16693. https://doi.org/10.1103/PhysRevB.49.16684.
37. Werner, W. S. M.; Glantschnig, K.; Ambrosch-Draxl, C. Optical Constants and Inelastic Electron-Scattering Data for 17 Elemental Metals. *J. Phys. Chem. Ref. Data* **2009**, *38*, 1013–1092. https://doi.org/10.1063/1.3243762.
38. Yubero, F.; Tougaard, S. Model for Quantitative Analysis of Reflection-Electron-Energy-Loss Spectra. *Phys. Rev. B* **1992**, *46*, 2486–2497. https://doi.org/10.1103/PhysRevB.46.2486.
39. Limandri, S. P.; Fadanelli, R. C.; Behar, M.; Nagamine, L.; Fernandez-Varea, J. M.; Abril, I.; Garcia-Molina, R.; Montanari, C.; Aguiar, J.; Mitnik, D.; Miraglia, J. E.; Arista, N. R. Stopping Cross Sections of TiO_2 for H and He Ions. *Eur. Phys. J. D* **2014**, *68*, 194. https://doi.org/10.1140/epjd/e2014-40782-6.
40. Denton, C. D.; Abril, I.; Garcia-Molina, R.; Moreno-Marín, J. C.; Heredia-Avalos, S. Influence of the Description of the Target Energy-Loss Function on the Energy Loss of Swift Projectiles. *Surf. Interface Anal.* **2008**, *40*, 1481. https://doi.org/10.1002/sia.2936.
41. Da, B.; Shinotsuka, H.; Yoshikawa, H.; Ding, Z. J.; Tanuma, S. Extended Mermin Method for Calculating the Electron Inelastic Mean Free Path. *Phys. Rev. Lett.* **2014**, *113*, 063201. https://doi.org/10.1103/physrevlett.113.063201.
42. Sabin, J. R.; Oddershede, J. Shell Corrections to Electronic Stopping Powers From Orbital Mean Excitation Energies. *Phys. Rev. A* **1982**, *26* (6), 3209–3219. https://doi.org/10.1103/physreva.26.3209.
43. Sabin, J. R.; Oddershede, J. An Analytical Representation of Shell Corrections for Stopping Power. *Int. J. Quantum Chem.* **2009**, *109* (13), 2933–2936. https://doi.org/10.1002/qua.22126.

44. Sauer, S. P. A.; Oddershede, J.; Sabin, J. Theory and Calculation of Stopping Cross Sections of Nucleobases for Swift Ions. In *Radiation Damage in Biological Systems*; Gómez-Tejedor, G. G., Fuss, M. C., Eds.; Springer: United States, 2012; pp. 191–200.
45. Lindhard, J.; Winther, A. Stopping Power of Electron Gas and Equipartition Rule. *Mat. Fys. Medd. Dan. Vid. Selsk.* **1964**, *34* (4), 1. http://gymarkiv.sdu.dk/MFM/kdvs/mfm%2030-39/mfm-34-4.pdf.
46. Lundqvist, B. I. Single-Particle Spectrum of the Degenerate Electron Gas. *Phys. Kondensierten Materie* **1967**, *6* (3), 206. https://doi.org/10.1007/bf02422717.
47. Ritchie, R. H.; Brandt, W.; Echenique, P. M. Wake Potential of Swift Ions in Solids. *Phys. Rev. B* **1976**, *14* (11), 4808–4812. https://doi.org/10.1103/physrevb.14.4808.

CHAPTER FOURTEEN

Hierarchical relaxation in frustrated systems

N.S. Sullivan[a,*], M. Pilla[b], E.B. Genio[c], J.A. Hamida[a], and K.A. Muttalib[a]
[a]Department of Physics, University of Florida, Gainesville, FL, United States
[b]Qorvo Inc., Richardson, TX, United States
[c]KETD East Qianjin Road Enterprise Science & Technology Park, Kunshan, China
*Corresponding author: e-mail address: sullivan@phys.ufl.edu

Contents

Abstract

We present a review of the role of hierarchical relaxation processes on the dynamics of geometrically frustrated systems and specifically the low temperature dynamics of the glass states of solid ortho-para hydrogen, orientational and quadrupolar glasses, and solid N_2–Ar mixtures. Comparison is made with the dynamics observed for the recently discovered Bose glass states.

1. Introduction

This contributed article is dedicated to *John R. Sabin* whose wisdom and sound common sense have been very influential on many occasions. Several years ago, researchers were struggling to understand the dynamical behavior observed in the quadrupolar glass state of solid ortho-para hydrogen mixtures for which the local ordering was that of an ensemble of quantum rotors. Jack's advice was to reduce the problem to its simplest terms and then look for the underlying physics. This advice resulted in treating the dynamics as a hierarchical process in which low energy states in a complex

Advances in Quantum Chemistry, Volume 85
ISSN 0065-3276
https://doi.org/10.1016/bs.aiq.2022.07.002
Copyright © 2022 Elsevier Inc.
All rights reserved.

configurational space must be accessed before higher energy states. This approach was able to explain a range of experimental data and was recently invoked to describe the anomalous frequency dependence of the dynamics of a frustrated Bose glass. In this review, we will sketch the progress in understanding the basic features of the relaxational dynamics of three frustrated molecular systems: the quantum rotor glass, the classical orientational glasses, and the simple Bose glass state.

2. The frustrated molecular glass

Of all the molecular glasses, the quadrupolar glass[1–12] formed at low temperatures by dilute mixtures of ortho- and para-hydrogen[13–18] is the most fascinating as the underlying physics is quantum mechanical. The homonuclear diatomic hydrogen can exist as one of two possible species, ortho and para, determined by symmetry. The molecular wave function is the product of a rotational wave function $\Phi_{orb}(J)$, determined by the angular momentum J, and a nuclear wave function $\chi_{nucl}(I)$, determined by the total nuclear spin I which can be 1 or 0. The full wave function $\Psi = \Phi_{orb}(J)\chi_{nucl}(I)$ must be totally antisymmetric. Consequently, ortho-H_2 must have odd values of J and para-H_2 even values of J, as shown in Table 1.

Because of the small moment of inertia, the rotational energies for a hydrogen molecule, $E_J = BJ(J+1)$ with $B = 85.6$ K[19] have a large separation: $E_{J=1} = 171$ K and $E_{J=2} = 510$ K compared to the ground state $J = 0$. In the solid state, we therefore need only consider $J = 1$ and $J = 0$. The anisotropic interactions that can mix the $J = 1$ and $J = 0$ states (principally electrostatic quadrupole–quadrupole interactions[20,21]) are very weak (typically ≈ 1 K) and J may be considered as a good quantum number. Furthermore, the conversion of the ortho species ($J = 1$) to the para species ($J = 0$) is doubly forbidden because two different symmetries (nuclear and rotational) must be broken simultaneously. The ortho-para conversion rate is therefore very slow ($\approx 1.9\%$ per hour[22,23]) and a solid mixture of ortho-para hydrogen will decay very slowly (over a few days) to the pure para state.

Table 1 Symmetry states of nuclear spin species of H_2.

	ortho-H_2	**para-H_2**
χ_{nucl}	Even, $I = 1$	Odd, $I = 0$
Φ_{orb}	Odd, $J = 1, 3, 5\ldots$	Even, $J = 0, 2, 4\ldots$

Only the ortho species ($J = 1$) has orientational degrees of freedom. In the absence of dipolar interactions the dipole moments J_α are expected to vanish for $\alpha = x, y, z$. The five remaining degrees of freedom are in general determined by the expectation values of the operator T_{lm}, which are the operator equivalents of the spherical harmonics Y_{lm} in the manifold $J = 1$. Five parameters are needed to fully specify all the degrees of freedom and we consider two quadrupolar order parameters

$$\sigma_i = \left(3J_z^2 - 2\right)_i \text{ and } \eta_i = \left(J_x^2 - J_y^2\right)_i \tag{1}$$

for a molecule at site i, with three principal axes x_i, y_i, z_i for the quadrupole moments.

At high ortho-H_2 concentrations, a first-order phase transition occurs from a rotationally disordered state with $\sigma_i = 0$ and $\eta_i = 0$ to an orientationally ordered state at low temperatures. In this orientationally ordered state, the molecules are arranged in an interpenetrating four-sublattice Pa3 structure for which the molecular alignment in each sublattice is along one of the diagonals of a face-centered cubic lattice. This results in a geometrically frustrated system as expected from an orientational ordering driven by electrostatic quadrupole–quadrupole interactions. (For a review see van de Bund and Ackland.[21])

The ordered state is geometrically frustrated because the geometrical alignment of nearest-neighbor molecules does not minimize the short-ranged quadrupole–quadruple interaction for an isolated pair of molecules. Diluting the ortho concentration weakens the orientational ordering and below a critical concentration of about 55%, long range ordering is lost, and the low temperature state is observed to be that of a broad distribution of local order parameters and local axes for the mean alignment at each site as illustrated in Fig. 1.

Some insight into the nature of this order–disorder transition can be obtained from simple molecular field theory.[18] If we assume axial symmetry and a given *mean* local order parameter

$$\sigma = \frac{1}{N}\sum_i \langle 3J_{zi}^2 - 2\rangle \tag{2}$$

we need to evaluate the free energy

$$F(\sigma) = E(\sigma) - TS(\sigma) \tag{3}$$

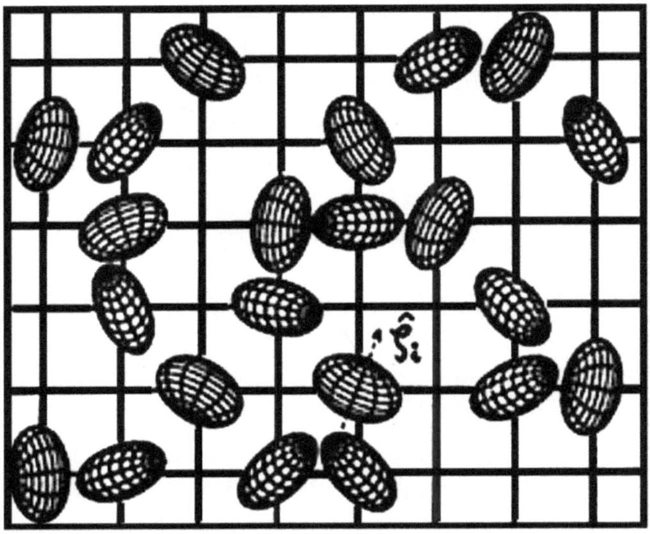

Fig. 1 Schematic representation of a quadrupolar glass as a distribution of prolate and oblate ellipsoids with random directions of principal axes. *After Fig. 4 of Sullivan, N. S.; Devoret, M.; Cowan, B. P.; Urbina, C. Evidence for Quadrupolar Glass Phases in Solid Hydrogen at Reduced Ortho Concentrations. Phys. Rev. B* ***1978****, 17 (12), 5016–5024, https://doi.org/10.1103/PhysRevB.17.5016.*

where S(σ) is the entropy for fixed σ. We then minimize $F(\sigma)$ with respect to σ to find the temperature dependence of σ.

The single particle density matrix can be written as

$$\rho = \frac{1}{3} I_3 + \sum_m T^{\dagger}_{2m} \langle T_{2m} \rangle \tag{4}$$

where I_3 is the unit 3 × 3 matrix and T_{2m} are the quadrupole operators. (The dipole terms vanish.) With the assumption of axial symmetry, ρ reduces to

$$\rho = \frac{1}{3} I_3 + T_{20} \langle T_{20} \rangle. \tag{5}$$

In the manifold $J = 1$, the ortho-normalized tensor operator is given by

$$T_{20} = \frac{1}{\sqrt{6}} (3J_z^2 - 2). \tag{6}$$

The density matrix becomes

$$\rho = \begin{bmatrix} \frac{1}{3}+\frac{\sigma}{6} & 0 & 0 \\ 0 & \frac{1}{3}-\frac{\sigma}{3} & 0 \\ 0 & 0 & \frac{1}{3}+\frac{\sigma}{6} \end{bmatrix}$$

where the rows and columns are for $J_z = 1, 0, -1$.
The entropy is given by

$$S(\sigma) = Tr(\rho \ln \rho) \tag{7}$$

which becomes

$$\begin{aligned} S(\sigma)) = & \frac{1}{3}[(2+\sigma)\ln(2+\sigma) + (1-\sigma)\ln(1-\sigma) \\ & -(2+\sigma)\ln 2 - 3\ln 3]. \end{aligned} \tag{8}$$

The intermolecular quadrupolar interaction between two molecules at sites i and j can be written as

$$\mathcal{H}_{ij}^{QQ} = \sum_{m,n} \Gamma_{ij}^{mn} T_i^{2m} T_j^{2n} \tag{9}$$

where the T^{2m} are the tensorial operators introduced previously and $\Gamma_{ij}^{m,n}$ are interaction constants that depend on the orientations of the local symmetry axes, which are well known for the Pa3 structure.

The quadrupolar interaction varies with molecule separation as r_{ij}^{-5} so we need only consider the mean energy for nearest neighbors. This energy is found to be

$$\langle \mathcal{H}_Q \rangle = 6X\gamma\sigma^2 \tag{10}$$

where γ measures the interaction strength and X is the ortho-H_2 fraction. The total free energy per particle is

$$F(\sigma) = \frac{1}{3}k_BT[(2+\sigma)\ln(2+\sigma) + (1-\sigma)\ln(1-\sigma) - \sigma\ln 2] + 6X\gamma\sigma^2. \tag{11}$$

$F(\sigma)$ is minimized for values of σ that satisfy

$$\frac{36X|\gamma|\sigma}{k_BT} = \ln\frac{(2+\sigma)}{(2-2\sigma)}. \tag{12}$$

For low temperatures the solution is $\sigma_0 = -2$, and as T increases, σ increases toward the value of -1 and then has a discontinuous jump to $\sigma = 0$ at a critical temperature

$$T_C = \frac{18|\gamma|X}{k_B \ln 2}. \tag{13}$$

While this mean-field approach demonstrates the first-order nature of the transition, it is only in qualitative agreement with the observed phase diagram (Fig. 2). It does not explain the origin of the critical concentration below which the glass state appears and the numerical values of T_c are too high.

The first attempt to provide an understanding of the physics responsible for the critical concentration was provided by Kirkwood's method[24,25] of restricted traces carried out to second order. Kirkwood's method is important for the case of quantum rotors because it includes the contributions of correlation terms of the form

$$\langle T_{2m} T_{2n} \rangle. \tag{14}$$

To first order, Kirkwood's method reproduces the mean field result. The second cumulant of Kirkwood's treatment limited to the isotropic terms m, $n = 0$ yields

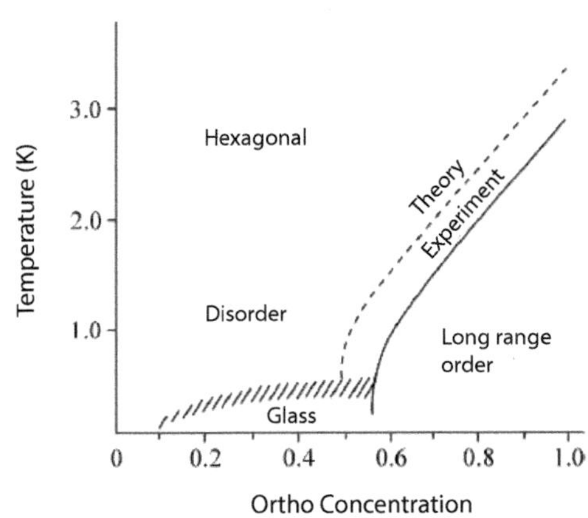

Fig. 2 Phase diagram observed for solid-ortho-para hydrogen mixtures. The hatched region indicates a smooth growth of orientational order into the quadrupolar glass state. *After Fig. 5 from Sullivan, N. S.; Hamida, J. A.; Pilla, S.; Muttalib, K. A.; Genio, E. Molecular Glasses: NMR and Dielectric Susceptibility Measurements. J. Struct. Chem.* ***2016****, 57 (2), 301–307, https://doi.org/10.1134/S0022476616020098.*

$$M_2 = \left\langle \mathcal{H}^2_{QQ} \right\rangle - \langle \mathcal{H}_{QQ} \rangle^2 \\ = 6N X^2 \gamma^2 (2 - \sigma - \sigma^2)[2 - \sigma - (2X - 1)\sigma^2]. \tag{15}$$

From this expression, we see that the correlation term changes sign for $X < 0.5$. Detailed calculations that include the anisotropic terms[25] put the critical concentration closer to 55%.

One can use a simple model of a broad distribution of interaction strengths to describe the quadrupolar glass state.[18] Consider the truncated version of Eq. (9)

$$\mathcal{H}_{ij} = \sum_{ij} \Gamma_{ij} T_{20}(i) T_{20}(j) \tag{16}$$

where there is a Gaussian distribution of possible values for the Γ_{ij} given by[26]

$$P(\Gamma_{ij}) = \left(\frac{1}{2\pi \widetilde{\Gamma}^2} \right)^{1/2} \exp\left[-(\Gamma_{ij} - \Gamma_0)^2 / 2\widetilde{\Gamma}^2 \right]. \tag{17}$$

Γ_0 is the mean value of the interaction and $\widetilde{\Gamma}^2$ is the variance. One can now follow the technique of Edwards and Anderson to calculate the free energy by taking an average over an ensemble of n replicas.

$$F = -k_b T \lim_{n \to 0} \left[\frac{1}{n} (\overline{Z^n} - 1) \right]. \tag{18}$$

The quadrupolar order parameter is defined as

$$Q = \overline{\langle (T_{20})_\mu \rangle \langle (T_{20})_\nu \rangle}. \tag{19}$$

μ and ν refer to different replicas and the bar represents a configurational average. The local molecular alignment is $\sigma = \overline{\langle T_{20} \rangle}$.

Minimizing F with respect to Q and S, yields

$$\sigma = \left(\frac{1}{2\pi} \right)^{1/2} \int e^{\frac{1}{2}y^2} F(\xi) dy \tag{20}$$

and

$$Q = 2 - \sigma - \left(\frac{1}{2\pi} \right)^{1/2} \int e^{\frac{1}{2}y^2} F'(\xi) dy \tag{21}$$

with

$$F(x) = \frac{2e^x - 2e^{-2x}}{2e^x + e^{-2x}} \tag{22}$$

and

$$\xi = \frac{1}{k_B T}\left[\widetilde{\Gamma}(zQ)^{1/2}y + \Gamma_0 z\sigma\right] \tag{23}$$

where z is the number of nearest neighbors.

If $\widetilde{\Gamma} = 0$, one recovers the mean field results

$$Q = 2 - \sigma \tag{24}$$

and

$$k_b T_c = 36\Gamma_0 / \left[\ln \frac{2 + \sigma}{2 + 2\sigma}\right]. \tag{25}$$

On the other hand, for $\widetilde{\Gamma} \gg \Gamma_0$

$$Q = \frac{5}{2} - \frac{27}{2} e^{B^2} erfc(B) \tag{26}$$

with

$$B = 3\widetilde{\Gamma}(zQ/2)^{1/2}. \tag{27}$$

On expanding the error function, one finds a nontrivial Q for

$$k_b T_g = \frac{5}{27}\sqrt{\left(\frac{5\pi z}{3}\right)}\widetilde{\Gamma} = 0.4z^{1/2}\widetilde{\Gamma}. \tag{28}$$

On estimating $\widetilde{\Gamma}$ by averaging over the angle dependent constants in the electrostatic quadrupole–quadrupole interaction constant one finds for the estimated glass transition[18]

$$T_g = 0.7x^{1/2} \tag{29}$$

which is quite close to the reported values.

The phase diagram was deduced from NMR studies[15,16,27,28] where the NMR line shapes provide a direct measurement of the local order parameters. This property results from the intramolecular nuclear dipole–dipole interaction which in the manifold $J = 1$ can be written as

$$\mathcal{H}_{DD} = hD(3I_Z^2 - 2)\sigma_i(3\cos^2\theta_i - 1) \tag{30}$$

where D is a constant, σ_i is the molecular order parameter, and (θ_i, ϕ_i) define the polar angles for the orientation of the applied magnetic field with respect to the local symmetry axes. (For simplicity we assume axial symmetry with $\eta_i = 0$ for all i.) Each molecule contributes a doublet to the NMR spectrum at frequencies

$$\Delta f_i = \pm 3D\sigma_i(3\cos^2\theta_i - 1). \tag{31}$$

For a fixed value of σ_i one can sum over the polar angles for a powder sample and one obtains the familiar Pake doublet line shape with two pronounced peaks separated by $6D\sigma_i$. The NMR results were very clear in showing that *both* the local order parameters σ_i and the local axes vary widely in the glass state as represented schematically in Fig. 1. At low temperatures ($T < 0.2$ K) and low ortho concentrations ($X < 0.4$), the observed NMR line shapes correspond approximately to a linear probability distribution of order parameters

$$P(\sigma) \propto \sigma. \tag{32}$$

This linear dependence at small σ is expected since the expression for the entropy is quadratic in σ for small values (Fig. 3).

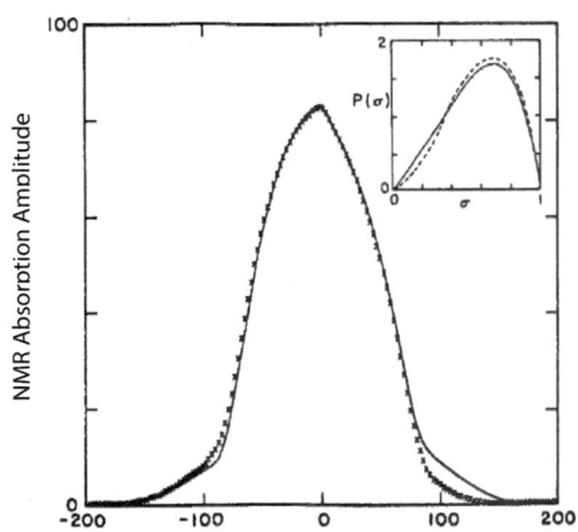

Fig. 3 Fit to the observed NMR line shape at very low temperatures ($T = 44$ mK) for a distribution $P(\sigma)$ of local order parameters σ. There is a small but distinct asymmetry to the line shape and this is due to the nonnegligible nuclear spin polarization (≈6%) at these temperatures. *After Fig. 8 from Edwards, C. M.; Zhou, D.; Lin, Y.; Sullivan, N. S. Local Ordering in Dilute Ortho-Para-Hydrogen Mixtures at Low Temperatures. J. Low Temp. Phys.* ***1988****, 72 (1/2), 1–24, https://doi.org/10.1007/BF00681725.*

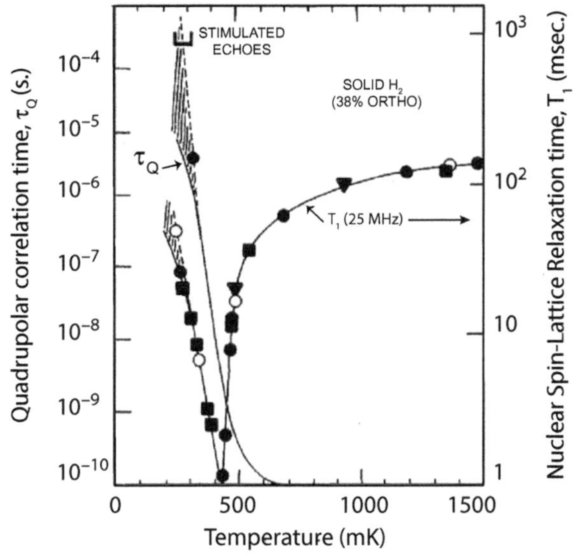

Fig. 4 Rapid variation of molecular relaxation dynamics in quadrupolar glass state of ortho-para hydrogen mixtures, near the transition to the glass state. *After Fig. 1 from Sullivan, N. S.; Estéve, D. Critical Slowing-Down in Spin Glasses: Quadrupolar Glass Phase of Solid H_2. Physica B+C* ***1981****, 107 (1), 189–190, https://doi.org/10.1016/0378-4363(81)90400-9.*

The hashed area in Fig. 2 designates a smooth evolution of the local order parameters from the disordered state to the glass state. There is no sharp discontinuity such as the cusp seen in dipolar spin glasses. There is, however, a very rapid change in the molecular dynamics in the neighborhood of the local glass ordering as shown in Fig. 4.[27,29] This behavior shows that the glass freezing is undoubtedly collective in nature and not simply a para-orientational freezing in fixed local crystal fields.

3. Hierarchical dynamics: Quantum rotor glass

With the continuous evolution of the local order parameters as samples of solid ortho-para hydrogen are cooled into the glass state, the challenge has been to understand the origin of the rapid change in dynamics over a relatively small temperature change. This challenge was addressed by Lin et al.[30,31] who invoked the behavior of the system in the presence of a complex configurational space with a very large number of energy states.

The most striking observation was how the decay of NMR stimulated echoes behaved in the glass state.[32] The decay of these echoes are determined by the spectral density of slow rotational motions (as opposed to the NMR spin-lattice relaxation times which depend on the fluctuations at the nuclear Larmor frequency).

The observed logarithmic decay (see Fig. 5) follows directly from the scaling model of Fisher and Huse.[33,34] The long-time correlations are dominated by the low energy excitations of coherently re-oriented molecules in clusters (or droplets). If E_B is the free energy barrier of a cluster, the tunneling rate to overcome a given barrier at temperature T is

$$W(E_B) = W_0 \exp\left(-\frac{E_B}{k_B T}\right) \tag{33}$$

where W_0 is the characteristic attempt frequency for a given cluster. In the limit of long times (which is the experimental case), W_0 is well defined because it depends on the cluster size. In hierarchical relaxation, the low energy barriers must be crossed first, before the larger barriers can be accessed. In a time t, the energy barriers overcome will therefore be limited to the range

$$0 < E_B < E_{max}(t) \quad \text{where} \quad E_{max}(t) = k_B T \ln(W_0 t). \tag{34}$$

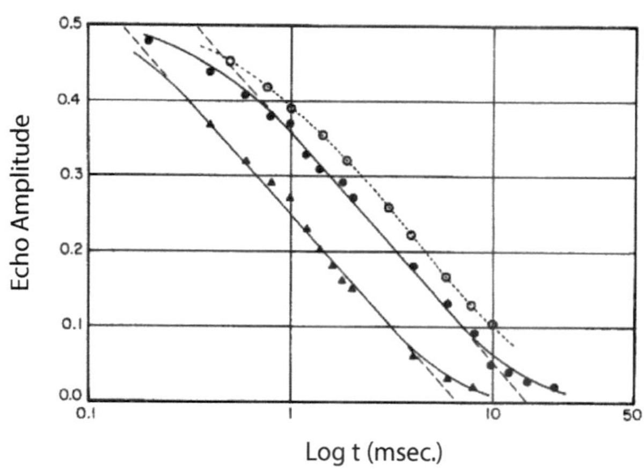

Fig. 5 Logarithmic decay of stimulated echoes in the glass state. The different symbols refer to different pulse periods τ and different samples (*solid circles:* $\tau = 1.7D^{-1}$; *triangles:* $\tau = 3.5D^{-1}, X = 0.54, T = 0.22$ K; *open circles:* $\tau = 7.7D^{-1}, X = 0.43, T = 0.15$ K.). *After Fig. 1 from Lin, Y.; Sullivan, N. S. Low-Frequency Dynamics of Orientational Glasses. Phys. Rev. B* ***1988****, 38 (7), 5158–5161, https://doi.org/10.1103/Phys-RevB.38.5158.*

When the barriers are crossed, the local order parameters change significantly and those clusters will therefore not contribute to the NMR stimulated echoes. As a result, the motion associated with barrier crossing erases the orientational *memory* of those clusters. In this scenario, the amplitude of the stimulated echo will therefore be given by

$$A(t) = C\left[1 - \int_0^{E_{max}(t)} \rho(E_B)dE_B\right] \tag{35}$$

where C is a constant dependent on the NMR parameters only, and $\rho(E_B)$ is density of states for rotational energy barriers. At low temperatures, only states with $E_B \approx 0$ need be considered and we have

$$A(t) = C\left[1 - k_B T\rho(0)\ln\left(\frac{t}{t_0}\right)\right] \tag{36}$$

where $t_0 = W_0^{-1}$.

This calculated logarithmic time dependence is not only in good agreement with the NMR results, but it also allows the comparison of the values deduced for the density of states $\rho(0)$ with those deduced from measurements of the heat capacities at low temperatures. For the local quantum rotors $J = 1$, the $J_z = \pm 1$ states are separated from the state $J_z = 0$ by a gap E_B. The broad distribution of low energy states $\rho(E_B)$ therefore leads to a heat capacity given by[14]

$$\frac{C_v}{NR} = \frac{2}{3}k_b TX\int_0^{\frac{3\Delta_0}{k_B T}} \frac{u^2 du}{(2e^{-u/2} + e^{u/2})^2}\rho(E) \tag{37}$$

where $u = \frac{3E}{k_B T}$. Fig. 6 shows an excellent agreement of the experimental measurements with the calculated values. At the lowest energies, the density of States $\rho(0) \approx 0.55$ which leads to a predicted stimulated echo decay of

$$A(T)^{calc} = const - 0.28\log_{10}\left(\frac{t}{t_0}\right) \tag{38}$$

which is in good quantitative agreement with the observed decay

$$A(T)^{obs} = const - 0.30\log_{10}\left(\frac{t}{t_0}\right) \tag{39}$$

for a sample with $X = 0.54$ at $T = 0.22$ K.[30–32]

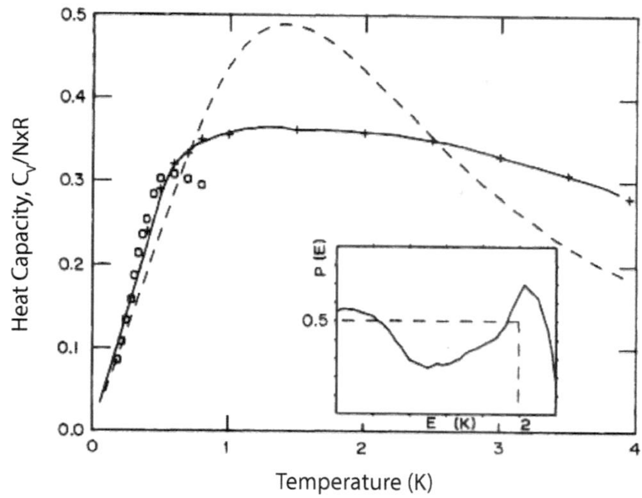

Fig. 6 Comparison of observed and calculated temperature variation of the heat capacity of solid ortho-para H_2 quadrupolar glass state. The symbols refer to experimental data (*crosses*: Engelsberg and Borges,[35] *circles*: Haase et al.[36]). The *broken line* is the prediction for a constant density of states and the *solid line* is the best fit as shown in the inset. *After Fig. 2 from Lin, Y.; Sullivan, N. S. Low-Frequency Dynamics of Orientational Glasses. Phys. Rev. B* ***1988****, 38 (7), 5158–5161, https://doi.org/10.1103/Phys-RevB.38.5158.*

4. Classical quadrupolar glasses

In addition to the quantum rotor glasses formed by solid ortho-para H_2 mixtures at low temperatures, classical rotors in the form of N_2 molecules also transit to orientational glass states at low temperatures when sufficiently diluted with argon.[37–43] A phase diagram very similar to that established for dilute ortho-para H_2 mixtures is observed in solid N_2–Ar mixtures. For high N_2 concentrations ($X > 77\%$), a first-order transition to a Pa3 structure occurs, driven by the quadrupole–quadrupole interactions between N_2 molecules. At lower N_2 concentrations the low temperature state is characterized by a broad distribution of local order parameters (as inferred from NMR line shapes of $^{15}N_2$ molecules) and the crystal lattice structure remains hcp.

At very low N_2 concentrations, $42\% < X < 57\%$, the lattice structure at low temperatures remains fcc and instead of a random distribution of local order parameters, the NMR line shapes are consistent with an approximately fixed value for the local orientational order parameter with the local symmetry axes remaining broadly distributed. The overall phase diagram is shown in Fig. 7.

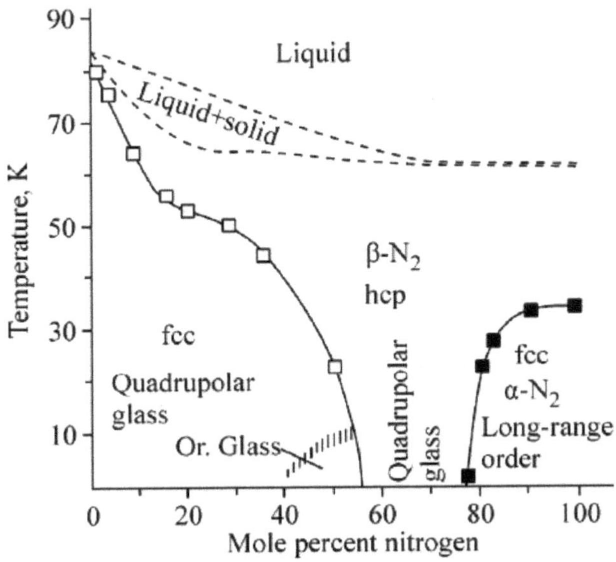

Fig. 7 Phase diagram for solid N_2–Ar mixtures. *After Fig. 4 from Pilla, S.; Hamida, J. A.; Muttalib, K. A.; Sullivan, N. S. Molecular Solid Glasses: New Insights Into Frustrated Systems. New J. Phys.* ***2001****, 3, 17, https://doi.org/10.1088/1367-2630/3/1/317.*

There is special interest in studying the local glass states in sold N_2–Ar mixtures because the order parameters can be probed at very low frequencies using high dielectric measurements of the dielectric susceptibility. This was carried out by Pilla et al.[41,44] using low frequency AC susceptibility measurements. The low frequency dynamics revealed by these measurements exhibit distinct hysteresis loops as shown in Fig. 8. In the long range ordered Pa3 phase, the hysteresis loops are closed and the glass state occurs in a narrow region between $X = 56\%$ and $X = 77\%$. Below $X = 56\%$ the hysteresis loops remain open, indicating that the glass state persists down to the lowest temperatures explored.

In the glass state one can apply a generalized fluctuation-dissipation theorem assuming a model of simple replica symmetry breaking.[45,46] One result of this theorem is that below the glass transition temperature T_G, the area of the hysteresis loop scales as

$$A(\Delta T) \propto \left(\frac{\Delta T}{T_G}\right)^2. \tag{40}$$

Pilla et al.[41,44] showed that the experimental results were in good agreement with this prediction.

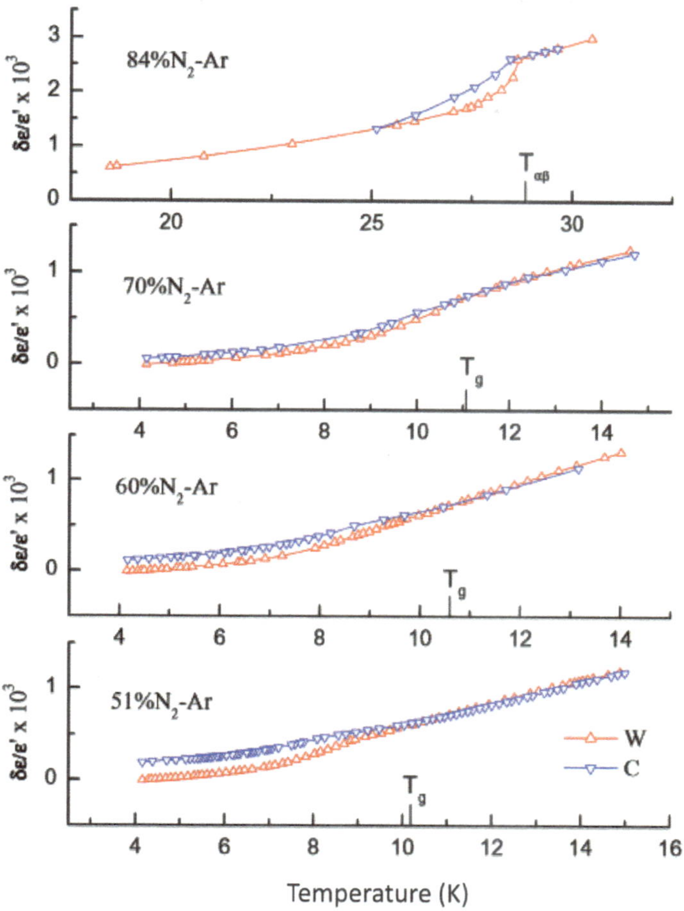

Fig. 8 Evidence for hysteresis in N_2–Ar solid mixtures from measurements of the electric susceptibility ϵ; *W*, warming; *C*, cooling. *After Fig. 1 from Pilla, S.; Hamida, J. A.; Muttalib, K. A.; Sullivan, N. S. Dielectric Response of N_2-Ar Solid Solutions in the Audio Frequency Range. Phys. Rev. B* ***2003****, 67 (17), 174204, https://doi.org/10.1103/PhysRevB.67.174204.*

5. Frustrated Bose glass dynamics

Dichloro-tetrakis thiourea-nickel [$NiCl_2$-4SC(NH_2)$_2$], often referred to as DTN, is a simple quantum molecular magnet exhibiting frustration. It forms a Bose glass state in the presence of disorder which occurs when a small percentage of Cl are replaced with Br.[47,48]

Pure DTN has a body-centered tetragonal crystal structure with the Ni-Cl-Cl-Ni linkage parallel to the *c*-axis. Thiourea molecules in the ab-plane are linked via hydrogen bonding and are polarizable. The Ni atoms form two interpenetrating tetragonal sub-lattices. Because of high ion anisotropy of the form DS_Z^2, the $S_Z = \pm 1$ states are separated from the $S_Z = 0$ state. Applying a strong magnetic field lowers the energy of the $S_z = +1$ state and when it reaches the energy of the $S_Z = 0$ state an anti-ferromagnetic ordering in the ab-plane occurs.

The magnetic properties of the Ni ions can be described as an ensemble of interacting bosons with $S^- = (2-n)b$, $S^+ = (2-n)b^\dagger$ and $S_Z = n - 1$, with $n = bb^\dagger$. b, $b^\dagger$ are triplet-boson operators and n can be 0,1, or 2. The bosonic interaction Hamiltonian is

$$\begin{aligned}\mathcal{H} &= -\sum_{(ij),c} \sqrt{1-\frac{n_i}{2}} b_i b_j^\dagger \sqrt{1-\frac{n_j}{2}} + h.c. \\ &+ \sum_{(ij),ab} \sqrt{1-\frac{n_i}{2}} b_i b_j^\dagger \sqrt{1-\frac{n_j}{2}} + h.c. \\ &+ \sum_i D_i (n_i - 1)^2 - g\mu_B H \sum_i n_i.\end{aligned} \tag{41}$$

In zero field the large value of D leads to a quantum paramagnetic state, corresponding to a Mott insulator with $n = 1$ particles per Ni atom. The energy gap between the $S_z = 0$ state and the next state is

$$\Delta = D - 2J_c - 4J_{ab}. \tag{42}$$

Applying a magnetic field of strength

$$H_{c1} = \Delta/\mu g \approx 2.1\text{T} \tag{43}$$

closes the gap. Increasing the field beyond H_{c1} leads to an increase in the number of bosons and it is these bosons that condense into a magnetic Bose–Einstein condensate. As the applied field is further increased, the spins cant away from the ab-plane and become saturated at about 12T.

When disorder is added by randomly replacing the Cl linker spins with larger Br spins, a Bose glass[49] is created. The observed phase diagram for 8% Br replacement of the Cl linkers is shown in Fig. 9

In the Bose condensation of the Ni spin excitations in DTN, one observes a strong magneto-electric effect.[50,51] The surprising feature of the Bose glass state is that the observed magneto-electric effect (although

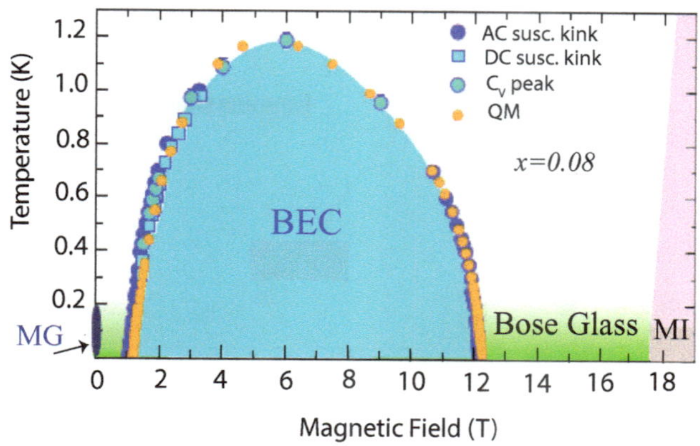

Fig. 9 Phase Diagram for the BEC and Bose glass states for Br (8%) doped DTN. *After Fig. 1 from Yin, L.; Xia, J. -S.; Sullivan, N. S.; Fry, J. N.; Cheng, H. -P.; Yazback, M.; Zapf, V. S.; Paduan-Filho, A. Anomalous Frequency Dependence of Magneto-Electric Effect in Doped DTN. Physica B* ***2021****, 608, 412875, https://doi.org/10.1016/j.physb.2021.412875.*

showing the expected quadratic dependence as a function of applied magnetic field) exhibited a strong peak as a function of frequency near 500 Hz followed by a power law decay as shown in Fig. 10.

As with the case of the frustrated quantum spins ($J = 1$) in the orientational glasses, the frustrated spins in the Bose glass state where there is also disorder, needs to be approached using hierarchical relaxation dynamics. The magnetic susceptibility is measured with an AC electric field applied at frequency f and the change in electric polarization is determined from the impedance change ΔZ in a carefully balanced capacitance bridge.[50] The impedance change at the bridge is

$$\Delta Z = -\frac{1}{2\pi f}\frac{\Delta C}{C_0^2} \tag{44}$$

where ΔC is the change in capacitance due to the sample's dielectric susceptibility and C_0 is the sample cell capacitance. The observed signal from the bridge (reported in Yin et al.[50]) is therefore

$$S = -Kf^{-1}\delta\chi(f) \tag{45}$$

where $\delta\chi(f)$ is the change in susceptibility of the sample for an excitation frequency f, and K is a constant that depends on the sample cell construction.

For hierarchical relaxation, the energy barriers crossed in a time $t = 1/f$ will satisfy

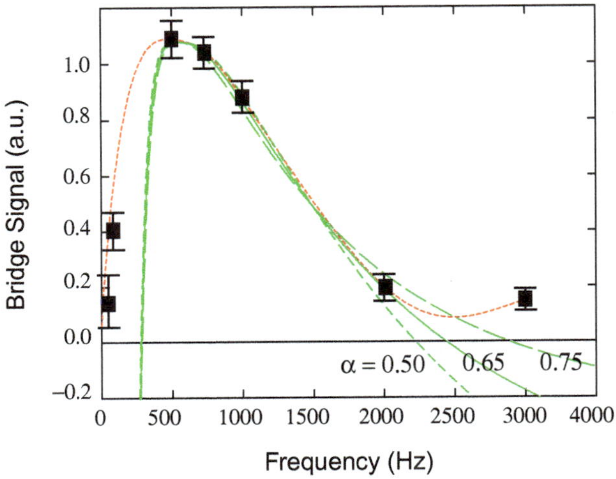

Fig. 10 Anomalous frequency dependence of the magneto-electric effect in the Bose glass state of DTN-Br. The *solid squares* are experimental points. The *dashed, solid,* and *broken green lines* are for different power laws with exponent $\alpha = 0.50, 0.65$, and 0.75, respectively. *After Fig. 3 in Yin, L.; Xia, J. -S.; Sullivan, N. S.; Fry, J. N.; Cheng, H. -P.; Yazback, M.; Zapf, V. S.; Paduan-Filho, A. Anomalous Frequency Dependence of Magneto-Electric Effect in Doped DTN. Physica B* ***2021****, 608, 412875, https://doi.org/10.1016/j.physb.2021.412875.*

$$0 < E < E_{max}(t) \tag{46}$$

where

$$E_{max}(t) = k_B T \ln(\Gamma_0 t). \tag{47}$$

Γ_0 is the attempt rate to cross energy barriers and needs to be calculated from the spin–spin interactions in the frustrated state. After the barriers are traversed, the electric polarization beforehand or the memory of the previous states will be erased. The resulting change in electric susceptibility will be

$$\Delta\chi = G\int_0^{E_{max}} \rho(E)dE. \tag{48}$$

Here $\rho(E)$ is the density of spin excitations and G is a constant. Assuming that the density of states is constant at low temperatures (as is the usual case for spin glasses), the frequency dependence of the change in susceptibility is then predicted to be

$$\Delta\chi = k_B T G \ln\left[\frac{\Gamma_0}{f}\right]. \tag{49}$$

The signal recorded by the bridge circuit will then be

$$\Delta Z = -G\frac{k_B T}{f}\left|\ln\frac{f}{\Gamma_0}\right|. \tag{50}$$

While this frequency dependence provided a qualitative fit to the experimental data, it does not account for the scaling arguments of Thill and Huse,[52] following the pioneering study of Fisher and Huse.[34] Those scaling arguments showed that the low energy excitations of Ising-like spin glass states that dominate the long-time correlations consist of clusters of "flipped" spins or "droplets" that scale with length as $L^{-\theta}$ where $0 < \theta \leq (d-1)/2$ where d is the effective dimensionality of the excitation system ($d = 2$ for DTN). Fisher and Huse[33,34] considered barrier heights to scale as $B \approx \Delta L^{\Psi}$ and they found that the spin–spin autocorrelation functions scaled as

$$C \approx T\left[\frac{\Delta}{T\ln(t/t_0)}\right]^{\frac{\theta}{\Psi}} \tag{51}$$

where $\theta \leq \Psi \leq 2$. As a result, the frequency-dependence expected for the magneto-electric measurement scales as

$$C(\omega) \approx \frac{k_b T}{\omega}\left|\ln\frac{\omega}{2\pi\Gamma_0}\right|^{\alpha} \tag{52}$$

where the exponent $0 < \alpha < 1$. In Fig. 10, we see that the experimental data is best described by an exponent $\alpha = 0.65$.

The power law behavior with exponent $\alpha < 1$ has also been observed in ordinary spin glasses. Joh et al.[53] showed that the dynamical correlation length for CuMn and $CdCr_{1.7}In_{0.3}S_4$ obeyed the form

$$\xi(t,T) \propto \left[\left(\frac{T}{T_G}\right)\ln(t/t_0)\right]^{1/\Psi} \tag{53}$$

where $1/t_0$ is an attempt frequency to cross energy barriers and T_G is the spin glass temperature. By varying the applied magnetic field, they showed that the states at the barrier height depended on the waiting time t_W with

$$\Delta(t_W, T) = k_B T\ln(t_W/\tau_0). \tag{54}$$

Kisker et al.[54] also reported a similar scaling law for the off-equilibrium dynamics of finite-dimensional spin-glass models.

6. Conclusion

We have considered a wide range of disparate physical systems that have a common feature: the interacting elements (electric or magnetic dipoles, or quadrupoles) are highly frustrated and in the presence of disorder, the landscape of different energy states is a complex array of energy minima and maxima. While some very general and approximate mean field treatments can explain the observed phase diagrams semi-quantitatively, the variation of the dynamics of the system, and in particular the changes as local ordering sets in, can be challenging. In this review we showed how taking Jack Sabin's principle of keeping the underlying driving physics as simple as possible, one can resolve the dynamical behavior and show how it is connected to thermodynamic properties such as the heat capacity. The underlying theme was to invoke hierarchical dynamics as the systems move through configuration space as a function of temperature: low energy barriers must be overcome before attempting to cross high energy barriers. Following this theme, we were able to understand the peculiar logarithmic decay of NMR echoes in the quadrupolar glass state of solid ortho-para hydrogen mixtures and their relation to the heat capacity of those glassy systems. One was also able to understand the behavior of the hysteresis observed for the temperature dependence of the electric susceptibility in solid N_2–Ar mixtures and its relation to replica symmetry breaking in classical orientational or quadrupolar glasses such as solid N_2–Ar mixtures. Finally the anomalous dynamics reported for Bose glass states of the quantum magnet DTN (when diluted with Br) could also be understood in terms of hierarchical relaxation.

Acknowledgments

This work was supported as part of the University of Florida's Center for Molecular Magnetic Quantum Materials, an Energy Frontier Research Center funded by the U.S. Department of Energy, Office of Science, Basic Energy Sciences Program under Award # DE-SC0019330.

References

1. Binder, K. Quadrupolar Glasses. In *Spin Glasses and Random Fields*; Young, A. P., Ed.; World Scientific: Oxford, 1997; pp. 99–108 (Chapter 4).
2. Michel, K. H. Quadrupolar Glasses. *Ferroelectrics* **1988**, *78* (1), 183–189. https://doi.org/10.1080/00150198808215904.
3. Novikov, S. V. Quadrupolar Glass as a Model for Charge Carrier Transport in Nonpolar Organic Materials. *Ann. Phys.* **2009**, *18* (12), 954–958. https://doi.org/10.1002/andp.200910395.

4. Tadic, B.; Pirc, R.; Blinc, R. Local Polarization Distribution in Quadrupolar Glasses. *Phys. Rev. B* **1997**, *55* (2), 816–823. https://doi.org/10.1103/PhysRevB.55.816.
5. Hu, Z.; Garland, C. W.; Wells, A. Elastic Behavior of $K_xNa_{(1-x)}CN$ Quadrupolar Glasses. *Phys. Rev. B* **1989**, *40* (8), 5757–5760. https://doi.org/10.1103/PhysRevB.40.5757.
6. Carmesin, H.-O.; Binder, K. Isotropic Edwards-Anderson Models for Quadrupolar Glasses: A Monte Carlo Simulation. *Europhys. Lett.* **1987**, *4* (3), 269–274. https://doi.org/10.1209/0295-5075/4/3/003.
7. Kokshenev, V. B. Orientational Quadrupolar Glass in Ortho-Para Hydrogen Mixtures. I. Mean-Field Microscopic Theory. *J. Low Temp. Phys.* **1996**, *104*, 1–23. https://doi.org/10.1007/BF00754087.
8. Gribova, N. V.; Tareyeva, E. E. Replica Symmetry Breaking in an Axial Model of Quadrupolar Glass. *Theor. Math. Phys.* **2002**, *131* (3), 852. https://doi.org/10.1023/A:1015983725317.
9. Schelkacheva, T. I. Quadrupolar Glass State in Para-Hydrogen and Ortho-Deuterium Under Pressure. *JETP Lett.* **2002**, *76*, 374–376. https://doi.org/10.1134/1.1525040.
10. Sellar, J. R.; Jeffrey, R.; Chen, C. Y. Metastability, Tetragonality and Quadrupolar Glass States in Cubic Stabilized Zirconia. *J. Am. Ceram. Soc.* **2020**, *103* (10), 5546–5553. https://doi.org/10.1111/jace.17334.
11. Müser, M. H. Computer Simulation of a Quadrupolar Glass. *J. Phys. Condens. Matter* **1996**, *8*, 913–925. https://doi.org/10.1088/0953-8984/8/8/004.
12. Loidl, A. Orientational Glasses. *Annu. Rev. Phys. Chem.* **1989**, *40* (1), 29–60. https://doi.org/10.1146/annurev.pc.40.100189.000333.
13. Mancini, F. P.; Sherrington, D. One-Step Replica Symmetry Breaking Solution of the Quadrupolar Glass Model. *J. Phys. A Math. Gen.* **2006**, *39* (43), 13393–13403. https://doi.org/10.1088/0305-4470/39/43/001.
14. Candela, D.; Buchman, S.; Vetterling, W. T.; Pound, R. V. NMR Lineshapes of the Solid Hydrogen Quadrupolar Glass. *Physica B+C* **1981**, *107* (1), 187–188. https://doi.org/10.1016/0378-4363(81)90399-5.
15. Sullivan, N. S.; Pound, R. V. Nuclear-Spin-Lattice Relaxation of Solid Hydrogen at Low Temperatures. *Phys. Rev. A* **1972**, *6* (3), 1102–1107. https://doi.org/10.1103/PhysRevA.6.1102.
16. Sullivan, N. S.; Devoret, M.; Cowan, B. P.; Urbina, C. Evidence for Quadrupolar Glass Phases in Solid Hydrogen at Reduced Ortho Concentrations. *Phys. Rev. B* **1978**, *17* (12), 5016–5024. https://doi.org/10.1103/PhysRevB.17.5016.
17. Lukierska-Walasek, K.; Walasek, K. Effect of the Zero-Point Rotational Motions on the Quadrupolar Glass Order Parameter in Solid Hydrogen. *Phys. Rev. B* **1994**, *50* (17), 12437–12440. https://doi.org/10.1103/PhysRevB.50.12437.
18. Sullivan, N. S. Orientational Order-Disorder Transitions in Solid Hydrogen. *Can. J. Chem.* **1988**, *66*, 908–914. https://doi.org/10.1139/v88-154.
19. Silvera, I. F. The Solid Molecular Hydrogens in the Condensed Phase: Fundamentals and Static Properties. *Rev. Mod. Phys.* **1980**, *52* (2), 393–452. https://doi.org/10.1103/RevModPhys.52.393.
20. Harris, A. B. Interactions Between Molecules in Solid Hydrogen. *Phys. Rev. B* **1970**, *1* (5), 1881–1902. https://doi.org/10.1103/PhysRevB.1.1881.
21. van de Bund, S.; Ackland, G. J. Quadrupole Arrangements and the Ground State of Solid Hydrogen. *Phys. Rev. B* **2020**, *101* (1), 014103. https://doi.org/10.1103/PhysRevB.101.014103.
22. Ahlers, G. Ortho-Para Conversion Rate in Solid Hydrogen. *J. Chem. Phys.* **1964**, *40*, 3123.
23. Berlinsky, A. J.; Hardy, W. N. Theory of Ortho-Para Conversion and its Effect on the NMR Spectrum of Ordered Solid Ortho-Hydrogen. *Phys. Rev. B* **1973**, *8* (11), 5013–5027. https://doi.org/10.1103/PhysRevB.8.5013.

24. Kirkwood, J. G. Statistical Mechanics of Fluid Mixtures. *J. Chem. Phys.* **1935**, *3* (5), 300–313. https://doi.org/10.1063/1.1749657.
25. Sullivan, N. S. Orientational Ordering in Solid hydrogen. *J. Phys.* **1976**, *37*, 981. https://doi.org/10.1051/jphys:01976003707-8098100.
26. Tareyeva, E. E.; Schelkacheva, T. I.; Chtchelkatchev, N. M. Orientational Glass: Full Replica Symmetry Breaking in Generalized Spin Glass-Like Models Without Reflection Symmetry. *Phys. Lett. A* **2013**, *377* (7), 507–512. https://doi.org/10.1016/j.physleta.2012.12.023.
27. Ishimoto, H.; Nagamine, K.; Kimura, Y.; Kumagai, H. Measurements of Nuclear Spin-Lattice Relaxation Time of Solid Hydrogen in the Ordered State. *J. Phys. Soc. Jpn.* **1974**, *37* (4), 956–960. https://doi.org/10.1143/JPSJ.37.956.
28. Cochran, W. T.; Gaines, J. R.; McCall, R. P.; Sokol, P. E.; Patton, B. R. Study of the Quadrupolar-Glass Region in Solid D_2 Via Proton Magnetic Resonance. *Phys. Rev. Lett.* **1980**, *45* (19), 1576–1580. https://doi.org/10.1103/PhysRevLett.45.1576.
29. Sullivan, N. S.; Estève, D. Critical Slowing-Down in Spin Glasses: Quadrupolar Glass Phase of Solid H_2. *Physica B+C* **1981**, *107* (1), 189–190. https://doi.org/10.1016/0378-4363(81)90400-9.
30. Lin, Y.; Sullivan, N. S. Low-Frequency Dynamics of Orientational Glasses. *Phys. Rev. B* **1988**, *38* (7), 5158–5161. https://doi.org/10.1103/PhysRevB.38.5158.
31. Sullivan, N. S.; Lin, Y. Relaxation and Large-Scale Excitations in the Orientational Glass State of Solid H_2. *Chem. Phys. Lett.* **1989**, *156* (2), 218–222. https://doi.org/10.1016/S0009-2614(89)87123-4.
32. Sullivan, N. S.; Esteve, D.; Devoret, M. NMR Pulse Studies of Molecular Solids: $^{15}N_2$ and H_2. II. Stimulated Echoes and Slow Rotational Motions. *J. Phys. C Solid State Phys.* **1982**, *15* (23), 4895–4911. https://doi.org/10.1088/0022-3719/15/23/024.
33. Fisher, D. S.; Huse, D. A. Nonequilibrium Dynamics of Spin Glasses. *Phys. Rev. B* **1988**, *38* (1), 373–385. https://doi.org/10.1103/PhysRevB.38.373.
34. Fisher, D. S.; Huse, D. A. Ordered Phase of Short-Range Ising Spin-Glasses. *Phys. Rev. Lett.* **1986**, *56* (15), 1601–1604. https://doi.org/10.1103/PhysRevLett.56.1601.
35. Engelsberg, M.; Borges, N. M. Nuclear Spin Stimulated Echoes in Glasses. *J. Phys. C Solid State Phys.* **1984**, *17* (20), 3633–3644. https://doi.org/10.1088/0022-3719/17/20/014.
36. Haase, D. G.; Perrell, L. R.; Saleh, A. M. Specific Heat, Pressure, and the Grüneisen Relation in Solid Hydrogen. *J. Low Temp. Phys.* **1984**, *55* (3), 283–296. https://doi.org/10.1007/BF00683447.
37. Hamida, J. A.; Genio, E. B.; Sullivan, N. S. NMR Studies of the Orientational Ordering in Dilute Solid N_2-Ar Mixtures. *J. Low Temp. Phys.* **1996**, *103* (1), 49–70. https://doi.org/10.1007/BF00754657.
38. Klee, H.; Carmesin, H. O.; Knorr, K. Quadrupolar Glass State of $(Ar)_{1-x}(N_2)_x$. *Phys. Rev. Lett.* **1988**, *61* (16), 1855–1858. https://doi.org/10.1103/PhysRevLett.61.1855.
39. Hamida, J. A.; Pilla, S.; Sullivan, N. S. The Orientational Ordering in Dilute Solid N_2-Ar Mixtures. *J. Low Temp. Phys.* **1998**, *111*, 365–370. https://doi.org/10.1023/A:1022287720183.
40. Manzhelii, V. G.; Bagatskii, M. I.; Minchina, I. Y.; Aleksandrovskii, A. N. Thermodynamic Properties of Molecular Orientational Glasses With Indirect Interaction. *J. Low Temp. Phys.* **1998**, *111* (3), 257–270. https://doi.org/10.1023/A:1022219232478.
41. Pilla, S.; Hamida, J. A.; Muttalib, K. A.; Sullivan, N. S. Molecular Solid Glasses: New Insights Into Frustrated Systems. *New J. Phys.* **2001**, *3*, 17. https://doi.org/10.1088/1367-2630/3/1/317.

42. Suan Li, M.; Gawron, T. R.; Cieplak, M. Scaling Properties of Quadrupolar and Octupolar Glasses. *Phys. Lett. A* **1992**, *164* (1), 120–125. https://doi.org/10.1016/0375-9601(92)90917-B.
43. Müser, M. H.; Löding, D.; Nielaba, P.; Binder, K. Computer Simulations of a Lennard-Jones Model for $Ar_{(1-x)}$ $(N_2)_x$: A Prototype System for Quadrupolar Glasses. *Ferroelectrics* **1998**, *208–209* (1), 293–327. https://doi.org/10.1080/00150199808014882.
44. Pilla, S.; Hamida, J. A.; Muttalib, K. A.; Sullivan, N. S. Dielectric Response of N_2-Ar Solid Solutions in the Audio Frequency Range. *Phys. Rev. B* **2003**, *67* (17), 174204. https://doi.org/10.1103/PhysRevB.67.174204.
45. Cugliandolo, L. F.; Kurchan, J. Analytical Solution of the Off-Equilibrium Dynamics of a Long-Range Spin-Glass Model. *Phys. Rev. Lett.* **1993**, *71* (1), 173–176. https://doi.org/10.1103/PhysRevLett.71.173.
46. Cugliandolo, L. F.; Kurchan, J. Mean-Field Theory of Temperature Cycling Experiments in Spin Glasses. *Phys. Rev. B* **1999**, *60* (2), 922–930. https://doi.org/10.1103/PhysRevB.60.922.
47. Yin, L.; Xia, J. S.; Zapf, V. S.; Sullivan, N. S.; Paduan-Filho, A. Direct Measurement of the Bose-Einstein Condensation Universality Class in $NiCl_2$-$4SC(NH_2)_2$ at Ultralow Temperatures. *Phys. Rev. Lett.* **2008**, *101* (18), 187205. https://doi.org/10.1103/PhysRevLett.101.187205.
48. Pillai, K.; Mun, E.; Wilcox, J.; Manson.; Scott, B.; Tobash, P.; Zapf, V. S. The Origin and Coupling Mechanism of the Magnetoelectric Effect in $TMCl_2$-$4SC(NH_2)_2$ (TM = Ni and Co). *Adv. Condens. Matter Phys.* **2014**, *2014*, 1–4. https://doi.org/10.1155/2014/512621.
49. Yu, R.; Yin, L.; Sullivan, N. S.; Xia, J. S.; Huan, C.; Paduan-Filho, A.; Oliveira, N. F.; Haas, S.; Steppke, A.; Miclea, C. F.; Weickert, F.; Movshovich, R.; Mun, E. D.; Scott, B. L.; Zapf, V. S.; Roscilde, T. Bose Glass and Mott Glass of Quasiparticles in a Doped Quantum Magnet. *Nature* **2012**, *489*, 379–384. https://doi.org/10.1038/nature11406.
50. Yin, L.; Xia, J.-S.; Sullivan, N. S.; Zapf, V. S.; Paduan-Filho, A. Magnetic Susceptibility Measurements at Ultra-Low Temperatures. *J. Low Temp. Phys.* **2010**, *158* (3), 710. https://doi.org/10.1007/s10909-009-0029-2.
51. Xia, J. S.; Yin, L.; Sullivan, N. S.; Zapf, V. S.; Paduan-Filho, A. Magneto-Electric Effect and Dielectric Susceptibility Measurement Technique at Very Low Temperature. *J. Low Temp. Phys.* **2017**, *187* (5), 627–632. https://doi.org/10.1007/s10909-016-1723-5.
52. Thill, M. J.; Huse, D. A. Equilibrium Behaviour of Quantum Ising Spin Glass. *Physica A* **1995**, *214* (3), 321–355. https://doi.org/10.1016/0378-4371(94)00247-Q.
53. Joh, Y. G.; Orbach, R.; Wood, G. G.; Hammann, J.; Vincent, E. Extraction of the Spin Glass Correlation Length. *Phys. Rev. Lett.* **1999**, *82* (2), 438–441. https://doi.org/10.1103/PhysRevLett.82.438.
54. Kisker, J.; Santen, L.; Schreckenberg, M.; Rieger, H. Off-Equilibrium Dynamics in Finite-Dimensional Spin-Glass Models. *Phys. Rev. B* **1996**, *53* (10), 6418–6428. https://doi.org/10.1103/PhysRevB.53.6418.

CHAPTER FIFTEEN

Electronic stopping from orbital mean excitation energies including both projectile and target electronic structure

Remigio Cabrera-Trujillo[a,b,*] and Jens Oddershede[c]
[a]Instituto de Ciencias Físicas, Universidad Nacional Autónoma de México, Cuernavaca, Morelos, Mexico
[b]Department of Physics, University of Gothenburg, Gothenburg, Sweden
[c]Department of Physics, Chemistry and Pharmacy, University of Southern Denmark, Campusvej, Denmark
*Corresponding author: e-mail address: trujillo@icf.unam.mx

Contents

Abstract

In this work, we revisit our Bethe theory for the energy deposition of swift charged particles with electronic structure when colliding with atomic targets [Phys. Rev. A, **55** 2864 (1997)]. As the projectile ion beam has a mixture of several projectile charge states, the electronic stopping cross section is the sum of contributions from all the Z_1 electrons. We use the independent particle approximation for the mean excitation energies such that the electronic stopping cross section becomes a sum of contributions from all the target electrons, N_2. We use an analytical expression of the generalized oscillator strength based on the Harmonic Oscillator approximation such that the stopping cross section becomes a function of only the orbital mean excitation energy, I_{0i}. These orbital mean excitation energies determine the atomic mean excitation energy I_0 via the Bragg's rule. The projectile electronic density for a given number of electrons $N_1 \leq Z_1$ is derived from the Local Plasma Approximation and we thereby obtain an analytic expression for the total stopping cross section for projectiles with structure in terms of the target orbital mean excitation energies within the First Born Approximation.

Advances in Quantum Chemistry, Volume 85
ISSN 0065-3276
https://doi.org/10.1016/bs.aiq.2022.05.004
Copyright © 2022 Elsevier Inc.
All rights reserved.

We compare the results of our calculations for several projectile–target combination to experiment showing an excellent agreement to the experimental data available in the literature.

1. Introduction

When a swift ion moves through matter it loses energy by collisions with the material target. Due to the complexity of the collision several processes take place with a probability of occurring, e.g., target excitation, ionization, and charge exchange. For fully stripped projectiles with velocity v much greater than the orbital velocity of the target electrons, v_e, the Bethe–Bloch theory[1,2] accounts for the electronic stopping cross section with good accuracy. According to Bethe's theory,[1] the electronic stopping cross section for a swift stripped ion with charge Z_1 is given by

$$S_e(v) = -\frac{1}{n_2}\frac{dE}{dx} = \frac{4\pi e^4}{m_e v^2} Z_1^2 Z_2 \ln\left(\frac{2mv^2}{I_0}\right), \tag{1}$$

where I_0 is the mean excitation energy of the target, $-dE/dx$ is the energy loss per unit path (stopping power), and n_2 is the number of target scatterer per volume. Here and below, the subscripts 1 and 2 refers to the projectile and target, respectively. In this treatment, the projectile has a fixed charge. However, while penetrating matter, the projectile ion may carry, at least temporarily, some bound electrons even if $v > v_e$.[3,4] Furthermore, capture and loss of electrons by the projectile are processes with a high probability that leads to important contributions to the energy loss, in particular as the projectile loses energy as it moves through the target.[5–8] Thus, for a given projectile velocity, the ion beam may consist of projectiles with varying number of electrons.[9,10]

The problem of the stopping power of screened projectiles was first studied by Dalgarno and Griffing[11] in the plane-wave Born approximation (PWBA) for the most simple collision system (H on H). More recently, Sigmund[12] has developed a model for the charge-dependent electronic stopping of swift heavy ions within the framework of Bohr's classical theory and Bethe–Bloch stopping theory. A simple model for the electronic energy loss as a function of the impact parameter for screened projectiles at high velocities based on the impact parameter approach is proposed by Azevedo et al.[13] For solid targets, the first attempt to include the influence of bound projectile electrons on the stopping cross section was performed by Ferrell and

Ritchie[14] in the dielectric-function formalism. The most elaborate method to include the projectile screening at low velocities has been developed by Artacho et al. by using the density functional theory (DFT).[15]

Within the spirit of the Bethe theory, there have been various efforts to account for the stopping of partially stripped ions. A scheme was suggested by Kim and Cheng[3] where the electronic stopping cross section S_e is obtained from the original Bethe formula by replacing the projectile nuclear charge Z_1, and target mean ionization energy I_0 by Z_{eff} and I_{eff}, respectively. Both Z_{eff} and I_{eff} are derived from properties of the projectile and target that can be calculated from first principles. Similarly, Arnau and Echenique[16] proposed a way to introduce the electronic structure of the projectile into the dielectric theory of stopping. Also, Moneta and Czerbniak[17] have used the First Born Approximation for estimating the stopping power of partially stripped ions based on the semiclassical impact parameter description. In their model, the projectile electrons are allowed to be excited or ionized but no electron exchange is considered.

In a previous work, Cabrera–Trujillo et al.[18] reported a First Born Approximation extension of the Bethe theory that accounts for partially stripped projectiles carrying bound electrons. The approach has successfully been applied to several different projectile-target systems by several authors.[13,19–32]

In this work, we propose to augment our previous study by considering two recent developments. First, we make use of the recent determination of the atomic mean excitation energies which may be used to calculate the contributions to the stopping cross section arising from individual electrons. Second, we use the implementation of the Harmonic oscillator approach to obtain an analytical expression of the stopping cross section.

The layout of our work is the following. In Section 2.1, we provide a review of the First Born Approximation theory of stopping power for an ion with electronic structure impinging on a target consisting of atoms. In Section 3 we discuss and compare our results with other theories and experiment. Finally, in Section 4 we present our summary and conclusions.

2. Stopping power

2.1 The Bethe theory for structured projectiles within the First Born Approximation

In a previous work,[18] we have reported the extension of the Bethe theory to the case of a projectile with electronic structure. The main results are briefly presented here.

Let us consider a collision of an ion moving with a velocity v (in the laboratory frame), mass M_1, nuclear charge Z_1e, and carrying N_1 bound electrons described by the electronic eigenstate $|n_0\rangle$, that collides with a stationary target with mass M_2 and N_2 bound electrons in an initial state denoted $|m_0\rangle$. The projectile is deflected into the solid-angle element $d\Omega$ along a direction with polar angle (θ, φ) measured in the laboratory frame. Suppose the projectile-target system undergoes a transition to final states $|n\rangle$ and $|m\rangle$, respectively, with energies E_n and E_m. Then the kinetic energy of the projectile, when electron transfer is not permitted, is reduced by $(E_n - E_{n_0}) + (E_m - E_{m_0})$ due to electronic excitations. In this case, the stopping cross section is given by,[18]

$$S_e(v) = S_{e,1}(v) + S_{e,2}(v) \tag{2}$$

with

$$\begin{aligned} S_{e,1}(v) &= \frac{2e^4}{m_e v^2} \sum_{n,m} \int_{q_{min}}^{q_{max}} \left| Z_2 \delta_{m_0 m} - {}_2M_{m_0 m} \right|^2 {}_1F_{n_0 n}(\mathbf{q}) \frac{d\mathbf{q}}{q^2}, \\ S_{e,2}(v) &= \frac{2e^4}{m_e v^2} \sum_{n,m} \int_{q_{min}}^{q_{max}} \left| Z_1 \delta_{n_0 n} - {}_1M_{n_0 n} \right|^2 {}_2F_{m_0 m}(\mathbf{q}) \frac{d\mathbf{q}}{q^2}, \end{aligned} \tag{3}$$

where

$$\begin{aligned} {}_1M_{n_0 n}(\mathbf{q}) &= \left\langle n_0 \right| \sum_{j=1}^{N_1} e^{-i\mathbf{q}\cdot\mathbf{x}_j} \left| n \right\rangle \quad ; \\ {}_2M_{m_0 m}(\mathbf{q}) &= \left\langle m_0 \right| \sum_{j=1}^{N_2} e^{-i\mathbf{q}\cdot\mathbf{y}_j} \left| m \right\rangle \end{aligned} \tag{4}$$

are the atomic form factors[33,34] for the electronic transition between the initial state and the final state of the projectile and target, respectively. Here, $\mathbf{q}$ is the momentum transferred during the collision from the projectile to the target. The generalized oscillators strengths (GOS) for projectile and target are defined as[1]

$${}_1F_{n_0 n}(\mathbf{q}) = \frac{2m_e}{\hbar q^2} \left| {}_1M_{n_0 n}(\mathbf{q}) \right|^2 w_{nn_0} \quad ; \quad {}_2F_{m_0 m}(\mathbf{q}) = \frac{2m_e}{\hbar q^2} \left| {}_2M_{m_0 m}(\mathbf{q}) \right|^2 w_{mm_0}. \tag{5}$$

where $\hbar w_{mm_0}$ is the energy difference between the two states.

The limits of the q-integration in Eq. (3) are determined by the kinematics of the collision process by conservation of momentum and energy. Therefore for heavy ions[35,36]

$$q_{min} = \frac{M_1 v}{\hbar}\left[1 - \sqrt{1 - \frac{2[(E_n - E_{n_0}) + (E_m - E_{m_0})]}{M_1 v^2}}\right],$$
$$q_{max} = \frac{M_1 v}{\hbar}\left[1 + \sqrt{1 - \frac{2[(E_n - E_{n_0}) + (E_m - E_{m_0})]}{M_1 v^2}}\right]. \tag{6}$$

The first term on the right-hand side of Eq. (3) contains the quantity ${}_1F_{n_0 n}$, which gives the contributions to the stopping cross section from the atomic transition of the projectile modulated by the electronic structure of the target. Similarly, the second term containing ${}_2F_{m_0 m}$, which describes the contributions from the electronic transitions in the target modulated by the electronic structure of the projectile. Knowing the atomic form factors and the GOS's for both projectile and target, it is a straightforward task to compute $S_e(v)$ from Eq. (3). However, in order to do so it is necessary to know the full excitation spectrum of both the projectile and the target, and in practice it is necessary to resort to approximate methods.

Our first approximation consists in assuming that the projectile remains mostly in the ground state with N_1 electrons. This assumption is justified by Bohr's adiabatic criteria,[37] which states that *the projectile electrons are stripped, if their orbital velocity v_e is less than the projectile velocity, while electrons with $v_e > v$ are kept since the collisions—seen, for example, from the system moving with the projectile—are adiabatic.* This dragging process leaves the projectile mostly with N_1 electrons in the ground state.[38] In this case the projectile GOS's are zero and the only contribution arises from the target excitation. Thus

$$S_e(v) = \frac{2e^4}{m_e v^2}\sum_m \int_{q_{min}}^{q_{max}} \left|Z_1 - {}_1M_{n_0 n_0}\right|^2 {}_2F_{m_0 m}(\mathbf{q})\frac{d\mathbf{q}}{q^2} \tag{7}$$

with

$$q_{min} = \frac{M_1 v}{\hbar}\left[1 - \sqrt{1 - \frac{2(E_m - E_{m_0})}{M_1 v^2}}\right];$$
$$q_{max} = \frac{M_1 v}{\hbar}\left[1 + \sqrt{1 - \frac{2(E_m - E_{m_0})}{M_1 v^2}}\right]. \tag{8}$$

Eq. (7) implies that the momentum transferred to the target contains a screened contribution from the projectile electronic structure through the ground state atomic form factor.

This is the expression we will use to determine the stopping cross section from a projectile with structure that collides with an atomic target.

2.2 Projectile atomic form factor

In order to determine the projectile atomic form factor ${}_1M_{n_0n_0}$, when $N_1 \leq Z_1$ electrons are bound to the projectile, we notice the following facts:

From the Local Plasma Approximation, one can determine the mean excitation energy from the target electronic density as

$$\ln I_0 = \frac{1}{N}\int \rho(\mathbf{r}) \ln\left[\hbar\sqrt{\frac{4\pi e^2\rho(\mathbf{r})}{m}}\right] d^3r. \tag{9}$$

as a function of the number of electrons N. Alternatively, the atomic density $\rho(\mathbf{r})$ can be determined from the same equation if we know the behavior of the mean excitation energy I_0 as a function of the atom number of electrons $N \leq Z$. In Ref. 39, the Thomas–Fermi–Amaldi theory together with the Thomas–Fermi–Dirac–Weizsäcker theory provides a guide to determine I_0 as a function of N. Furthermore, we notice that Sauer et al.[40,41] have calculated the mean excitation energy for all ionic charge of the atoms with $Z < 39$ by means of the Random Phase Approximation. Interestingly, the plot of I_0/Z vs N/Z shows a universal structure, as observed in Fig. 1. Guided by this, we find that

$$I_0 = \frac{20Z}{\left(40\left(\frac{N}{Z}\right)^{3/2} + 1\right)} H_r, \tag{10}$$

which provides a good description of the results of Sauer et al.[40,41] for I_0 as a function of the number of electrons N. Here $H_r = 27.2114$ eV is the Hartree energy (1 a.u.).

We propose to use a functional for the electronic density of the form

$$\rho(r) = \frac{N\alpha^3}{8\pi} e^{-\alpha r}, \tag{11}$$

such that it reproduces Eq. (10). Eq. (11) provides an averaged description of the electronic structure of an atom with N electrons very similar to the Thomas–Fermi electronic density with the advantage of having a Slater-like behavior. We find that

$$\alpha = \left(\frac{Z}{N}\right)^{1/3} \frac{\beta Z^{1/3}}{\left(40\left(\frac{N}{Z}\right)^{3/2} + 1\right)^{2/3} a_0}, \tag{12}$$

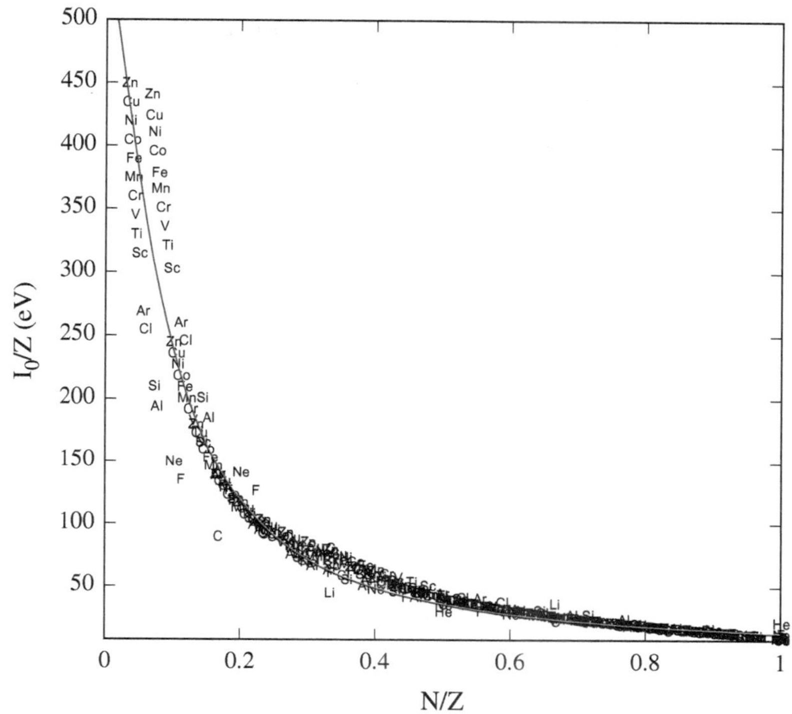

Fig. 1 Total mean excitation energy, I_0, as calculated by Sauer et al. at the RPA level of theory.[40,41] The *purple solid line* is the result of Eq. (10). See text for discussion.

where $\beta = 25.23429 = (20\sqrt{2})^{2/3} e$ with e Euler number. When Eq. (12) is used in Eq. (9) it reproduces Eq. (10).

Consequently, we can now determine the projectile atomic form factor. Due to the assumed spherical symmetry of the ground state $|0\rangle$ of the projectile, Eq. (4) becomes

$$ {}_1M_{00}(q) = 4\pi \int \frac{\rho(r) \sin(qr)}{qr} r^2 dr, \tag{13} $$

where $\rho(r)$ is the projectile ground state electronic density as obtained from Eq. (11). Introducing Eq. (11) in Eq. (13), we get

$$ {}_1M_{00}(q) = \frac{N_1 \alpha^4}{(\alpha^2 + q^2)^2} \tag{14} $$

with α given by Eq. (12). A more complex expression is reported in Ref. 18 where a Thomas-Fermi approach was used. The advantage of Eq. (14) is that it describes any cation with $N \leq Z$ and it is much simpler.

2.3 Harmonically bound target electrons

For the description of the target excitation, we consider a target electronic description within the Harmonic Oscillator approach.[42] We have found that the GOS has an analytical expression

$$F_{mm_0}(q) = \frac{1}{(m-1)!}\left(\frac{q^2}{2\gamma^2}\right)^{m-1} e^{-q^2/2\gamma}, \tag{15}$$

where $\gamma = m_e\omega_0/\hbar$ with ω_0 the oscillator frequency. In this description, the mean excitation energy for the target becomes $I_0 = \hbar\omega_0$. With this, we can determine the stopping cross section from Eq. (7). If we define $x = q^2/2\gamma^2$, then Eq. (7) becomes

$$S_e(v) = \frac{\pi S_0 Z_1^2}{\epsilon I_0}\sum_{m=1}^{m_{max}}\frac{1}{(m-1)!}\int_{x_{min}}^{x_{max}}\left(1-\frac{\xi_1}{(1+x\sigma)^2}\right)^2 e^{-x}x^{m-2}dx \tag{16}$$

where $\xi_1 = N_1/Z_1$, $m_{max} = [E_p/I_0]$, $S_0 = 0.76199564 \times 10^{-15}$ eV cm^2, and

$$x_{\substack{max\\min}} = \left(\frac{M_1}{m_e}\right)^2 \epsilon\left[1 \pm \sqrt{1-\frac{m_e}{M_1}\frac{v}{\epsilon}}\right]^2. \tag{17}$$

Here, $\epsilon = m_eE_p/M_pI_0$ is the reduced projectile collision energy.

This expression is straightforward to implement and it gives the stopping cross section calculated for a projectile with nuclear charge Z_1 and a number of bound electrons N_1 that collides with a target with mean orbital excitation energies I_0.

Our next assumption is to assume that the target consists of N_2 electrons such that each electron contributes to the electronic stopping cross section through a Bragg-like rule. This assumption is justified by the independent particle approach[42] where

$$S_e(v) = \sum_{i=1}^{N_2} S_{e,i}(v) \tag{18}$$

such that we use Eq. (16) for each orbital contribution, i.e.,

$$S_{e,i}(v) = \frac{\pi S_0 Z_1^2}{\epsilon_i I_{0,i}}\sum_{m=1}^{m_{max}}\frac{1}{(m-1)!}\int_{x_{min}}^{x_{max}}\left(1-\frac{\xi_1}{(1+x\sigma)^2}\right)^2 e^{-x}x^{m-2}dx. \tag{19}$$

Here $m_{max} = [E_p/I_{0,i}]$ and

$$x_{\substack{max\\min}} = \left(\frac{M_1}{m_e}\right)^2 \epsilon_i \left[1 \pm \sqrt{1 - \frac{m_e}{M_1}\frac{\nu}{\epsilon_i}}\right]^2, \tag{20}$$

where $\epsilon_i = m_e E_p / M_p I_{0,i}$ for each orbital mean excitation energy.

Eq. (18) gives the stopping cross section for a projectile with $N_1 \leq Z_1$ electrons that collides with a target with N_2 electrons, each with an orbital mean excitation energy $I_{0,i}$.

2.4 Projectile beam fraction

Experimentally, it is known that a beam contains all the charge states as a function of the beam energy.[9] Hence, the beam will consist of all the projectile charge fraction f_i with a given probability distribution. Consequently, Allison et al.[9,10,43] suggest that the final electronic stopping cross section for each charge fraction in the beam is given by

$$S_e(\nu) = \sum_{i=0}^{N_1} f_i S_e^i(\nu) \tag{21}$$

where f_i is the fraction of the beam ion with i electrons. Thus $\sum_i f_i = 1$ and $S_e^i(\nu)$ is the electronic stopping cross section for that projectile number of electrons, as given by Eq. (18).

In the case of a hydrogen beam the charge fractions are given by[10]

$$\begin{aligned} f_0 &= \frac{\sigma_{01}}{\sigma_{01} + \sigma_{10}}, \\ f_1 &= \frac{\sigma_{10}}{\sigma_{01} + \sigma_{10}}, \end{aligned} \tag{22}$$

where σ_{if} is the charge exchange cross section for the projectile to go from an initial charge state i to a final charge state f. Thus, according to the charge exchange process, a projectile will have a unique charge fraction for a given projectile collision energy when colliding with the target, as shown by Allison et al.[9] From Eqs. (21) and (22), we notice that a projectile colliding with different targets may have a different beam fractions as this is a consequence of the charge exchange process.

In order to simplify our approach, we implement a further approximation. For the case of a hydrogen beam, Montenegro et al.[44] recognized the probabilistic character of the beam charge fraction. They assume that the hydrogen beam fraction for a proton to capture an electron, as a function of the projectile velocity v, can be approximated analytically by $f_0 \approx e^{-u}$, where $u = v/v_0$ and v_0 is Bohr's velocity. Consequently $f_1 = 1 - e^{-u}$.

Thus, for the case of hydrogen beams, the electronic stopping cross section is given by

$$S_e(v) = S_e^0(v)(1 - e^{-v/v_0}) + S_e^1(v)e^{-v/v_0}, \tag{23}$$

where S_e^1 corresponds to the collision of a neutral hydrogen projectile with a given target and S_e^0 corresponds to the case of a bare proton. These results corresponds to a weighted electronic stopping cross section over projectile charge fractions. This approximation assumes that σ_{if} has the same behavior for hydrogen ions on all the targets which we know is an approximation. However, it is justified statistically by the effective charge approach of Betz.[45] Of course, it would be more proper to determine the correct charge exchange process, e.g., as done in the Electron-Nuclear Dynamics approach.[46]

3. Analysis and discussion

As examples of our approach, we calculate the stopping cross section for a hydrogen beam colliding with H, He, C, F, Ne, Al, Si, Cl, and Ar targets which we compare to available experimental data.[51–56] The orbital mean excitation energies used in Eqs. (19) and (21) are those reported by Sauer et al.[40,41] for the same set of targets. Our results are shown in Fig. 2 by the purple solid line. For comparison, we show the results obtained through the Electron-Nuclear Dynamics approach which includes charge exchange at low collision energy and shows the universal behavior of the energy deposition (red long dashed-dot line).[46] In the same figure, we show the results of SRIM (Stopping and Range In Matter) code by Ziegler.[48] We observe a good agreement between our results and the experimental data from the maximum of the stopping power curve towards high energies. Because our treatment does not include many of the processes that contribute to low energy stopping, we do not reproduce the experimental stopping power curve below the maximum. However, since we are not using the Bethe approximation our results extend towards small collision energies and do not diverge.

In Fig. 3, we show the electronic stopping cross section for a hydrogen beam colliding with Sc, Ti, V, Cr, Mn, Fe, Co, Ni, and Cu targets. We observe that for projectile collision energies from the maximum to higher

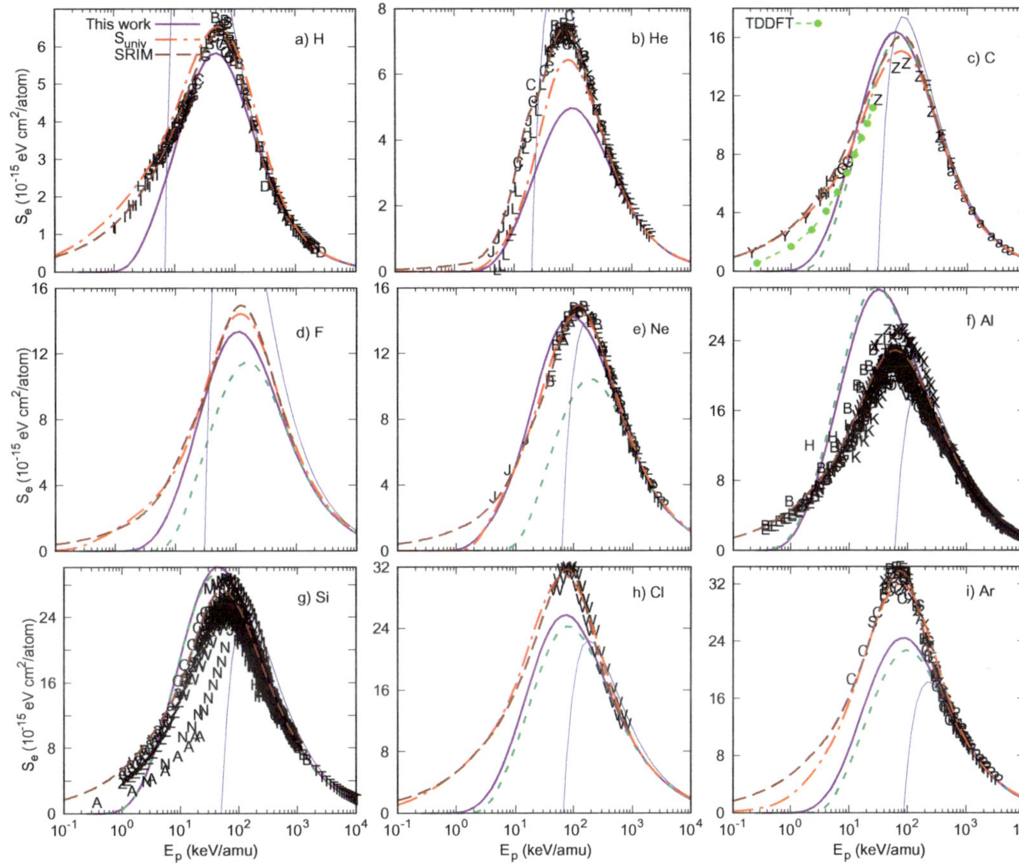

Fig. 2 Electronic stopping cross section for a hydrogen beam colliding with H, He, C, F, Ne, Al, Si, Cl, and Ar atomic targets. The *purple solid line* is the results of Eq. (21) when using the orbital mean excitation energies of Sauer et al.[40,41] The *green dashed lines* are results of Eq. (21) when using the orbital mean excitation energies of Oddershede and Sabin.[47] For comparison purposes, the *red dash-dot line* is the universal expression as reported in Ref. 46, while the *brown dashed lines* are the SRIM results[48] and the *blue thin line* is the standard Bethe expression for a bare projectile, Eq. (1). The experimental data are taken from H. Paul's database[49,50] and various other experimental sources.[51–56]

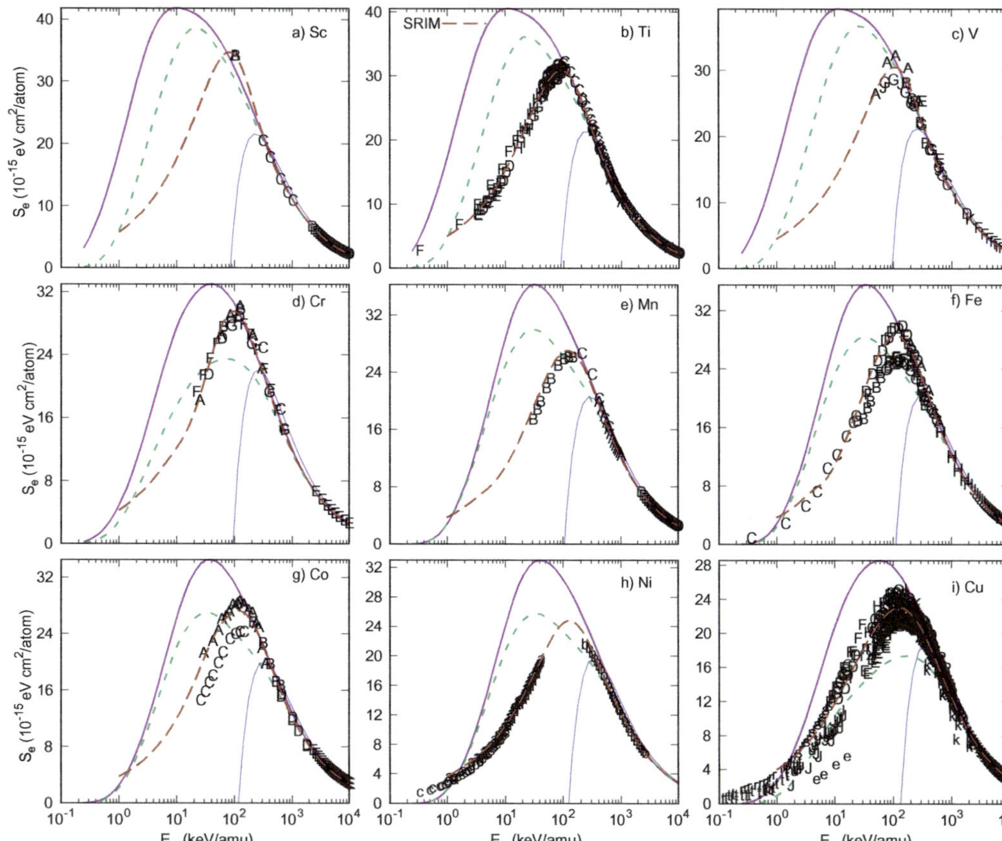

Fig. 3 The same as Fig. 2, but showing results for the electronic stopping cross section for a hydrogen beam colliding with Sc, Ti, V, Cr, Mn, Fe, Co, Ni, and Cu atomic targets. See text for discussion.

collision energies, our approach agrees very well with the experimental data, although we should stress that the experimental data is for condensed phase materials when our theoretical approach is developed for gas phase targets. Also, in the same figures we plot the standard Bethe term (Eq. 1 for a bare projectile).

4. Summary

We have described a method to calculate the electronic stopping power of swift ions with $N_1 \leq Z_1$ bound electrons colliding with an ionic target which includes the effects of the projectile bound electrons as well as the target electronic excitations. The method applies many of the same assumptions as are involved in the standard Bethe theory of the stopping power in the First Born Approximation. The approach only requires the knowledge of the projectile number of electrons and target orbital mean excitation energies. The total stopping power takes into account the contribution of the electronic structure of the projectile plus the contribution of the electronic structure of the target.

Acknowledgments

It is a pleasure to dedicate this paper to Jack Sabin. We would not have been active in this field had it not been for Jack's influence on our scientific carriers. The content of the paper is inspired by our long association to Jack Sabin, both in terms of the subject and presentation. We would like to take this opportunity to thank Jack for scientific inspiration, wonderful discussions, and friendship over many years. R.C.-T. acknowledges support from grant UNAM-DGAPA-PAPIIT IN-111-820 and LANCAD-UNAM-DGTIC-228.

References

1. Bethe, H. Zur Theorie des Durchgangs schneller Korpuskularstrahlen durch Materie. *Ann. Phys.* **1930**, *397* (3), 325–400. https://doi.org/10.1002/andp.19303970303.
2. Bloch, F. Zur Bremsung Rasch Bewegter Teilchen Beim Durchgang Durch Materie. *Ann. Phys. (Leipzig)* **1933**, *16*, 285.
3. Kim, Y. K.; Cheng, K. Stopping Power for Partially Stripped Ions. *Phys. Rev. A* **1980**, *22*, 61.
4. Kaneko, T. Stopping Power for Hydrogenlike and Heliumlike Particles: Bethe Theory. *Phys. Rev. A* **1991**, *43*, 4780.
5. Bohr, N. The Penetration of Atomic Particles Through Matter. *Mat. Fys. Medd. Dan. Vid. Selsk.* **1948**, *18* (8).
6. Cabrera-Trujillo, R.; Sabin, J. R.; Öhrn, Y.; Deumens, E. Charge Exchange and Threshold Effect in the Energy Loss of Slow Projectiles. *Phys. Rev. Lett.* **2000**, *84* (23), 5300–5303. https://doi.org/10.1103/PhysRevLett.84.5300.

7. Cabrera-Trujillo, R.; Öhrn, Y.; Deumens, E.; Sabin, J. R. Stopping Cross Section in the Low- to Intermediate-Energy Range: Study of Proton and Hydrogen Atom Collisions With Atomic N, O, and F. *Phys. Rev. A* **2000**, *62* (5), 052714. https://doi.org/10.1103/PhysRevA.62.052714.
8. Cabrera-Trujillo, R.; Sabin, J. R.; Öhrn, Y.; Deumens, E. Stopping Cross Section for $N^{4+} \rightarrow H$ at Low Projectile Velocity. *Phys. Rev. A* **2002**, *66*, 022706.
9. Allison, S. K.; Cuevas, J.; Garcia-Munoz, M. Partial Atomic Stopping Power of Gaseous Hydrogen for Hydrogen Beams. I. *Phys. Rev.* **1962**, *127*, 792.
10. Cuevas, J.; Garcia-Munoz, M.; Torres, P.; Allison, S. K. Partial Atomic and Ionic Stopping Powers of Gaseous Hydrogen for Helium and Hydrogen Beams. *Phys. Rev.* **1964**, *135*, A335.
11. Dalgarno, A.; Griffing, G. W.; Bates, D. R. Energy Loss of Protons Passing Through Hydrogen. *Proc. R. Soc. Lond. A Math. Phys. Sci.* **1955**, *232* (1190), 423–434. https://doi.org/10.1098/rspa.1955.0228.
12. Sigmund, P. Charge-Dependent Electronic Stopping of Swift Nonrelativistic Heavy Ions. *Phys. Rev. A* **1997**, *56* (5), 3781–3793. https://doi.org/10.1103/PhysRevA.56.3781.
13. de Azevedo, G. M.; Grande, P. L.; Schiwietz, G. Impact-Parameter Dependent Energy Loss of Screened Ions. *Nucl. Instrum. Methods Phys. Res. B* **2000**, *164–165*, 203–211. https://doi.org/10.1016/S0168-583X(99)01074-5.
14. Ferrell, T. L.; Ritchie, R. H. Energy Losses by Slow Ions and Atoms to Electronic Excitation in Solids. *Phys. Rev. B* **1977**, *16* (1), 115–123. https://doi.org/10.1103/PhysRevB.16.115.
15. Ullah, R.; Artacho, E.; Correa, A. A. Core Electrons in the Electronic Stopping of Heavy Ions. *Phys. Rev. Lett.* **2018**, *121* (11), 116401. https://doi.org/10.1103/PhysRevLett.121.116401.
16. Arnau, A.; Echenique, P. M. Stopping Power of an Electron Gas for Partially Stripped Ions. *Nucl. Instrum. Methods Phys. Res. B* **1989**, *42* (2), 165–170. https://doi.org/10.1016/0168-583X(89)90702-7.
17. Moneta, M.; Czerbniak, J. Electronic Stopping Power of Atoms for Ions. *Nucl. Instrum. Methods* **1994**, *B90*, 67.
18. Cabrera-Trujillo, R.; Cruz, S. A.; Oddershede, J.; Sabin, J. R. Bethe Theory of Stopping Incorporating Electronic Excitations of Partially Stripped Projectiles. *Phys. Rev. A* **1997**, *55*, 2864.
19. Stevanović, N.; Nikezić, D. Stopping Power. Projectile and Target Modeled as Oscillators. *Phys. Lett. A* **2005**, *340* (1), 290–298. https://doi.org/10.1016/j.physleta.2005.03.070.
20. Stevanović, N.; Nikezić, D. Calculation of Stopping Power for Partially Stripped Ions Using an Oscillator Model. *Eur. Phys. J. D* **2007**, *42*, 397–406. https://doi.org/10.1140/epjd/e2007-00128-3.
21. Matveev, V. I.; Sidorov, D. B. Effective Stopping of Fast Heavy Highly Charged Structure Ions in Collisions With Complex Atoms. *JETP Lett.* **2006**, *84*, 243–248. https://doi.org/10.1134/S0021364006170036.
22. Matveev, V. I.; Sidorov, D. B. Theory of the Stopping of Fast Heavy Highly Charged Structured Ions in Their Collisions With Complex Atoms. *J. Exp. Theor. Phys.* **2007**, *105*, 502–510. https://doi.org/10.1134/S1063776107090051.
23. Matveev, V. I.; Sidorov, D. B. Energy Losses of Fast Heavy Multiply Charged Structural Ions in Collisions With Complex Atoms. *Tech. Phys.* **2007**, *52*, 839–844. https://doi.org/10.1134/S1063784207070031.
24. Çağatay Tufan, M.; Kabadayı, Önder; Gümüş, H. Stopping Power Calculations of Molecules for Swift Projectiles. *Radiat. Phys. Chem.* **2007**, *76* (3), 631–635. https://doi.org/10.1016/j.radphyschem.2006.01.034. Proceedings of the 3rd International Conference on Elementary Processes in Atomic Systems.

25. Gusarevich, E. S. Excitation of Electron Shells in Fast Structural Ions and Their Contribution to Ion Energy Losses. *Tech. Phys.* **2017**, *62*, 965–967. https://doi.org/10.1134/S1063784217060123.
26. Gusarevich, E. S. Effect of the Nuclear Charge of a Fast Structural Ion on Its Internal Effective Stopping in Collisions With Atoms. *J. Exp. Theor. Phys.* **2017**, *124*, 196–201. https://doi.org/10.1134/S1063776117010137.
27. Gusarevich, E. S. Stopping of Fast Structural Ions in a Substance by Means of Excitation of Their Own Electrons. *Russ. Phys. J.* **2017**, *60*, 1255–1261. https://doi.org/10.1007/s11182-017-1203-7.
28. Usta, M.; Çağatay Tufan, M. Stopping Power and Range Calculations in Human Tissues by Using the Hartree-Fock-Roothaan Wave Functions. *Radiat. Phys. Chem.* **2017**, *140*, 43–50. https://doi.org/10.1016/j.radphyschem.2017.03.005. Second International Conference on Dosimetry and its Applications (ICDA-2) University of Surrey, Guildford, United Kingdom, 3–8 July 2016.
29. Usta, M.; Çağatay Tufan, M.; Aydın, G.; Bozkurt, A. Stopping Power and Dose Calculations With Analytical and Monte Carlo Methods for Protons and Prompt Gamma Range Verification. *Nucl. Instrum. Methods Phys. Res. A* **2018**, *897*, 106–113. https://doi.org/10.1016/j.nima.2018.04.045.
30. Usta, M. The Calculation of Stopping Power and Range for Radium, Thorium and Uranium Using New Electronic Potential Energy Function. *Appl. Radiat. Isot.* **2019**, *152*, 193–199. https://doi.org/10.1016/j.apradiso.2019.06.005.
31. GÜMÜŞ, H. Calculation of stopping power and range for heavy ions at intermediate energies by using Tietz screening function. *Nucl. Instrum. Methods Phys. Res. B* **2019**, *461*, 93–97. https://doi.org/10.1016/j.nimb.2019.09.024.
32. Gil, J. M.; Beltrán, P. R.; Rodríguez, R.; Espinosa, G.; Barriga-Carrasco, M. D.; González-Gallego, L. Bound Electron Stopping Power Model of Partially Stripped Ions in Partially Ionized Plasmas. *X-Ray Spectrometry* **2020**, *49* (1), 234–238. https://doi.org/10.1002/xrs.3116.
33. Fano, U.; Rau, A. R. P. *Atomic Collisions and Spectra*; Academic Press: New York, 1986.
34. James, R. W. The Atomic Scattering Factor. In *The Optical Principles of the Diffraction of X-rays*; Bragg, L., Ed.; G. Bell and Sons Ltd.: London, 1967; p. 109.
35. Bethe, H. A.; Jackiw, R. *Intermediate Quantum Mechanics*; Benjamin/Cummings: New York, 1986.
36. Inokuti, M. Inelastic Collisions of Fast Charged Particles With Atoms and Molecules—The Bethe Theory Revisited. *Rev. Mod. Phys.* **1971**, *43* (3), 297–347. https://doi.org/10.1103/RevModPhys.43.297.
37. Bohr, N. Velocity-Range Relation for Fission Fragments. *Phys. Rev.* **1941**, *59*, 270.
38. Yarlagadda, Y. S.; Robinson, J. E.; Brandt, W. Effective-Charge Theory and the Electronic Stopping Power of Solids. *Phys. Rev. B* **1978**, *17*, 3473.
39. Cabrera-Trujillo, R.; Cruz, S. A. Chapter Seven–Mean Total and Orbital Excitation Energies of Atomic Ions in Two Approaches of the Thomas-Fermi Theory. In *A Tribute to John R. Sabin*; Oddershede, J., Brändas, E., Eds.; Advances in Quantum Chemistry, Vol. 85; Academic Press, 2022.
40. Sauer, S. P. A.; Oddershede, J.; Sabin, J. R. Chapter Three–The Mean Excitation Energy of Atomic Ions. In *Concepts of Mathematical Physics in Chemistry: A Tribute to Frank E. Harris–Part A*; Sabin, J. R., Cabrera-Trujillo, R., Eds.; Advances in Quantum Chemistry, Vol. 71; Academic Press, 2015; pp. 29–40. https://doi.org/10.1016/bs.aiq.2015.02.001.
41. Sauer, S. P. A.; Sabin, J. R.; Oddershede, J. Calculation of Mean Excitation Energies of 3d-Elements and Their Cations. *Mol. Phys.* **2021**, *119* (5), e1823508. https://doi.org/10.1080/00268976.2020.1823508.
42. Cabrera-Trujillo, R. Stopping Power in the Independent-Particle Model: Harmonic Oscillator Results. *Phys. Rev. A* **1999**, *60* (4), 3044–3052. https://doi.org/10.1103/PhysRevA.60.3044.

43. Allison, S. K.; Garcia-Munoz, M. Electron Capture and Loss at High Energies. In *Atomic and Molecular Processes*; Bates, D. R., Ed.; Pure and Applied Physics, Vol. 13; Academic Press: New York, 1962; p. 721.
44. Montenegro, E. C.; Cruz, S. A.; Vargas-Aburto, C. A Universal Equation for the Electronic Stopping of Ions in Solids. *Phys. Lett. A* **1982**, *92* (4), 195–202. https://doi.org/10.1016/0375-9601(82)90532-1.
45. Betz, H. D. Charge States and Charge-Changing Cross Sections of Fast Heavy Ions Penetrating Through Gaseous and Solid Media. *Rev. Mod. Phys.* **1972**, *44* (3), 465–539. https://doi.org/10.1103/RevModPhys.44.465.
46. Cabrera-Trujillo, R. Analytical Expression for the Electronic Stopping Cross Section of Atomic Gas Targets for Hydrogen Projectiles. *Phys. Rev. A* **2021**, *103* (3), 032812. https://doi.org/10.1103/PhysRevA.103.032812.
47. Oddershede, J.; Sabin, J. R. Orbital and Whole-Atom Proton Stopping Power and Shell Corrections for Atoms With $Z \leq 36$. *At. Data Nucl. Data Tables* **1984**, *31* (2), 275–297. https://doi.org/10.1016/0092-640X(84)90024-X.
48. Ziegler, J. F.; Biersack, J. P.; Littmark, U. *The Stopping and Range of Ions in Solids*; Pergamon Press: New York, 1985. See http://www.srim.org.
49. IAEA. Electronic Stopping Power of Matter for Ions: Graphs, Data, Comments, and Programs. https://www-nds.iaea.org/stopping, (accessed 10 February 2022).
50. Montanari, C. C.; Dimitriou, P. The IAEA Stopping Power Database, Following the Trends in Stopping Power of Ions in Matter. *Nucl. Instrum. Methods Phys. Res. B* **2017**, *408*, 50–55. https://doi.org/10.1016/j.nimb.2017.03.138.
51. Northcliffe, L. C.; Schilling, R. F. Range and Stopping Power Tables for Heavy Ions. *At. Data Nucl. Data Tables* **1970**, 7, 233.
52. Santry, D. C.; Werner, R. D. Stopping Power Measurements of C, Al, Si, Ti, Ni, Ag, Au and Mylar Using Radioactive Alpha Sources. *Nucl. Instrum. Methods* **1984**, *B1*, 13.
53. Santry, D. C.; Werner, R. D. Stopping Power Values of Be, C, Al and Si for ^{4}He Ions. *Nucl. Instrum. Methods* **1980**, *178*, 523.
54. Andersen, H. H.; Bak, J. F.; Knudsen, H.; et al. Stopping Power of Al, Cu, Ag, and Au for MeV Hydrogen, Helium, and Lithium Ions. Z_1^3 and Z_1^4 Proportional Deviations From the Bethe Formula. *Phys. Rev. A* **1977**, *16*, 1929.
55. Lin, H. H.; Li, L. W.; Norbeck, E. Stopping Powers of C, Al, Ni, Cu, In, Sn, Ag and Au for ^{7}Li Ions of 1.0–4.7 MeV. *Nucl. Instrum. Methods* **1986**, *B17*, 91.
56. Booth, W.; Grant, I. S. The Energy Loss of Oxygen and Chlorine Ions in Solids. *Nucl. Phys.* **1965**, *63*, 481.

CHAPTER SIXTEEN

The propensity of terpenes to invoke concerted reactions in their biosynthesis

B. Andes Hess Jr.*
Department of Chemistry, Vanderbilt University, Nashville, TN, United States
*Corresponding author: e-mail address: b.andes.hess@vanderbilt.edu

To Jack: My friend, a fellow Ephman of the Class of 1962, Williams College.

Contents

Abstract

Recent advances in the understanding of mechanistic aspects of terpene biosynthesis is primarily due to computational chemistry ranging from DFT to QMMM MD. In such a way, it has been determined that terpene biosynthetic pathways make extensive use of multi-step concerted reactions that involve carbocation intermediates, including cyclizations, alkyl and [1,2]hydride shifts. The rearrangements that occur in the mechanism of the bisabolyl carbocation to form a number of terpene products involve a [1,2]hydride shift followed by two cyclizations each of which produces a new cyclopentane rings. It was found earlier, using the B3LYP functional, that while two rings were formed via a typical, concerted reaction, the hydride shift did not, but instead an intermediate was produced along the pathway. Being aware of the deficiencies in treating the formation of carbocycles with the B3LYP functional, the two pathways of conformers of the bisabolyl carbocation were revisited here, applying the M062X functional which includes dispersion. In both analyzed cases the [1,2]hydride shift occurred in concert with the two subsequent cyclizations.

Advances in Quantum Chemistry, Volume 85
ISSN 0065-3276
https://doi.org/10.1016/bs.aiq.2022.05.002
Copyright © 2022 Elsevier Inc.
All rights reserved.

1. Introduction

In 1926 Heilbron suggested that squalene is a possible precursor to cholesterol, one of the first examples of a mechanism of the biosynthesis of terpenes.[1] This complex transformation involves at first the formation of a carbocation by protonation of the epoxide and its ring opening.[2] This is followed by the formation of four rings, the backbone of all steroids.[3,4] However, details of the mechanism were controversial for many years,[5] in particular, the formation of the pentacyclic ring D from the intermediate tertiary carbocation on ring C. It was generally believed that this D-ring formation was occurring via ring-expansion of the C-ring to a hexacyclic secondary carbocation even though this was in violation of Markovnikov's rule, followed by ring closure of the D-ring to give the 6665 tetracyclic ring system.[6,7] In 2002, based on DFT calculations for a model system, it was shown that the Markovnikov rule in fact need not be violated in the case of the formation of the D-ring. Indeed, the C-ring expansion and formation of the D-ring might be realized with a concerted reaction,[8] which would preserve Markovnikov's rule. Almost exactly 1 year later, the same idea was repeated in a QM/MM study[9] without referencing the original paper, reference 8. Subsequent theoretical studies on this tetracyclic ring formation demonstrated that in the conversion of oxido-squalene to the protosterol carbocation, the precursor to all steroids, the 665 tricyclic ring system was formed in a concerted fashion followed by a concerted C-ring expansion and D-ring formation.[5] Further QM/MM MD calculations, which included the enzyme, confirmed this biosynthetic mechanism.[10]

In the past 20 years in many studies it has been shown that concerted reactions play a key role in the biosynthesis of terpenes. Much of this work has been performed by Tantillo.[11–14] Interestingly, 1,2-hydride shifts are not very common in terpene biosynthesis, however Tantillo reported two such cases in 2009,[11] neither of which led to participation in a concerted reaction in subsequent steps. This present work revisits these two cases that used the pseudopotential B3LYP, as illustrated below in Fig. 1. Two examples of carbocation rearrangements that involve 1,2-hydride shifts studied, which included dispersion, have been recently reported.[15,16]

When analyzing this figure, it is seen that the reactions of both conformers (**1** and **4**) involve intermediates (**2** and **5**) produced by [1,2]hydride shifts. They subsequently undergo concerted formation of two pentacylic rings to give products **3** and **6**.

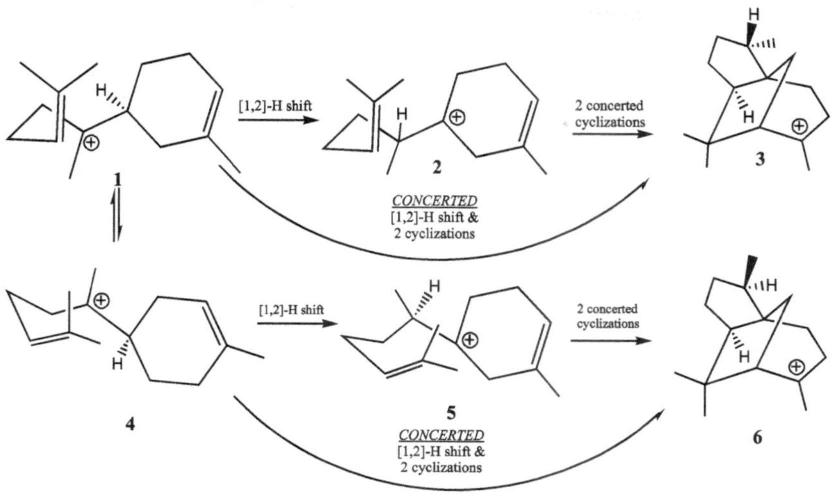

Fig. 1 Conversions of the two conformers (**1** and **4**) to diasteromeric products (**3** and **6**) via non-concerted pathways and concerted pathways.

2. Methods

All calculations have been performed applying the density functional theory (DFT) with the pseudopotential M062X and the basis set of 6-31G*, via the Gaussian 09 program suite.[17] In this way the geometry optimizations, vibrational frequencies and intrinsic reaction coordinate calculations (IRC) were calculated.[18,19]

3. Results and discussion

It has been well documented that the use of the B3YLP DFT functional[20,21] in the calculations of cyclic systems performs poorly, particularly in such cases where rings are formed during the course of a reaction. In addition to poor thermodynamic results, the application of this particular pseudopotential often leads to false minima during the course of reaction pathways that involve formation of multiple rings.[22–24] Intrigued by Tantillo's results mentioned above, obtained with the B3LYP functional, we decided to repeat these calculations with Truhlar's M062X functional, which takes into account some dispersion.

In Fig. 2 the new transition structure (**7**) generated with the M062X potential is compared with Tantillo's first transition structure (**8**) that

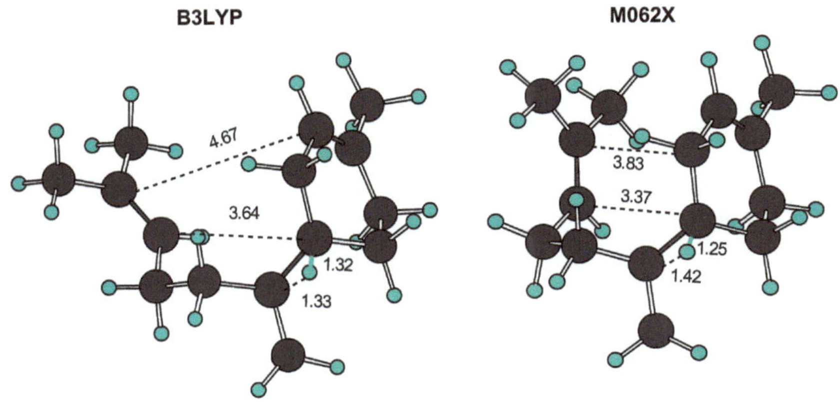

8 7

Fig. 2 The two transition structures obtained with the B3LYP and M062X functionals. The two C—C bond distances shown (in Ångstroms) are those forming the two pentacycles.

connect reactant (**1**) with intermediate (**2**) in Fig. 1. Note that the distances between the carbons of the developing carbon-carbon single bonds (3.64 and 4.67 Å in **8**) are significantly smaller in **7** (3.37 and 3.83 Å). Hence the M062X functional allows a stronger interaction of the first reacting double bond with the developing tertiary carbocation on the carbon from which the hydride is migrating. Further comparison of these two transition structures shows that in the B3LYP TS the migrating hydrogen is equidistant from the two carbons of the migration. At the same time however, in the case of the M062X TS the migrating hydrogen has moved to a lesser extent in its migration (1.25 vs 1.32 Å). The difference in these two transition structures in Fig. 2 indicates that ring formation has already begun in the TS, as seen in (**7**).

Intrinsic reaction coordinate calculations were undertaken in order to follow the reaction pathway, and their results are presented in Fig. 3. Note that the pathway clearly shows that the three parts of the reaction are highly asynchronous but concerted (no minimum was located between the reactant and product). The overall change in the reaction energy (HF+ZPE) is −33 kcal/mol. The B3LYP calculations by Tantillo, which were done with a much larger basis set than here, shows only −12 kcal/mol as a heat of reaction. This is further evidence for the rather poor reproduction of the thermodynamics in the case of carbocycle ring formation when the B3LYP pseudopotential is applied.[23–26]

In Fig. 4 it is clearly seen that as the hydrogen begins to migrate (C13-H21 and C5-H21 lines), the C—C bond distance (C5-C11 line) of

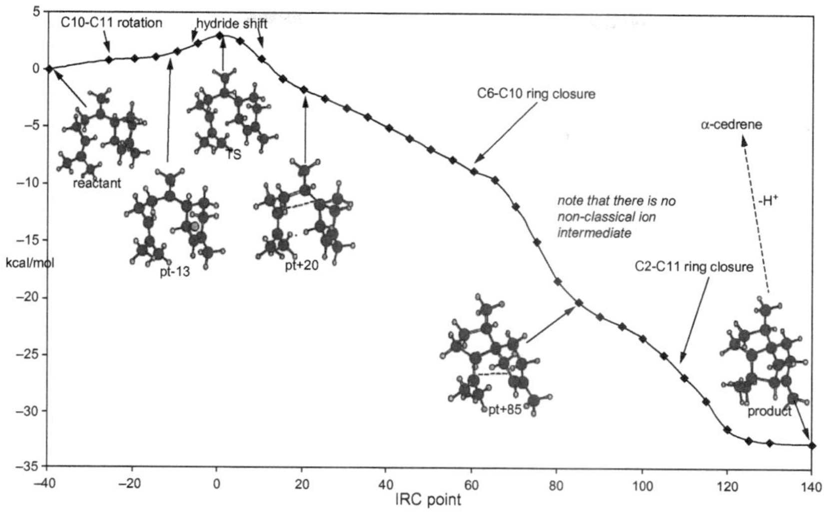

Fig. 3 IRC points (pt) of the M062X pathway, which shows the overall concerted nature of the hydride shift with two pentacyclics formed. The IRC energies are the DFT/M062X energies without zero point energy corrections.

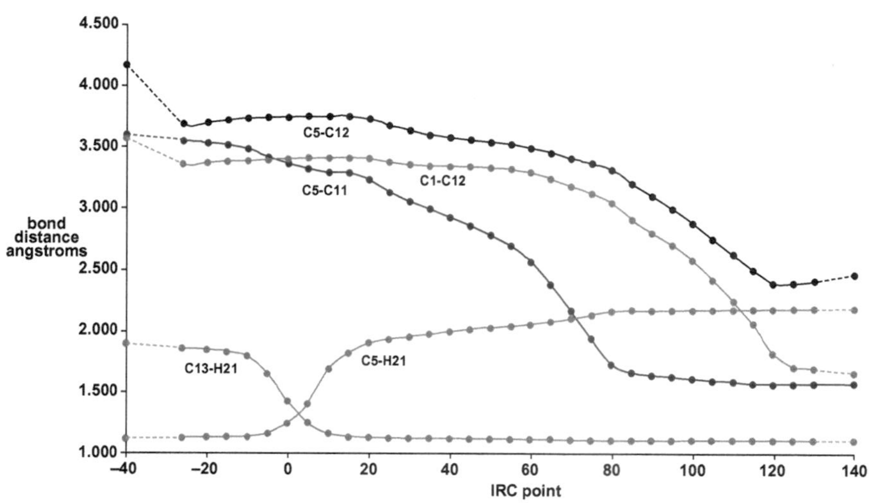

Fig. 4 Change in the key C—C and C—H bond distances as the concerted reaction proceeds. C13-H21 and C5-H21 show the shifting hydrogen. C5-C11 is the bond forming the first pentacycle. C1–12 is the bond forming the second pentacycle. C2-C12 is the distance of the adjacent sp2 hybridized carbon to the carbon forming the second pentacyle.

the first formed carbocycle has begun to decrease showing the concerted nature of the reaction. The C1-C12 line shows the C—C distance of the second forming carbocycle. Note that it has begun to approach the developing positive charged carbon in the latter stages of the formation of the first pentacycle. The C2-C12 line shows that the second carbon of the C═C, which forms the second carbocycle, roughly parallels the C1-C12 line. This indicates that the C═C approaches the developing positive charge in a "symmetric way" until quite late in in the formation of the second pentacycle.

In order to verify the improvement of the above description of the reaction pathway with application of the M062X vs B3LYP another system was studied. Namely, calculations have been performed for the second reaction depicted in Fig. 1 in which reactant **4** is converted to product **6.** A transition structure (**9**) was easily located and is compared with the B3LYP structure (**10**) in Fig. 5.

As in the previous case the two transition structures are quite different. Again the interaction of the forming carbocation on the cyclohexane ring with the carbon that will form the first pentacycle is significantly shorter with the M062X functional. The structure is such that the forming carbocation in the acyclic chain also moves toward the double bond in the cyclohexene, which will eventually form the second pentacycle. An IRC was performed and as in the previous case above it was found that this reaction also involves a [1,2]hydride shift and the formation of two pentacycles in a concerted but highly asynchronous reaction.

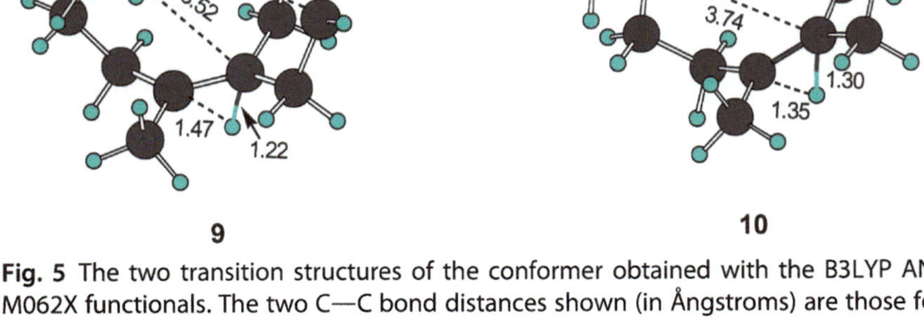

Fig. 5 The two transition structures of the conformer obtained with the B3LYP AND M062X functionals. The two C—C bond distances shown (in Ångstroms) are those forming the two pentacycles.

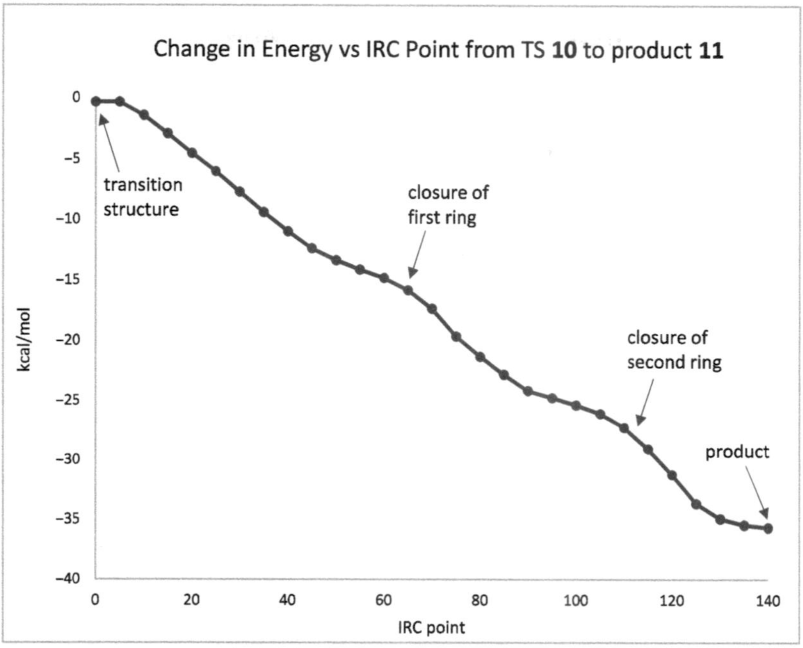

Fig. 6 Change in energy along the pathway as the concerted reaction proceeds. The energies are DFT/M062X and do not include zero point energy corrections (ZPE).

The IRC plot (Fig. 6) of the energy vs IRC point show the conversion of the transition structure **9** to the final tricyclic product to be a concerted reaction.

4. Conclusions

1. These two examples provide evidence that concerted reactions in the biosynthesis of terpenes can include [1,2]hydride shifts as well as the more common cyclizations. These hydride shifts add to the diversity of terpenes in Nature. It is also apparent that the earlier concerns about the reliability of B3LYP are confirmed, since to obtain an accurate description of these types of reactions, dispersion must be included in the potential applied for the calculations.

5. Postscript

I met Jack in September of 1958 during our first day at Williams College as freshmen. We shared a common entrance to our dorm. A picture of the students in our "entry" taken several days later interestingly shows the absence of Jack. Was he already in the Chemistry Lab? To fill "this gap" I inserted his photograph to make the group of freshmen complete.

We both were chemistry majors, and I remember many nights we spent in the Thompson Lab our senior year performing honors research. This was the year when I really got to know Jack. We went our own ways after graduation, but we kept in contact over the years at conferences and my two visits to the QTP at Florida State. I remember especially warmly one visit when Jack hosted for dinner Professor Geerd Diercksen and his wife, from Giessen. That evening Jack really demonstrated his talents as a physical chemist when serving us perfectly grilled pork loin prepared by himself!

References

1. Heibron, I. M.; Kamm, E. D.; Owens, W. M. CCXIII.—The Unsaponifiable Matter From the Oils Elasmobranch Fish. Part I. A Contribution to the Study of the Constitution of Squalene (Spinacene). *J. Chem. Soc.* **1926**, *126*, 1630.
2. Langdon, R. G.; Bloch, K. The Biosynthesis of Squalene and Cholesterol. *J. Am. Chem. Soc.* **1952**, *74*, 1869.
3. Corey, E. J.; Russey, W. E.; de Montellano, P. R. O. Metabolic Fate of 10,11-Dihydrosqualene in Sterol-Producing Rat Liver Homogenate. *J. Am. Chem. Soc.* **1966**, *88*, 4750.
4. van Tamelen, E. E.; Willett, J. D.; Clayton, R. B. The Mechanism of Lanosterol Biosynthesis From Squalene 2,3-oxide. *J. Am. Chem. Soc* **1967**, *89*, 3371.
5. Hess, B. A., Jr. Computational Studies on the Biosynthesis of Squalene to the Steroids and Hopennes. *Org. Biomol. Chem.* **2017**, *15*, 2133.
6. Abe, L.; Rohmer, M.; Restwich, G. D. Enzymatic Cyclization of Squalene and Oxidosqualene to Sterols and Triterpenes. *Chem. Rev.* **1993**, *93*, 2189.

7. Wendt, K. U.; Schulz, G. E.; Corey, E. J.; Liu, D. Enzyme Mechanisms for Polycyclic Triterpene Formation. *Angew. Chem. Int. Ed.* **2000**, *39*, 2812.
8. Hess, B. A., Jr. Concomitant C-Ring Expansion and D-Ring Formation in Lanosterol Biosynthesis From Squalene Without Violation of Markovnikov's Rule. *J. Am. Chem. Soc.* **2002**, *124*, 10286.
9. Rajamani, R.; Gao, J. Balancing Kinetic and Thermodynamic Control: The Mechanism of Carbocation Cyclization by Squalene Cyclase. *J. Am. Chem. Soc.* **2003**, *125*, 12768.
10. Chen, N.; Wang, S.; Smentek, L.; Hess, B. A., Jr.; Wu, R. Biosynthetic Mechanism of Lanosterol: Cyclization. *Angew. Chem. Int. Ed.* **2015**, *54*, 8693.
11. Hong, Y. J.; Tantillo, D. J. Consequences of Conformational Preorganization in Sesquiterpene Biosynthesis: Theoretical Studies on the Formation of the Bisabolene, Curcumene, Acoradiene, Zizaene, Cedrene, Duprezianene, and Sesquithuriferol Sesquiterpenes. *J. Am. Chem. Soc.* **2009**, *131*, 7999.
12. Tantillo, D. J. Biosynthesis *via* Carbocations: Theoretical Studies on Terpene Formation. *Nat. Prod. Rep.* **2011**, *28, 1035.*
13. Hare, S. R.; Tantillo, D. J. Dynamic Behavior of Rearranging Carbocations—Implications for Terpen Biosynthesis. *Beilstein J. Org. Chem.* **2016**, *12*, 377.
14. Tantillo, D. J. The Carbocation Continuum in Terpene Biosynthesis—Where Are The Secondary Cations? *Chem. Soc. Rev.* **2010**, **39**, 2847.
15. Levi, S.; Zhang, Q.; Major, D. T. Thermodynamic and Kinetic Control Determine the Sesquiterpene Reaction Pathways Inside Nanocapsules. *ACS Catal.* **2020**, *10*, 6843.
16. Zhang, Q.; Rinkel, J.; Goldfuss, B.; Dickschat, J. S.; Tiefenbacher, K. Sesquiterpene Cyclisations Catalysed Inside the Resorcinarene Capsule and Application in the Short Synthesis of Isolongifolene and Isolongifolenone. *Nat. Catal.* **2018**, *1*, 609.
17. Frisch, M. J. et. al., Gaussian Inc., Wallingford CT, 2009.
18. Zhao, Y.; Truhlar, D. G. Construction of a Generalized Gradien Approximation by Restoring the Density-Gradient Expansion and Enforcing a Tight Lieb-Oxford Bound. *Theor. Chem. Acc.* **2008**, *120*, 215.
19. Hariharan, P. C.; Pople, J. A. The Influence of Polarization Functions on Molecular Orbital Hydrogenation Energies. *Theor. Chim. Acta* **1973**, *28*, 213.
20. Becke, A. D. Density-Functional Thermochemistry. III The Role of Exact Exchange. *J. Chem. Phys.* **1993**, *98*, 5648.
21. Lee, C.; Yang, W.; Parr, R. G. Development of the Colle-Salvetti Correlation-Energy Formula Into a Functional of the Electron Density. *Phys. Rev. B* **1988**, *37*, 785.
22. Walker, M.; Harvey, A. J. A.; Sen, A.; Dessent, C. E. H. Performance of M06, M06-2X, and M06-HF Density Functionals for Conformationally Flexible Anionic Clusters: M06 Functionals Perform Better Than B3LYP for a Model System With Dispersion and Ionic Hydrogen-Bonding Interactions. *J. Phys. Chem. A.* **2013**, *117*, 12590.
23. Matsuda, S. P. T.; Wilson, W. K.; Ziong, Q. Mechanistic Insights Into Triterpene Synthesis From Quantum Mechanical Calculations. Detection of Systematic Errors in B3LYP Cyclization Energies. *Org. Biomol. Chem.* **2006**, *4*, 530.
24. Weitman, M.; Major, D. T. Challenges Posed to Bornyl Diphosphate Synthase: Diverging Reaction Mechanisms in Monoterpenes. *J. Am. Chem. Soc.* **2010**, *132*, 6349.
25. Zev, S.; Gupta, P. K.; Pahima, E.; Major, D. T. Benchmark Study of QM and QM/MM Methods for Carbocation Chemistry. *J. Chem. Theory Comput.* **2022**, *18*, 167.
26. Raz, K.; Levi, S.; Gupta, P. K.; Major, D. T. Enzymatic Control of Product Distribution in Terpene Synthases: Insights From Multiscale Simulations. *Curr. Opin. Biotech.* **2020**, *65*, 248.

CHAPTER SEVENTEEN

An ionic Hamiltonian for transition metal atoms: Kondo resonances and tunneling currents

F. Flores[a,*], M.S. Tacca[b], and E.C. Goldberg[b]

[a]Departamento Física Teórica de la Materia Condensada and Condensed Matter Physics Center (IFIMAC) Universidad Autónoma de Madrid, Madrid, Spain
[b]Instituto de Física del Litoral, CONICET-UNL, Santa Fe, Argentine
*Corresponding author: e-mail address: fernando.flores@uam.es

Contents

Abstract

An extended ionic Hamiltonian for d^5-transition metal atoms, that incorporates both the effect of the first Hund rule and the effect of the orbital degeneracy in the ground state, is presented. We also discuss a Green-function method to analyze that Hamiltonian, solved by means of the Equation of Motion method. Results for different cases in a V^2 and aV^4-order are given and, in particular, Kondo frequencies are calculated for atoms with an orbital singlet state or with a multi-orbital degeneracy. The effect of the multi-orbital degeneracy is shown to be important.

1. Introduction

Kondo physics has been a fashionable topic in condensed matter physics for more than 50 years.[1] In addition to representing a very attractive model to analyze theoretically many-body effects in solids,[2] it has been introduced for understanding the behavior of many systems of applied interest. In the earliest

Advances in Quantum Chemistry, Volume 85
ISSN 0065-3276
https://doi.org/10.1016/bs.aiq.2022.05.003
Copyright © 2022 Elsevier Inc.
All rights reserved.

theoretical approaches, an Anderson-like Hamiltonian[3] was introduced to model atomic impurities, by means of an intrasite Coulomb interaction associated with a spin degenerate single level. This model has been extensively analyzed using many different techniques[2] and has also been extended to understanding the behavior of more complex systems like molecular magnets, quantum dots or aggregates of transition metal atoms,[4–13] systems that can also be contemplated as possible operational components of quantum computers. It should also be mentioned that the Kondo physics and the associated Mott phenomena are crucial to understanding magnetic materials[14] and high temperature superconductors.[15]

Most works on the Kondo effect use very idealistic models, with a poor description of the inner structure of the magnetic atom. A few works[16–18] have stressed, however, the importance of taking into account both, orbital and spin degeneracies of the atom. We should also mention also that the strong Hund's rule coupling limit in the ionic Hamiltonian has been used previously in many approaches, generally leading to a correct description of the analyzed systems.[19–27]

In this work we deepen the extension of an ionic Hamiltonian introduced to study the spectral density and the tunneling current for a Co atom adsorbed in graphene and copper surfaces.[28,29] This Hamiltonian includes two main effects operating in the atom: (a) the first Hund rule and (b) the degenerate orbital states. In the d^5-transition metal atoms, the electron exchange interaction is large enough to make the first Hund rule operative, with the electrons filling the levels with one spin, up to the half-filled d^5-shell; this behavior can be distorted by the crystal field effect which can split the d^5-levels into several independent groups that keep, however, their total initial spin. On the other hand, while the electrons fill the atomic levels following the first Hund's rule, they also fill those atomic levels following some rules with respect to the degeneracy of the spacial orbital states, typically filling the orbitals that have a specific angular momentum. In Section 2 of this paper, we present our ionic Hamiltonian and discuss how to include in this Hamiltonian both the first-Hund rule and the effects associated with the spacial orbital degeneracy. In Section 3, we discuss how to obtain the properties of that ionic Hamiltonian introducing appropriate Green-functions for it and using an Equation of Motion method (EOM) for analyzing their properties. In particular, we consider in the same section some specific examples, Mn and Ti+, and calculate their Kondo energies, including the case of Co adsorbed on a layer of CuN deposited on Cu(100),[9] whose tunneling currents are also discussed. Finally, in Section 4, we present our conclusions.

2. The ionic Hamiltonian and multiorbital degeneration

Our initial Hamiltonian reads as follows:

$$\hat{H} = \sum_{m,\sigma} \varepsilon_m \hat{n}_{m\sigma} + \sum_m U \hat{n}_{m\uparrow} \hat{n}_{m\downarrow} + \frac{1}{2} \sum_{m \neq m',\sigma} U' \hat{n}_{m\sigma} \hat{n}_{m'-\sigma}$$
$$+ \frac{1}{2} \sum_{m \neq m',\sigma} (U' - J) \hat{n}_{m\sigma} \hat{n}_{m'\sigma} - \frac{1}{2} \sum_{m \neq m',\sigma} J \hat{c}^{+}_{m\sigma} \hat{c}_{m-\sigma} \hat{c}^{+}_{m'-\sigma} \hat{c}_{m'\sigma}$$
$$+ \sum_{k,\sigma} \varepsilon_k \hat{n}_{k\sigma} + \sum_{k,m,\sigma} \left(V_{km} \hat{c}^{+}_{k\sigma} c_{m\sigma} + c.c. \right) \tag{1}$$

The first five terms describe the many electron atom. Here $\hat{c}^{+}_{m\sigma}(\hat{c}_{m\sigma})$ are the fermionic operators creating (annihilating) an electron with spin projection σ in the orbital m and $\hat{n}_{m\sigma} = \hat{c}^{+}_{m\sigma}\hat{c}_{m\sigma}$; the intra-atomic Coulomb repulsions in the same orbital, U, and between different orbitals U', as well as the intra-atomic exchange interaction J, are assumed to be constants independent of the m-orbital index. The fifth term, related to spin-flip processes, restores the invariance under rotation in spin space if $U = U' + J$. The two last terms describe the solid with k-states of energies ε_k and the charge exchange between the atom and the solid through a one electron tunneling mechanism $(V_{km}\hat{c}^{+}_{k\sigma}c_{m\sigma})$, respectively.

We assume initially the crystal field splitting of the d levels negligible, and present in Fig. 1 the atomic levels given by the atomic part of Hamiltonian (1) taking into account the first Hund's rule.

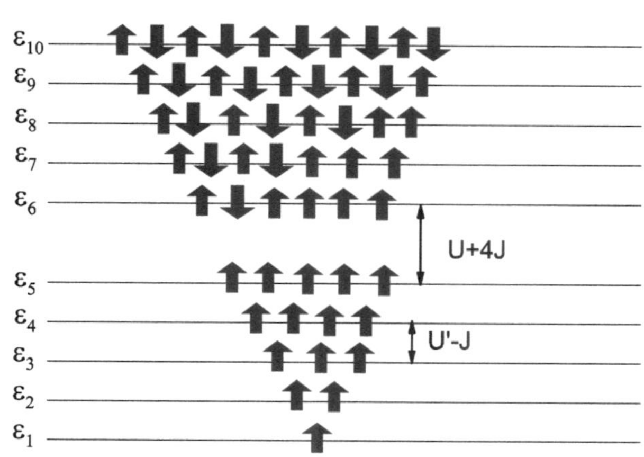

Fig. 1 Shows the atomic energy levels of a d^5-shell for a transition metal atom.

The energy levels in Fig. 1 are calculated as the difference between the total electron energies with N and N-1 electrons: $\varepsilon_i = E(N=i) - E(N=i-1)$. This yields $\varepsilon_1 = E(N=1) - E(N=0) = \varepsilon_d$ and in general $\varepsilon_i - \varepsilon_{i-1} = U' - J$ except for $\varepsilon_6 - \varepsilon_5 = U + 4J$.

For each level we find degeneracy associated with the orbitals and the spin. We start discussing the orbital degeneracy by assuming that all electrons have spin up, and for the sake of simplicity we consider now a strong crystal field introducing a splitting of the d levels into two groups: t_{2g} (d_{xy}, d_{xz}, d_{yz}) and e_g (d_z, $d_{x2\text{-}y2}$). Let us assume that the contribution from e_g-states can be neglected and concentrate our discussion on the t_{2g} orbitals. This is the case of an octahedral symmetry and a strong ligand field (see[30]). In this case, due to the first Hund's rule, we find the following states with maximum spin and different occupations ($M=3$; $N \leq 3$):

$$
\begin{aligned}
N &= 0; \quad |\phi_0\rangle = |000\rangle \quad \text{(one state)} \\
N &= 1; \quad |\phi_1\rangle = |\uparrow 00\rangle, |0 \uparrow 0\rangle, |00 \uparrow\rangle \ \text{(three states)} \\
N &= 2; \quad |\phi_2\rangle = |\uparrow\uparrow 0\rangle, |\uparrow 0 \uparrow\rangle, |0 \uparrow\uparrow\rangle \ \text{(three states)} \\
N &= 3; \quad |\phi_0\rangle = |\uparrow\uparrow\uparrow\rangle \quad \text{(one state)}
\end{aligned}
$$

The complete configurational space of atomic orbitals for this case can be obtained by rotating the spin in each of the previous states,

$$
\begin{aligned}
N=1;&\quad |\uparrow 00\rangle, |\downarrow 00\rangle; |0 \uparrow 0\rangle, |0 \downarrow 0\rangle; |00 \uparrow\rangle, |00 \downarrow\rangle \\
N=2;&\quad \begin{cases} |\uparrow\uparrow 0\rangle, \dfrac{1}{\sqrt{2}}(|\uparrow\downarrow 0\rangle + |\downarrow\uparrow 0\rangle), |\downarrow\downarrow 0\rangle; \\ |\uparrow 0 \uparrow\rangle, \dfrac{1}{\sqrt{2}}(|\uparrow 0 \downarrow\rangle + |\downarrow 0 \uparrow\rangle), |\downarrow 0 \downarrow\rangle; \\ |0 \uparrow\uparrow\rangle, \dfrac{1}{\sqrt{2}}(|0 \uparrow\downarrow\rangle + |0 \downarrow\uparrow\rangle), |0 \downarrow\downarrow\rangle \end{cases} \\
N=3;&\quad |\uparrow\uparrow\uparrow\rangle, \frac{1}{\sqrt{3}}(|\downarrow\uparrow\uparrow\rangle + |\uparrow\downarrow\uparrow\rangle + |\uparrow\uparrow\downarrow\rangle), \\
&\quad \frac{1}{\sqrt{3}}(|\downarrow\downarrow\uparrow\rangle + |\downarrow\uparrow\downarrow\rangle + |\uparrow\downarrow\downarrow\rangle), |\downarrow\downarrow\downarrow\rangle
\end{aligned}
$$

It is interesting to see how the different atomic states can be connected by the term $V_{km}\hat{c}^{+}_{k\sigma}c_{m\sigma}$ of Hamiltonian (1) which describes how one electron is transferred from the m orbital of the atom to the k-state of the metal. For instance, the state $|\uparrow\uparrow\uparrow\rangle$ can jump into the states $|0\uparrow\uparrow\rangle$, $|\uparrow 0 \uparrow\rangle$ and $|\uparrow\uparrow 0\rangle$ depending on the d orbital (channel) involved: d_{xy}, d_{xz} or d_{yz}.

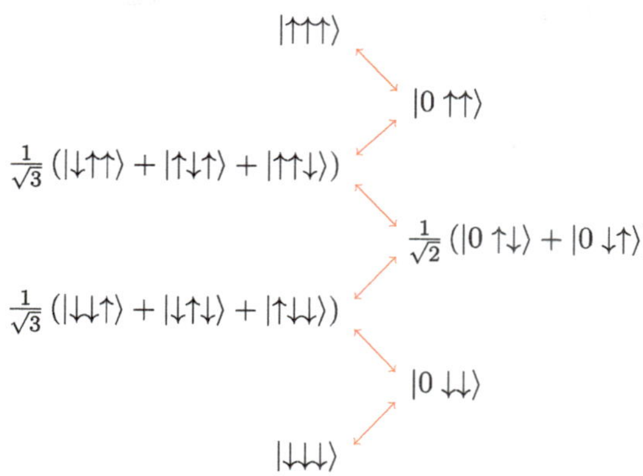

Fig. 2 States $|\uparrow\uparrow\uparrow\rangle$ and $|0\uparrow\uparrow\rangle$ are connected by a k_{dxy}-momentum.

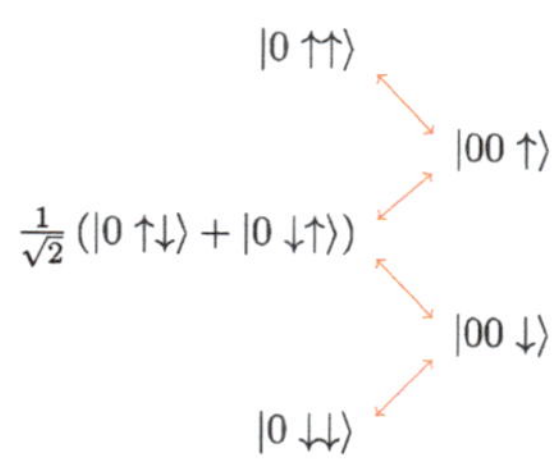

Fig. 3 States $|0\uparrow\uparrow\rangle$ and $|00\uparrow\rangle$ are connected by a k_{dxz}-channel.

In Fig. 2 we show the possible connections between the different states through the channel d_{xy}.

The lines linking the atomic states in the different transitions are related to the appropriate interaction V_{km}, in this case $V_{kd_{xy}}$. In a similar way we have the connections through the channels d_{xz} and d_{yz}. The state $|\uparrow\uparrow\uparrow\rangle$ is the same for the three channels while the set of states $|0\uparrow\uparrow\rangle$, $|\uparrow 0\uparrow\rangle$ and $|\uparrow\uparrow 0\rangle$ corresponds to a 3-fold degeneracy (with an angular momentum symmetry).

In the same way, we can obtain how the set of states $|\uparrow\uparrow 0\rangle$, $|\uparrow 0\uparrow\rangle$, … or $|0\uparrow\downarrow\rangle$ can be connected with the states $|\uparrow 00\rangle$, $|0\uparrow 0\rangle$, … or $|00\downarrow\rangle$ through the different channels, as it is shown in Fig. 3 for the channel d_{xz}.

Similar connections involving the other channels can be obtained in a similar way, showing that there is a threefold degeneracy for the set of states $|\uparrow\uparrow 0\rangle$, $|\uparrow 0\uparrow\rangle$ and $|0\uparrow\uparrow\rangle$, which are also connected to the threefold degenerate states $|\uparrow 00\rangle$,$|0\uparrow 0\rangle$ and $|00\uparrow\rangle$.

Finally, the states $|\uparrow 00\rangle$, $|0\uparrow 0\rangle$ and $|00\uparrow\rangle$ are connected with the vacuum $|000\rangle$ through the three channels d_{xy}, d_{xz} and d_{yz}. In this case a 3-fold degenerate set of states, with the angular momentum L = 1, is connected to a singlet state.

We can describe the previous results in a compact way by using the notation $|S, M\rangle$ for the atomic states, as shown below:

$$|3/2,3/2\rangle \equiv |\uparrow\uparrow\uparrow\rangle;|3/2,1/2\rangle = \frac{1}{\sqrt{3}}(|\downarrow\uparrow\uparrow\rangle + |\uparrow\downarrow\uparrow\rangle + |\uparrow\uparrow\downarrow\rangle);\cdots;|3/2,-3/2\rangle = |\downarrow\downarrow\downarrow\rangle$$
$$|1,1\rangle_1 = |\uparrow\uparrow 0\rangle;|1,1\rangle_2 = |\uparrow 0 \uparrow\rangle;|1,1\rangle_3 = |0\uparrow\uparrow\rangle;\cdots;|1-1\rangle_3 = |0\downarrow\downarrow\rangle$$
$$|1/2,1/2\rangle_1 = |\uparrow 00\rangle;|1/2,1/2\rangle_2 = |0\uparrow 0\rangle;|1/2,1/2\rangle_3 = |00\uparrow\rangle;\cdots;|1/2,-1/2\rangle_3 = |00\downarrow\rangle$$

Then, the previous diagrams in Figs. 2 and 3, take the form shown in Fig. 4.

Then, the Hamiltonian (1) which describes an adsorbed magnetic atom with total spin S with the charge fluctuating from N to N-1 electrons and the spin fluctuating from S to S-1/2, can be written as the following multiorbital ionic Hamiltonian by using the projection operators $|S, M\rangle_p\langle S, M|_p$ and $|S-1/2, M-\sigma\rangle_q\langle S-1/2, M-\sigma|_q$[28]:

$$\hat{H} = \sum_{k,\sigma} \varepsilon_k \hat{n}_{k\sigma} + \sum_{p,M} E_M^{S(p)} |S, M\rangle_p \langle S, M|_p + \sum_{q,m} E_m^{S-1/2(p)} |S-1/2, m\rangle_q \langle S-1/2, m)|_q$$
$$+ \sum_{k,\sigma,p,q,M} \left[V_{k,M,\sigma}^{S,p,q} c_{k\sigma}^{+} |S-1/2, M-\sigma\rangle_q \langle S, M)|_p + c.c. \right] \tag{2}$$

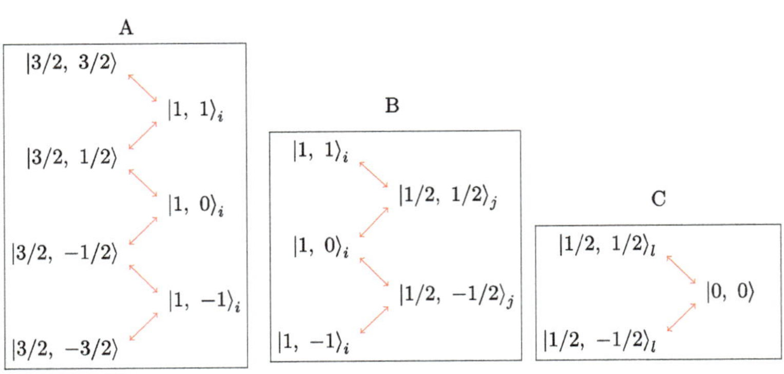

Fig. 4 (A) States $|3/2,M\rangle$ and $|1,m\rangle_i$ are connected by the k-tunneling mechanism. The index i indicates the channel involved. (B) The same as in (A) for states $|1,M\rangle_i$ and $|1/2,m\rangle_j$. (C) The same as in A for states $|1/2,m\rangle_l$ and $|0,0\rangle$.

In Eq. (2) the interaction between $|S, M\rangle_p$ and $|S-1/2, M-\sigma\rangle_q$ is given by:

$$V_{k,M,\sigma}^{S,p,q} = \langle S - 1/2,\, M - \sigma|_q \hat{c}_{d(p,\, q),\sigma} |S,\, M\rangle_p V_{kd(p,\, q)} \tag{3}$$

with $V_{kd(p,q)}$being the coupling between the k band state and the *d(p,q)* orbital involved in the fluctuation between both configurations, and its coefficient is:

$$\langle S - 1/2M - \sigma|_p \hat{c}_{d(pq),\sigma} |SM\rangle_q = \zeta_{pq} \sqrt{\frac{S + 2\sigma M}{2S}} \tag{4}$$

where $\zeta_{pq}$is 1 (-1) if an even (odd) number of permutations are necessary to annihilate the electron (hole) in the *d(p, q)* orbital, and it is equal to 0 if this *d* orbital is empty in the $|S, M\rangle_p$configuration. The indexes *p* and *q* indicate the different configurations depending on the occupied orbitals of the valence shell in each case (see Figs. 2, 3 and 4).[28,29]

Up to this point we have assumed that N, the number of electrons in the atom for the t_{2g}-levels, is $N \leq 3$, meaning that this shell is less than half-filled occupied. The case $N > 3$ can be treated as holes, and one can recover for this case the symmetric degeneracy of the cases discussed above for $N \leq 3$, due to the electron-hole symmetry of the problem. For example, the fluctuations between 4 and 5 electrons in that t_{2g}-shell can be described as fluctuations between 2 and 1 holes, with a similar interaction to the one given for the fluctuations between 1 and 2 electrons.

The case d^5 (with $n = 5$ orbitals) can be analyzed in a similar way to the case d^3. For the sake of brevity we will analyze only a couple of cases: (a) first, we will consider that the number of electrons fluctuate in the atom between $N = 5$ and $N = 4$.We also find for this case the same interacting ionic Hamiltonian given above, Eq. (2), with $S = M = 5/2$; moreover, in this particular case there is no degeneracy for the state with $S = 5/2$, and a five-fold degeneracy for $S = 2$. Mn^{2+} in some metals is well described by this case. (b) The second case we consider corresponds to Ti^+ with 1-electron in the d^5-shell: in this case, the state $S = 1/2$ has a 2×5-fold degeneracy, fluctuating to a singlet state with $S = 0$. We shall discuss in the next section these two cases from the point of view of a possible Kondo resonance.

Notice that the d^5-degeneracy can be lifted in a variety of ways by the crystal field. We have just mentioned that in a cubic symmetry $d^5 \rightarrow t_{2g} + e_g$; in an octahedral symmetry t_{2g} is lower in energy than the e_g-levels, while in a tetrahedral symmetry is the other way around. It is interesting to realize that for the octahedral symmetry, 3 electrons define by the first Hund's rule a half-filled t_{2g}-shell yielding an orbital singlet state. With 8 electrons, six electrons can fill the d^3-shell of t_{2g}, and 2 electrons fill the first levels of the e_g-shell yielding also an orbital singlet state. In the case of a tetrahedral symmetry, things are similar but with 2 electrons filling first the e_g-shell, or with 7 electrons filling the e_g-shell (with 4 electrons) and half-filling the d^3-shell with the remaining 3 electrons. These particular cases define an orbital singlet state for the d^5-shell. In many other cases, we find a multiorbital and degenerate ground state for the d^5-shell.

If a d^5-atom is adsorbed on a surface, the broken degeneracy of the states depends on the symmetry of the adsorption site of the atom. In the case of Co adsorbed on either graphene or Cu(111) surface,[28,29] the favorable hollow site leads to a C_6 symmetry that splits the five d orbitals into the groups E1 (d_{xz},d_{yz}), E2 (d_{xy}, $d_{x2\text{-}y2}$) and A1 (d_{z2}). A case of interest is Fe or Co on CuN[5,9]: in this case the site symmetry is so low that all the d^5-levels have a different energy and their atomic occupancies are all different. For example, DFT-calculations for Fe show that the atom fluctuates between 6 and 7 electrons, and that the $d_{x2\text{-}y2}$ and d_{z2} orbitals have occupancies for the minority spin of 0.72 and 0.49 electrons, respectively.[31] This indicates that the atom charge is fluctuating between the states $S=3/2$ (with 3 holes and the minority spins of $d_{x2\text{-}y2}$ and d_{z2} occupied) and $S=2$ (with only the minority spin level of d_{z2} occupied). Accordingly, we will discuss this case below considering the fluctuations from $S=3/2$ to $S=1$ without any degeneracy, because it corresponds to fluctuations in the occupancies of the $d_{x2\text{-}y2}$-level. We anticipate that in these particular cases, for Fe and Co, there appears another atomic term associated with an effective anisotropy interaction, $\Lambda_{ij}S_iS_j$, which comes from a second-order calculation of the atomic spin-orbit coupling $\lambda L.\ S$. That interaction is small but has an important effect on the behavior of the Kondo resonance for those transition metal atoms. In particular, $\Lambda_{ij}S_iS_j$ lifts partially the degeneracy of the different M-states appearing in the $S=3/2$ to $S=1$ fluctuations.

3. Green functions, EOM, tunneling currents

3.1 Green functions and EOM solution

In this section we present the equation of motion of the appropriate Green functions for the case of a system described by Hamiltonian (2).

$$G_{|SM_0\rangle_{p_0}\langle S-1/2M_0-\sigma|_{q_0}} = i\Theta(t'-t)\Big\langle\Big\{|SM_0\rangle_{p_0} \Big\langle S-1/2',M_0-\sigma\Big|_{q_0 t'};|S-1/2M_0-\sigma\rangle_{q_0}\langle SM_0|_{p_0 t}\Big\}\Big\rangle \quad (5)$$

In Eq. (5) we use the Heisenberg representation, and take $\{\hat{A};\hat{B}\} = \hat{A}\hat{B} + \hat{B}\hat{A}$. The operator $|S-1/2, M_0-\sigma\rangle_{q_0}\langle S, M_0|$ plays the role of an annihilation operator and expression (7) is equivalent to the conventional one $i\Theta(t'-t)\langle\{\hat{a}^+(t');\hat{a}(t)\}\rangle$defined with the fermionic creation ($\hat{a}^+$) and annihilation operators ($\hat{a}$), respectively.

A brief description of the equation of motion (EOM) method to calculate the Green functions (5) is given below. The details can be found in the Appendix. Let us write the expression (5) in the compact form,

$$G_{i'j',t'}\left(|j\rangle\langle i|_t\right) = i\Theta(t'-t)\left\langle\left\{|i'\rangle\langle j'|_{t'};|j\rangle\langle i|_t\right\}\right\rangle \tag{6}$$

and introduce the following time derivative:

$$\begin{aligned}\frac{d}{dt}G_{i'j',t'}\left(|j\rangle\langle i|_t\right) &= -i\delta(t'-t)\left\langle\left\{|i'\rangle\langle j'|_{t'};|j\rangle\langle i|_t\right\}\right\rangle \\ &+ i\Theta(t'-t)\left\langle\left\{|i'\rangle\langle j'|_{t'};\frac{d}{dt}|j\rangle\langle i|_t\right\}\right\rangle\end{aligned} \tag{7}$$

Using $\frac{d}{dt}\hat{O} = -i[\hat{H},\hat{O}] = -i(\hat{H}\hat{O}-\hat{O}\hat{H})$, and Fourier transforming Eq. (7) in time, we obtain:

$$\omega G_{i'j'}(|j\rangle\langle i|) = \langle\{|i'\rangle\langle j'|;|j\rangle\langle i|\}\rangle + G_{i'j'}\left(\left[|j\rangle\langle i|,\ \hat{H}\right]\right) \tag{8a}$$

Or equivalently,

$$\left(\omega - E_i + E_j\right)G_{i'j',t'}\left(|j\rangle\langle i|_t\right) = \langle\{|i'\rangle\langle j'|;\ |j\rangle\langle i|\}\rangle + G_{i'j'}\left(\left[|j\rangle\langle i|,\hat{H}_{\text{int}}\right]\right) \tag{8b}$$

where we have assumed $\hat{H} = \hat{H}_0 + \hat{H}_{\text{int}}$ and $\hat{H}_0|i\rangle = E_i|i\rangle$, $\hat{H}_0|j\rangle = E_j|j\rangle$. Eq. (8a) is the starting point for the EOM-method; from Eq. (8a) a second order Green function is obtained, by means of the equation:

$$\omega G_{i'j'}\left(\left[|j\rangle\langle i|,\hat{H}\right]\right) = \left\langle\left\{|i'\rangle\langle j'|;\ \left[|j\rangle\langle i|,\hat{H}\right]\right\}\right\rangle + G_{i'j'}\left(\left[\left[|j\rangle\langle i|,\hat{H}\right],\hat{H}\right]\right) \tag{9}$$

which is equivalent to Eq. (8a) replacing $|j\rangle\langle i|$ by $\left[|j\rangle\langle i|,\ \hat{H}\right]$.

Likewise, higher order Green-functions,

$G_{i'j'}\left(\left[\left[|j\rangle\langle i|,\hat{H}\right],\hat{H}\right]\right)$, $G_{i'j'}\left(\left[\left[\left[|j\rangle\langle i|,\hat{H}\right],\hat{H}\right],\hat{H}\right]\right)$, etc

are obtained by taking higher derivatives, and a closed system of equations is obtained by contraction of the operators $\hat{c}_k^+\hat{c}_k$ or $\hat{c}_k\hat{c}_k^+$ in the corresponding Green functions. See the Appendix for more details.

In the case of a degenerate set of orbitals involved in the transitions, $E^{S(p_0)}_{M_0-\sigma'+\sigma''} = E^S$, $E^{S-1/2(q_0)}_{M_0-\sigma} = E^{S-1/2}$, $\varepsilon_d = E^S - E^{S-1/2}$ and $\langle n_{k\sigma}\rangle \equiv \langle n_k\rangle$, the equations of motion (A11) take the following expression in the lowest V^2-order approximation[28,29]

$$\left[\begin{array}{l}\omega-\varepsilon_d \\ -\sum_{k,q'}\frac{\left|\zeta_{p_0q'}V_{kd(p_0q')}\right|^2\left\langle 1-n_k\right\rangle}{\left(\omega-\varepsilon_k\right)}-\frac{2S+1}{2S}\sum_{k,p'}\frac{\left|\zeta_{p'q_0}V_{kd(p'q_0)}\right|^2\left\langle n_k\right\rangle}{\left(\omega-\varepsilon_k\right)}\end{array}\right]G_{|SM_0\rangle_{p_0}\langle S-1/2,M_0-\sigma|_{q_0}}$$

$$=\left\langle\left|SM_0\right\rangle_{p_0}\left\langle SM_0\right|_{p_0}\right\rangle+\left\langle\left|S-1/2,M_0-\sigma\right\rangle_{q_0}\left\langle S-1/2,M_0-\sigma\right|_{q_0}\right\rangle$$

$$+\sum_{k,\sigma',q'}\frac{V^{S,p_0q'*}_{k,M_0,\sigma'}\left\langle\left|SM_0\right\rangle_{p_0}\left\langle S-1/2,M_0-\sigma'\right|_{q'}c_{k\sigma'}\right\rangle}{\left(\omega-\varepsilon_k\right)}$$

$$-\sum_{k,\sigma',p'}\frac{V^{S,p'q_0*}_{k,M_0-\sigma+\sigma',\sigma'}\left\langle\left|SM_0-\sigma+\sigma'\right\rangle_{p'}\left\langle S-1/2,M_0-\sigma\right|_{q_0}c_{k\sigma'}\right\rangle}{\left(\omega-\varepsilon_k\right)} \tag{10}$$

And for the fourth order in V, Eq. (A16) is reduced to the following one:

$$\left[\begin{array}{l}\omega-\varepsilon_d \\ -\sum_{k,q'}\frac{\left|\zeta_{p_0q'}V_{kd(p_0q')}\right|^2\langle 1-n_k\rangle}{\left(\omega-\varepsilon_k-\Xi^{S-1/2,(q'q_0)<}(k\omega)\right)}-\frac{2S+1}{2S}\sum_{k,p'}\frac{\left|\zeta_{p'q_0}V_{k'd(p'q_0)}\right|^2\langle n_k\rangle}{\left(\omega-\varepsilon_k-\Xi\wedge S-1/2,(p'p_0)>(k\omega)\right)}\end{array}\right]$$

$$G_{|SM_0\rangle_{p_0}\langle S-1/2,M_0-\sigma|_{q_0}}=\left\langle|SM_0\rangle_{p_0}\langle SM_0|_{p_0}\right\rangle+\left\langle|S-1/2,M_0-\sigma\rangle_{q_0}\langle S-1/2,M_0-\sigma|_{q_0}\right\rangle$$

$$+\sum_{k,\sigma',q'}\frac{V^{S,p_0q'*}_{k,M_0,\sigma'}\left\langle|SM_0\rangle_{p_0}\left\langle S-1/2,M_0-\sigma'\right|_{q'}c_{k\sigma'}\right\rangle}{\left(\omega-\varepsilon_k-\Xi^{S-1/2,(q'q_0)<}(k\omega)\right)}$$

$$-\sum_{k,\sigma',p'}\frac{V^{S,p'q_0*}_{k,M_0-\sigma+\sigma',\sigma'}\left\langle\left|SM_0-\sigma+\sigma'\right\rangle_{p'}\langle S-1/2,M_0-\sigma|_{q_0}c_{k\sigma'}\right\rangle}{\left(\omega-\varepsilon_k-\Xi S-1/2,(p'p_0)>(k\omega)\right)} \tag{11}$$

where,

$$\begin{aligned}\Xi^{S-1/2,(q'q_0)<}(k\omega)&=\frac{2S+1}{2S}\sum_{k',p''}\frac{\left|\zeta_{p''q'}V_{k'd(p'',q')}\right|^2\langle n_{k'}\rangle}{\omega-\varepsilon_k+\varepsilon_{k'}-\varepsilon_d}\\&+\frac{2S+1}{2S}\sum_{k',p''}\frac{\left|\zeta_{p''q_0}V_{k'd(p''q_0)}\right|^2\langle n_{k'}\rangle}{\omega-\varepsilon_k-\varepsilon_{k'}+\varepsilon_d}\\ \Xi\wedge S-1/2(p'p_0)>(k\omega)&=\sum_{k',q''}\frac{\left|\zeta_{p'q''}V_{k'd(p',q'')}\right|^2\langle 1-n_{k'}\rangle}{(\omega-\varepsilon_k+\varepsilon_{k'}-\varepsilon_d)}\\&+\sum_{k',q''}\frac{\left|\zeta_{p_0q''}V_{k'd(p_0q'')}\right|^2\langle 1-n_{k'}\rangle}{(\omega-\varepsilon_k-\varepsilon_{k'}+\varepsilon_d)}\end{aligned} \tag{12}$$

The sum over the spin index in the expressions below has been done by taking into account Eqs. (3) and (4)[28]:

$$\sum_{\sigma''}\left|V^{S,pq''}_{k',M,\sigma''}\right|^2 = \left|\zeta_{pq''}V_{k'd(p,q'')}\right|^2$$

$$\sum_{\sigma''}\left|V^{S,pq''}_{k',M+\sigma'',\sigma''}\right|^2 = \frac{2S+1}{2S}\left|\zeta_{pq''}V_{k'd(p,q'')}\right|^2$$

Notice that Eq. (11), $O(V^4)$, is reduced to Eq. (10), $O(V^2)$, by eliminating the self-energies $\Xi^<$ and $\Xi^>$ that represent the typical $O(V^4)$ contribution.

3.2 Kondo resonance for a d^5-shell with $S=5/2$ and $S=1/2$

1. *A singlet ground state with $S=5/2$ and n degenerate orbitals involved in the transition to $S=2$ (n channels).*

In this case, there is no degeneracy for the state with $S=5/2$, and a 5-fold degeneracy for $S=2$.

The feasibility of a Kondo resonance is studied through the analysis of the pole of the Green function for $\omega \to \varepsilon_F (\varepsilon_F = 0$is the Fermi energy). In the second order approximation (Eq. 10) we look for the zero of $(V_{kd(p,q'')} = V_{kd})$:

$$\omega_K - \varepsilon_d - \sum_{k,q'}\frac{\left|\zeta_{pq''}V_{kd}\right|^2\langle 1-n_k\rangle}{(\omega_K-\varepsilon_k)} - \frac{6}{5}\sum_{k,p'}\frac{\left|\zeta_{pq''}V_{kd}\right|^2\langle n_k\rangle}{(\omega_K-\varepsilon_k)} = 0 \quad (13)$$

The sum over p' has only one possibility ($|\uparrow\uparrow\uparrow\uparrow\uparrow\rangle$) while q' has n possibilities depending on the number of active channels ($|0\uparrow\uparrow\uparrow\uparrow\rangle$, $|\uparrow 0\uparrow\uparrow\uparrow\rangle$; ...). Then, Eq. (13) reads as:

$$\omega_K - \varepsilon_d - n\sum_{k}\frac{\left|V_{kd}\right|^2\langle 1-n_k\rangle}{(\omega_K-\varepsilon_k)} - \frac{6}{5}\sum_{k}\frac{\left|V_{kd}\right|^2\langle n_k\rangle}{(\omega_K-\varepsilon_k)} = 0 \quad (14)$$

From Eq. (14) we can observe that a Kondo resonance *only* exists for $n=1$ (one channel), and in this case $\omega_K = \Lambda\exp(-5\pi|\varepsilon_d|/\Gamma)$.[19]

This was found in the Co/graphene and Co/Cu(111) systems[28,29] where the second order in V predicts a Kondo resonance only when the channel involved in the fluctuation is a nondegenerated one ($A1(d_{z2})$), while Kondo resonances disappear when the active channel is an orbital from the 2-fold degenerated groups E1 (d_{xz},d_{yz}), E2 (d_{xy}, d_{x2-y2}).[32] However, in the fourth order approximation (Eq. 11), the poles are obtained from:

$$\omega_K - \varepsilon_d - n\sum_k \frac{|V_{kd}|^2 \langle 1-n_k \rangle}{\left(\omega_K - \varepsilon_k - \Xi^{S-1/2<}(k\omega)\right)} - \frac{6}{5}\sum_k \frac{|V_{k'd}|^2 \langle n_k \rangle}{\left(\omega_K - \varepsilon_k - \Xi^{S-1/2>}(k\omega)\right)} = 0 \quad (15)$$

For $\omega \approx \omega_K$, the relevant contributions from ε_k to the self-energies $\Xi^<$ and $\Xi^>$ are close to $\omega_K(\varepsilon_k \approx \omega_K)$, and the real parts of the two terms of these quantities cancel each other. However, in $\Xi^<$ the denominators of these two terms can be zero for $\varepsilon_{k'} = \varepsilon_d$, giving an imaginary contribution to $\Xi^<$ that takes the value of $(6/5)i\Gamma$; for $\Xi^>$ this is not the case and we find that $\Xi^> \approx 0$. Then, from Eq. (15) we find a Kondo resonance *for any number of degenerate orbitals and active channels*, with $\omega_K = \Lambda \exp(-5\pi|\varepsilon_d|/6\Gamma)$.

2. *S = 1/2 and n ≤ 5 degenerate orbitals involved in the transitions (n channels).* The state with $S = 1/2$ is 2×5-fold degenerate while the state with spin 0 is a singlet state. In this case p has $2n$ ($n \leq 5$) possibilities ($|\uparrow 0000\rangle$; $|\downarrow 000\rangle$; $|0 \uparrow 000\rangle$; ...) while q has only one, $|00000\rangle$. Then, in second order approximation,

$$\omega_K - \varepsilon_d - \sum_k \frac{|V_{kd}|^2 \langle 1 - n_k \rangle}{(\omega_K - \varepsilon_k)} - 2n \sum_k \frac{|V_{kd}|^2 \langle n_k \rangle}{(\omega_K - \varepsilon_k)} = 0, \qquad (16)$$

and the Kondo resonance appears for the following energy:

$$\omega_K = \Lambda \exp\left(-\pi|\varepsilon_d|/(2n-1)\Gamma\right), \qquad (17)$$

while in the fourth order approximation,

$$\omega_K - \varepsilon_d - \sum_k \frac{|V_{kd}|^2 \langle 1-n_k \rangle}{\left(\omega_K - \varepsilon_k - \Xi^{S-1/2<}(k\omega)\right)} - 2n\sum_k \frac{|V_{kd}|^2 \langle n_k \rangle}{\left(\omega_K - \varepsilon_k - \Xi^{S-1/2>}(k\omega)\right)} = 0 \quad (18)$$

By taking into account that near the Fermi level $\Xi^< \approx 2i\Gamma$ and $\Xi^> \approx 0$, we find $\omega_K = \Lambda \exp(-\pi|\varepsilon_d|/(2n)\Gamma)$ in agreement with the n-fold degenerate model of Hewson.[2]

3.3 Co on CuN, anisotropy interaction, Kondo resonance

In this section we consider the case of Co adsorbed on CuN, with the magnetic atom fluctuating from $|3/2, M\rangle$ to the $|S-1/2, M-\sigma\rangle$ state. As mentioned above, the low symmetry of the CuN adsorption site lifts completely the degeneracy of the d^5-levels, so that in Hamiltonian (2) we find for this case $p = q = 1$ with no orbital degeneracy, but still keeping the spin degeneracy, fourfold for $S = 3/2$ and threefold for $S = 1$. However, this degeneracy is lifted partially by an effective anisotropy interaction $\Lambda_{ij}S_iS_j$ that takes the form $D\hat{S}_z^2 + E\left(\hat{S}_x^2 - \hat{S}_y^2\right)$ with the z-direction of the uniaxial interaction D being parallel to the surface. For Co, $D=2.8$ meV and $E=0$[6,9]; then, we find the following eigenstates and eigenvalues for $D\hat{S}_z^2 + E\left(\hat{S}_x^2 - \hat{S}_y^2\right)$: for $S=3/2$, the eigenstates $|3/2,\pm 3/2\rangle$ have the energy $E_{3/2,\pm 3/2}=9D/4$, while for $|3/2,\pm 1/2\rangle$ the energy is $E_{3/2,\pm 1/2}=D/4$; for $S=1$, we find the eigenstate $|1,0\rangle$ with $E_{1,0}=0$, and the eigenstates $|1,\pm 1\rangle$ with $E_{1,\pm 1}=D$. Thus, we still have the same eigenstates that we had with no anisotropy interaction, but with different eigenvalues.

These results show that we can also introduce, to analyze this case, the Green-functions, and the EOM method to obtain them both in $O(V^2)$ and in $O(V^4)$, in the same way as we did above; the only difference to consider is associated with the new atom-levels appearing in both the initial equation for the Green-function, and in the denominators of the self-energies, $\Xi^<$ and $\Xi^>$. Notice that these changes yield three different Green functions:

$$G_{|3/2,1/2\rangle\langle 1,1|},\ G_{|3/2,-1/2\rangle\langle 1,0|},\ G_{|3/2,-3/2\rangle\langle 1,-1|}.$$

These Green-functions have been published in Advances of Quantum Chemistry,[33] and the reader is referred to that reference for their detailed expressions. Here, we only mention that these results can be used to calculate the tunneling currents, which have been measured by Otte et al.,[9] by means of the equation:

$$G_\sigma = \frac{d}{dt} I_\sigma = \frac{2e}{h} \sum_{M,\sigma} \Gamma^S_{t,M,\sigma} \mathrm{Im}\, G_{|SM\rangle\langle S-1/2 M-\sigma|} \tag{19}$$

where $\Gamma^S_{t,M,\sigma}(\omega) = \pi \sum_{kt} \left| V^S_{kt,M,\sigma} \right|^2 \delta(\hbar\omega - \varepsilon_k)$ is the broadening associated with the tip-atom coupling $V^S_{kt,M,\sigma}$ and the tip density of states $\sum_k \delta(\hbar\omega - \varepsilon_k)$. For Co, $|S,M\rangle = |3/2,M\rangle$ and $|S-1/2,M-\sigma\rangle = |1,M-\sigma\rangle$. Eq. (19) is a second order contribution in $V^S_{kt,M,\sigma}$, obtained assuming that we are in tunneling regime and that the tip-atom interaction is very small. Notice that Eq. (19) is similar to the well-known Bardeen's equation[34]: $dI/dV = \frac{4\pi e}{h} |V_{ta}|^2 \rho_t \mathrm{Im}\, G_a(\omega)$, where ρ_t is the tip density of states, and $G_a(\omega)$ the Green function of the atom. Fig. 5 in Advances in Quantum Chemistry[33] shows $dI/dV = dI_\uparrow/dV + dI_\downarrow/dV$ as a function of V for calculations of the Green function, in both, $O(V^2)$ and $O(V^4)$ approximations. Both calculations show a Kondo resonance although the $O(V^4)$ yields a better and reasonable agreement with the experiment. This difference in $O(V^2)$ and $O(V^4)$ is well appreciated in the calculation of the Kondo energy in both approximations. As discussed in that reference, the values of ω_K for those two cases are:

$$\omega_K = \Lambda \exp\left(-3\pi|\varepsilon_d|/2\Gamma\right) \rightarrow \text{for}\, O\left(V^2\right).$$
$$\omega_K = \Lambda \exp\left(-3\pi|\varepsilon_d|/4\Gamma\right) \rightarrow \text{for}\, O\left(V^4\right).$$

indicating that the $O(V^4)$ yields a much broader Kondo resonance.

4. Discussion and conclusions

We have first presented in this paper a detailed discussion of the spin and orbital degeneracies for a t_{2g}-shell, with a 3-fold degeneracy in the atomic levels, ε_d, assuming that the first Hund's rule is operative in the magnetic atom configuration. This particular case allows us to present an extended ionic Hamiltonian, whereby the interaction between the atom and a metal is described including both the effect of the spin degeneracy associated with that first Hund-rule, and also with the contributions associated with the orbital degeneracy of the atomic states. In this ionic Hamiltonian, Eq. (2), the atom jumps from an initial state, $|S, M\rangle_p$, to a final one, $|S-1/2, M-\sigma\rangle_q$, due to an electron channel i, as described by the interaction potential $V_{k,M,\sigma}^{S,p,q}$ of Eq. (3). The spin degeneracy is described by the (S,M)-parameters, and the orbital degeneracy by the p and q indexes. It should be mentioned that this orbital degeneracy is associated with the angular momentum of the atomic states; in particular, in the diagrams of Fig. 4A and B, the indices i, with values 1, 2 or 3, correspond to states with the three components of an angular momentum equal to 1.

In a second step, we present in Section 3 how to analyze that Hamiltonian using a Green-function method adapted to the atomic states $|S,M\rangle_p$ and $|S-1/2, M-\sigma\rangle_q$, and an EOM-method to calculate those Green-functions. These equations are solved in order V^2 and V^4, where V represents the parameter of expansion used to solve the EOMs.

Finally, in Section 4 results for some specific d^5-shells are presented. The case $S=5/2$ is representative of a magnetic atom having a singlet ground state; we have also considered similar cases such as: 3 electrons in a t_{2g}-shell; 2 electrons in a e_g-shell; or 1 electron in a non-degenerate level. It is worth mentioning that in those cases we have obtained the following Kondo energies:

$$\omega_K = \Lambda \exp\left(-3\pi|\varepsilon_d|/4\Gamma\right) \rightarrow \text{for } t_{2g} - \text{shell } O\left(V^4\right).$$
$$\omega_K = \Lambda \exp\left(-2\pi|\varepsilon_d|/3\Gamma\right) \rightarrow \text{for } e_g - \text{shell } O\left(V^4\right).$$

$$\omega_K = \Lambda \exp\left(-\pi|\varepsilon_d|/2\Gamma\right) \rightarrow \text{for } 1 - \text{level } O\left(V^4\right),$$

to be compared with the d^5-case:

$$\omega_K = \Lambda \exp\left(-5\pi|\varepsilon_d|/6\Gamma\right) \rightarrow \text{for } d^5 - \text{shell } O\left(V^4\right).$$

Notice that the result for the t_{2g}-shell coincides with the one obtained for Co on CuN. This is due to having in both cases the same spin and a singlet ground state for the atom. We also mention that in the 1-level case, we find the exponent of the Haldane Kondo frequency.[35] Notice also that the Kondo frequency, obtained for one level also coincides with the one calculated in Section 3.2 for the case of $S=1/2$ and $n \leq 5$ degenerate orbitals, if we take $n=1$.

It should be emphasized that in some cases the introduction of the V^4-order is mandatory for obtaining appropriate results. The case $S=5/2$ for d^5 is a good example: as discussed in Section 3.2, for $n=5$ the Kondo resonance disappears for a V^2-calculation while a V^4-order yields an appropriate Kondo energy. In other cases not so extreme, the effect of a fourth order in V calculation is to broaden the Kondo resonance in a better agreement with the experiment: the case of Co on CuN is a good example, since in order V^2 we find $\omega_K=\Lambda\exp(-3\pi|\varepsilon_d|/2\Gamma)$, and in order V^4, $\omega_K=\Lambda\exp(-3\pi|\varepsilon_d|/4\Gamma)$.

In conclusion, we have presented an extended ionic Hamiltonian for d^5-transition metal atoms that incorporates the effect of the first Hund's rule in the spin degeneracy, and the effect of the orbital degeneracy. We have also discussed how to solve that Hamiltonian using an EOM-method up to $O(V^2)$ and $O(V^4)$. The results for the Kondo resonance of different cases suggest that this approach offers a convenient way for treating d^5-magnetic atoms.

Acknowledgments

This paper has been written to honor Jack Sabin for his distinguished achievements to the field of modern Quantum Chemistry, and for his invaluable contribution as Serial Editor of Advances in Quantum Chemistry.

FF acknowledges financial support from the Spanish Ministry of Science and Innovation through projects No. MAT2017-88258-R and No. CEX2018-000805-M (Maria de Maeztu Programme for units of Excellence in R&D).

ECG and MST acknowledge financial support from Consejo Nacional de Investigaciones Científicas y Técnicas (CONICET) through PIP Grant No. 11220150100546CO, Universidad Nacional del Litoral (UNL) through CAI+D No. 50620190100034LI.

Appendix

The Hamiltonian which describes a magnetic atom adsorbed on a surface which fluctuates from a total spin S to S-1/2 due to the interaction with the surface is given by the expression:

$$H=\sum_{k,\sigma}\varepsilon_k n_{k\sigma}+H_{atom}+\sum_{k,\sigma,p,q,M}\left[V^{S,p,q}_{k,M,\sigma}c^{+}_{k\sigma}|S-1/2,M-\sigma\rangle_q\langle S,M)|_p+c.c.\right] \quad (A1)$$

By introducing spin anisotropy effects and the presence of an external magnetic field B, the atomic part of Hamiltonian (A1) can be written as follows,

$$\hat{H}_{atom}=\sum_{m,r}E^{s,r}_m|s,m\rangle_r\langle s,m|_r+D\hat{s}^2_z+\mathrm{E}\left(\hat{s}^2_x-\hat{s}^2_y\right)+\mu_B\hat{B}.g\hat{s} \quad (A2)$$

where D and E define the uniaxial and transversal anisotropy parameters and g the gyromagnetic factor.

The eigenstates of this atomic Hamiltonian written in the basis set of the different orbital configurations $|S,M\rangle_\beta$ and $|S-1/2,m\rangle_\alpha$ are:

$$\left|\Psi^{S(\beta)}_i\right\rangle=\sum_M A^{i,S(\beta)}_M|S,M\rangle_\beta;\quad \left|\Psi^{S-1/2(\alpha)}_j\right\rangle=\sum_m A^{j,S-1/2(\alpha)}_m|S-1/2,m\rangle_\alpha$$

And the Hamiltonian (A1) in this basis set takes the form:

$$\begin{aligned}H=&\sum_{k,\sigma}\varepsilon_k n_{k\sigma}+\sum_{i,\beta}\mathrm{E}^{S(\beta)}_i\left|\Psi^{S(\beta)}_i\right\rangle\left\langle\Psi^{S(\beta)}_i\right|+\sum_{j,\alpha}\mathrm{E}^{S-1/2(\alpha)}_j\left|\Psi^{S-1/2(\alpha)}_j\right\rangle\left\langle\Psi^{S-1/2(\alpha)}_j\right|\\&+\sum_{\substack{k,\sigma,i,j\\\beta,\alpha}}\left[T^{\sigma,S(\beta\alpha)}_{k,i,j}c^{+}_{k\sigma}\left|\Psi^{S-1/2(\alpha)}_j\right\rangle\left\langle\Psi^{S(\beta)}_i\right|+T^{\sigma,S(\beta\alpha)*}_{k,i,j}\left|\Psi^{S(\beta)}_i\right\rangle\left\langle\Psi^{S-1/2(\alpha)}_j\right|c_{k\sigma}\right]\end{aligned} \quad (A3)$$

where $T^{\sigma,S(p,q)}_{k,i,j}=\sum_M V^{S,p,q}_{k,M,\sigma}A^{j,S-1/2(q)}_{M-\sigma}A^{i,S(p)}_M$

The appropriate Green functions for calculating the spectral density and the tunneling current between a tip and the adsorbed atom on the surface in a system described by the Hamiltonian (A3) are:

$$G_{np,qm}^{(\beta,\,\alpha)} = i\Theta(t'-t)\left\langle\left\{\left|\Psi_n^{S(\beta)}\right\rangle\left\langle\Psi_p^{S-1/2(\alpha)}\right|_{t'};\left|\Psi_m^{S-1/2(\alpha)}\right\rangle\left\langle\Psi_q^{S(\beta)}\right|_t\right\}\right\rangle \tag{A4}$$

The conductance ($G = dI/dV$), measured with the tip (t) over the adsorbed atom (s), in the limit of low temperatures and small bias voltages, can be approximated by the expression (G_0 being the conductance quantum):

$$G/G_0 = \sum_{i,j,\beta,\alpha,\sigma}\Gamma_{ij}^{\sigma,S(\beta,\,\alpha)}\mathrm{Im}\;G_{ij,ij}^{(\beta,\,\alpha)}(eV) \tag{A5}$$

where $\Gamma_{ij}^{\sigma,S(\beta,\alpha)} = 2\Gamma_{t,ij}^{\sigma,S(\beta,\alpha)}\,\Gamma_{s,ij}^{\sigma,S(\beta,\alpha)}/(\Gamma_{t,ij}^{\sigma,S(\beta,\alpha)} + \Gamma_{s,ij}^{\sigma,S(\beta,\alpha)})$ is the effective broadening defined by both interactions: atom-tip ($\Gamma_{t,ij}^{\sigma,S(\beta,\alpha)}$) and atom-surface ($\Gamma_{s,ij}^{\sigma,S(\beta,\alpha)}$). These broadenings are given by $\Gamma_{n,ij}^{\sigma,S(\beta,\,\alpha)}(\varepsilon) = \pi\sum_k\left|T_{(n)k,i,j}^{\sigma,S(\beta,\alpha)}\right|^2\delta(\varepsilon-\varepsilon_k)$.

The equations of motion of (A4) result:

$$\begin{aligned}\left(\omega - E_q^{S(\beta)} + E_m^{S-1/2(\alpha)}\right)G_{np,qm}^{(\beta\alpha)} &= \left\langle\left|\Psi_n^{S(\beta)}\right\rangle\left\langle\Psi_q^{S(\beta)}\right| + \left|\Psi_m^{S-1/2(\alpha)}\right\rangle\left\langle\Psi_p^{S-1/2(\alpha)}\right|\right\rangle\\ &+\sum_{k,j,\alpha',\sigma}T_{k,q,j}^{\sigma,S(\beta\alpha')*}\,G_{np}^{(\beta\alpha)}\left(\left|\Psi_m^{S-1/2(\alpha)}\right\rangle\left\langle\Psi_j^{S-1/2(\alpha')}\right|c_{k\sigma}\right)\\ &+\sum_{k,i,\beta',\sigma}T_{k,i,m}^{\sigma,S(\beta'\alpha)*}\,G_{np}^{(\beta\alpha)}\left(\left|\Psi_i^{S(\beta')}\right\rangle\left\langle\Psi_q^{S(\beta)}\right|c_{k\sigma}\right)\end{aligned} \tag{A6}$$

New Green functions appeared in (A6),

$$\begin{aligned}&G_{np}^{(\beta\alpha)}\left(\left|\Psi_m^{S-1/2(\alpha)}\right\rangle\left\langle\Psi_j^{S-1/2(\alpha')}\right|c_{k\sigma}\right)\\ &\quad= i\Theta(t'-t)\left\langle\left\{\left|\Psi_n^{S(\beta)}\right\rangle\left\langle\Psi_p^{S-1/2(\alpha)}\right|_{t'};\left|\Psi_m^{S-1/2(\alpha)}\right\rangle\left\langle\Psi_j^{S-1/2(\alpha')}\right|c_{k\sigma}\right\}\right\rangle\end{aligned}$$

$$G_{np}^{(\beta\alpha)}\left(\left|\Psi_i^{S(\beta')}\right\rangle\left\langle\Psi_q^{S(\beta)}\right|c_{k\sigma}\right) = i\Theta(t'-t)\left\langle\left\{\left|\Psi_n^{S_N(\beta)}\right\rangle\left\langle\Psi_p^{S_N-1/2(\alpha)}\right|_{t'};\left|\Psi_i^{S(\beta')}\right\rangle\left\langle\Psi_q^{S(\beta)}\right|c_{k\sigma}\right\}\right\rangle$$

and their respective equations of motion are,

$$\begin{aligned}
&\left(\omega-\varepsilon_k-E_j^{S-1/2(\alpha')}+E_m^{S-1/2(\alpha)}\right)G_{np}^{(\beta\alpha)}\left(\left|\Psi_m^{S-1/2(\alpha)}\right\rangle\left\langle\Psi_j^{S-1/2(\alpha')}\right|c_{k\sigma}\right)\\
&=+\sum_{r,\beta'}T_{k,r,j}^{\sigma,S(\beta'\alpha)}\left\langle c_{k\sigma}c_{k\sigma}^{+}\right\rangle G_{np}^{(\beta\alpha)}\left(\left|\Psi_m^{S-1/2(\alpha)}\right\rangle\left\langle\Psi_r^{S(\beta')}\right|\right)\\
&\quad+\sum_{k'\neq k,\sigma'\neq\sigma,r,\beta'}T_{k',r,j}^{\sigma',S(\beta'\alpha)}G_{np}^{N(\beta\alpha)}\left(\left|\Psi_m^{S-1/2(\alpha)}\right\rangle\left\langle\Psi_r^{S(\beta')}\right|c_{k\sigma}c_{k'\sigma'}^{+}\right)\\
&\quad-\sum_{k'\neq k,\sigma'\neq\sigma,r,\beta'}T_{k',r,m}^{\sigma',S(\beta',\alpha')}G_{np}^{(\beta\alpha)}\left(\left|\Psi_r^{S(\beta')}\right\rangle\left\langle\Psi_j^{S-1/2(\alpha')}\right|c_{k'\sigma'}c_{k\sigma}\right)\\
&\quad+\delta_{pm}\left\langle\left|\Psi_n^{S(\beta)}\right\rangle\left\langle\Psi_j^{S-1/2(\alpha')}\right|c_{k\sigma}\right\rangle
\end{aligned}\tag{A7}$$

$$\begin{aligned}
&\left(\omega-\varepsilon_k-E_q^{S(\beta)}+E_i^{S(\beta')}\right)G_{np}^{(\beta\alpha)}\left(\left|\Psi_i^{S_N(\beta')}\right\rangle\left\langle\Psi_q^{S_N(\beta)}\right|c_{k\sigma}\right)=\delta_{qn}\left\langle\left|\Psi_i^{S(\beta')}\right\rangle\left\langle\Psi_p^{S-1/2(\alpha)}\right|c_{k\sigma}\right\rangle\\
&\quad+\sum_{u,\alpha''}T_{k,i,u}^{\sigma,S(\beta',\alpha'')}\left\langle c_{k\sigma}^{+}c_{k\sigma}\right\rangle G_{np}^{N(\beta\alpha)}\left(\left|\Psi_u^{S-1/2(\alpha'')}\right\rangle\left\langle\Psi_q^{S(\beta)}\right|\right)\\
&\quad+\sum_{k'\neq k,\sigma',u,\alpha''}T_{k',i,u}^{\sigma',S(\beta',\alpha'')}G_{np}^{N(\beta\alpha)}\left(\left|\Psi_u^{S-1/2(\alpha'')}\right\rangle\left\langle\Psi_q^{S(\beta)}\right|c_{k'\sigma'}^{+}c_{k\sigma}\right)\\
&\quad-\sum_{k'\neq k,\sigma',u,\alpha''}T_{k',q,u}^{\sigma',S(\beta\alpha'')}G_{np}^{(\beta\alpha)}\left(\left|\Psi_i^{S(\beta')}\right\rangle\left\langle\Psi_u^{S-1/2(\alpha'')}\right|c_{k\sigma}c_{k'\sigma'}\right)
\end{aligned}\tag{A8}$$

The second order in V_{kd} approximation is obtained by disregarding the last two terms in both (A7) and (A8):

$$\begin{aligned}
G_{np}^{(\beta\alpha)}\left(\left|\Psi_m^{S-1/2(\alpha)}\right\rangle\left\langle\Psi_j^{S-1/2(\alpha')}\right|c_{k\sigma}\right)&=\delta_{pm}\frac{\left\langle\left|\Psi_n^{S(\beta)}\right\rangle\left\langle\Psi_j^{S-1/2(\alpha')}\right|c_{k\sigma}\right\rangle}{\left(\omega-\varepsilon_k-E_j^{S_N-1/2(\alpha')}+E_m^{S_N-1/2(\alpha)}\right)}\\
&\quad+\sum_{r,\beta'}\frac{T_{k,r,j}^{\sigma,S(\beta'\alpha)}\left\langle c_{k\sigma}c_{k\sigma}^{+}\right\rangle}{\left(\omega-\varepsilon_k-E_j^{S_N-1/2(\alpha')}+E_m^{S_N-1/2(\alpha)}\right)}G_{np}^{(\beta\alpha)}\left(\left|\Psi_m^{S-1/2(\alpha)}\right\rangle\left\langle\Psi_r^{S(\beta')}\right|\right)
\end{aligned}\tag{A9}$$

$$G_{np}^{(\beta\alpha)}\left(\left|\Psi_i^{S_N(\beta')}\right\rangle\left\langle\Psi_q^{S_N(\beta)}\right|c_{k\sigma}\right)=\frac{\delta_{qn}\left\langle\left|\Psi_i^{S(\beta')}\right\rangle\left\langle\Psi_p^{S-1/2(\alpha)}\right|c_{k\sigma}\right\rangle}{\left(\omega-\varepsilon_k-E_q^{S(\beta)}+E_i^{S(\beta')}\right)}$$
$$+\sum_{u,\alpha''}\frac{T_{k,i,u}^{\sigma,S(\beta',\alpha'')}\left\langle c_{k\sigma}^{+}c_{k\sigma}\right\rangle}{\left(\omega-\varepsilon_k-E_q^{S(\beta)}+E_i^{S(\beta')}\right)}G_{np}^{N(\beta\alpha)}\left(\left|\Psi_u^{S-1/2(\alpha'')}\right\rangle\left\langle\Psi_q^{S(\beta)}\right|\right) \tag{A10}$$

By replacing Eqs. (A9)–(A10) in the Eq. (A6) and only conserving the diagonal Green functions,

$$G_{qm,qm}^{(\beta,\alpha)}=i\Theta(t'-t)\left\langle\left\{\left|\Psi_q^{S_N(\beta)}\right\rangle\left\langle\Psi_m^{S-1/2(\alpha)}\right|_{t'};\left|\Psi_m^{S-1/2(\alpha)}\right\rangle\left\langle\Psi_q^{S(\beta)}\right|_t\right\}\right\rangle,$$

we finally obtain for the second order in the atom-surface coupling the following expression:

$$\left(\begin{array}{l}\omega-\mathrm{E}_q^{S(\beta)}+\mathrm{E}_m^{S-1/2(\alpha)}\\[4pt] -\displaystyle\sum_{k,j,\alpha',\sigma}\frac{\left|T_{k,q,j}^{\sigma,S(\beta\alpha')*}\right|^2\left\langle c_{k\sigma}c_{k\sigma}^{+}\right\rangle}{\left(\omega-\varepsilon_k-E_j^{S-1/2(\alpha')}+E_m^{S-1/2(\alpha)}\right)}-\sum_{k,i,\beta',\sigma}\frac{\left|T_{k,i,m}^{\sigma,S(\beta'\alpha)*}\right|^2\left\langle c_{k\sigma}^{+}c_{k\sigma}\right\rangle}{\left(\omega-\varepsilon_k-E_q^{S(\beta)}+E_i^{S(\beta')}\right)}\end{array}\right)G_{qm,qm}^{(\beta\alpha)}$$
$$=\left\langle\left|\Psi_q^{S_N(\beta)}\right\rangle\left\langle\Psi_q^{S_N(\beta)}\right|+\left|\Psi_m^{S_N-1/2(\beta)}\right\rangle\left\langle\Psi_m^{S_N-1/2(\beta)}\right|\right\rangle$$
$$+\sum_{k,j,\alpha',\sigma}\frac{T_{k,q,j}^{\sigma,S(\beta\alpha')*}}{\left(\omega-\varepsilon_k-E_j^{S-1/2(\alpha')}+E_m^{S-1/2(\alpha)}\right)}\left\langle\left|\Psi_q^{S(\beta)}\right\rangle\left\langle\Psi_j^{S-1/2(\alpha')}\right|c_{k\sigma}\right\rangle$$
$$+\sum_{k,i,\beta',\sigma}\frac{T_{k,i,m}^{\sigma,S(\beta'\alpha)*}}{\left(\omega-\varepsilon_k-E_q^{S(\beta)}+E_i^{S(\beta')}\right)}\left\langle\left|\Psi_i^{S(\beta')}\right\rangle\left\langle\Psi_m^{S_N-1/2(\alpha)}\right|c_{k\sigma}\right\rangle \tag{A11}$$

If we continue with the equations of motion of the new Green functions appearing in the last two terms of expressions (A7) and (A8) and close them performing appropriate mean-field approximations, we obtain the following expressions[33]:

$$\left(\omega - \varepsilon_k + \varepsilon_{k'} - \mathrm{E}_r^{S_N(\beta\prime)} + \mathrm{E}_m^{S_N-1/2(\alpha)}\right) G_{np}^{(\beta\alpha)} \left(\left|\Psi_m^{S_N-1/2(\alpha)}\right\rangle \left\langle \Psi_r^{S_N(\beta\prime)}\right| c_{k\sigma} c_{k'\sigma'}^{+}\right)$$

$$= + \sum_{u,\beta''} T_{k',r,u}^{\sigma\prime, S_N(\beta\prime\prime\alpha)*} \langle n_{k'\sigma'}\rangle G_{np}^{N(\beta\alpha)} \left(\left|\Psi_m^{S_N-1/2(\alpha)}\right\rangle \left\langle \Psi_u^{S_N-1/2(\beta\prime\prime)}\right| c_{k\sigma}\right)$$

$$- \sum_{t,\beta''} T_{k',t,m}^{\sigma\prime, S_N(\beta\prime\prime\alpha)*} \left\langle c_{k'\sigma'} c_{k'\sigma'}^{+}\right\rangle G_{np}^{N(\beta\alpha)} \left(\left|\Psi_t^{S_N(\beta\prime\prime)}\right\rangle \left\langle \Psi_r^{S_N(\beta\prime)}\right| c_{k\sigma}\right) \quad \text{(A12)}$$

$$\left(\omega - \varepsilon_{k'} - \varepsilon_k - \mathrm{E}_j^{S-1/2(\alpha\prime)} + \mathrm{E}_r^{S(\beta\prime)}\right) G_{np}^{(\beta,\ \alpha)} \left(\left|\Psi_r^{S(\beta\prime)}\right\rangle \left\langle \Psi_j^{S-1/2(\alpha\prime)}\right| c_{k'\sigma'} c_{k\sigma}\right)$$

$$\approx - \sum_{u,\alpha''} T_{k',r,u}^{\sigma\prime, S(\beta\prime,\ \alpha\prime\prime)} \left\langle c_{k'\sigma'}^{+} c_{k'\sigma'}\right\rangle G_{np}^{(\beta,\ \alpha)} \left(\left|\Psi_u^{S-1/2(\alpha\prime\prime)}\right\rangle \left\langle \Psi_j^{S-1/2(\alpha\prime)}\right| c_{k\sigma}\right)$$

$$+ \sum_{t,\beta''} T_{k',t,j}^{\sigma\prime, S(\beta\prime\prime,\ \alpha\prime)} \left\langle c_{k'\sigma'} c_{k'\sigma'}^{+}\right\rangle G_{np}^{(\beta,\ \alpha)} \left(\left|\Psi_r^{S(\beta\prime)}\right\rangle \left\langle \Psi_t^{S(\beta\prime\prime)}\right| c_{k\sigma}\right) \quad \text{(A13)}$$

$$\left(\omega - \varepsilon_k + \varepsilon_{k'} - \mathrm{E}_q^{S(\alpha)} + \mathrm{E}_u^{S-1/2(\alpha\prime\prime)}\right) G_{np}^{N(\beta)} \left(\left|\Psi_u^{S-1/2(\alpha\prime\prime)}\right\rangle \left\langle \Psi_q^{S(\beta)}\right| c_{k\sigma} c_{k'\sigma'}^{+}\right)$$

$$= + \sum_{r,\alpha'''} T_{k',q,r}^{\sigma\prime, S(\beta\alpha\prime\prime\prime)*} \langle n_{k'\sigma'}\rangle G_{np}^{(\beta\alpha)} \left(\left|\Psi_u^{S_N-1/2(\alpha\prime\prime)}\right\rangle \left\langle \Psi_r^{S_N-1/2(\alpha\prime\prime\prime)}\right| c_{k\sigma}\right)$$

$$- \sum_{t,\beta'} T_{k',t,u}^{\sigma\prime, S(\beta\prime\alpha\prime\prime)*} \left\langle c_{k'\sigma'} c_{k'\sigma'}^{+}\right\rangle G_{np}^{N(\beta)} \left(\left|\Psi_t^{S_N(\beta\prime)}\right\rangle \left\langle \Psi_q^{S_N(\beta)}\right| c_{k\sigma}\right) \quad \text{(A14)}$$

$$\left(\omega - \varepsilon_{k'} - \varepsilon_k - \mathrm{E}_u^{S-1/2(\alpha\prime\prime)} + \mathrm{E}_i^{S(\beta\prime)}\right) G_{np}^{(\beta,\ \alpha)} \left(\left|\Psi_i^{S(\beta\prime)}\right\rangle \left\langle \Psi_u^{S-1/2(\alpha\prime\prime)}\right| c_{k'\sigma'} c_{k\sigma}\right)$$

$$\approx - \sum_{r,\alpha'''} T_{k',i,r}^{\sigma\prime, S(\beta\prime,\ \alpha\prime\prime\prime)} \left\langle c_{k'\sigma'}^{+} c_{k'\sigma'}\right\rangle G_{np}^{(\beta,\ \alpha)} \left(\left|\Psi_r^{S-1/2(\alpha\prime\prime\prime)}\right\rangle \left\langle \Psi_u^{S-1/2(\alpha\prime\prime)}\right| c_{k\sigma}\right)$$

$$+ \sum_{t,\beta''} T_{k',t,u}^{\sigma\prime, S(\beta\prime\prime,\ \alpha\prime\prime)} \left\langle c_{k'\sigma'} c_{k'\sigma'}^{+}\right\rangle G_{np}^{(\beta,\ \alpha)} \left(\left|\Psi_i^{S(\beta\prime)}\right\rangle \left\langle \Psi_t^{S(\beta\prime\prime)}\right| c_{k\sigma}\right) \quad \text{(A15)}$$

By replacing Eqs. (A12)–(A15) in the Eqs. (A7)–(A8), then the obtained result in Eq. (A6), and conserving only diagonal Green functions, we obtain the final result for the fourth order in the atom-surface coupling:

$$
\begin{aligned}
&\left(\omega - \mathrm{E}_q^{S(\beta)} + \mathrm{E}_m^{S-1/2(\alpha)} - \sum_{k,j,\alpha',\sigma} \frac{\left|T_{k,q,j}^{\sigma,S(\beta\alpha')*}\right|^2 \left\langle c_{k\sigma} c_{k\sigma}^{+} \right\rangle}{\left(\omega - \varepsilon_k - E_j^{S-1/2(\alpha')} + E_m^{S-1/2(\alpha)} - \Xi_{mj}^{S(\alpha\alpha')<}(k\omega)\right)} - \sum_{k,i,\beta',\sigma} \frac{\left|T_{k,i,m}^{\sigma,S(\beta'\alpha)*}\right|^2 \left\langle c_{k\sigma}^{+} c_{k\sigma} \right\rangle}{\left(\omega - \varepsilon_k - E_q^{S(\beta)} + E_i^{S(\beta')} - \Xi_{iq}^{S(\beta'\beta)>}(k\omega)\right)}\right) \mathrm{G}_{qm,qm}^{(\beta\alpha)} \\
&\quad = \left\langle \left|\boldsymbol{\Psi}_q^{S_N(\beta)}\right\rangle \left\langle \boldsymbol{\Psi}_q^{S_N(\beta)}\right| + \left|\boldsymbol{\Psi}_m^{S_N-1/2(\beta)}\right\rangle \left\langle \boldsymbol{\Psi}_m^{S_N-1/2(\beta)}\right| \right\rangle \\
&\quad + \sum_{k,j,\alpha',\sigma} \frac{T_{k,q,j}^{\sigma,S(\beta\alpha')*}}{\left(\omega - \varepsilon_k - E_j^{S-1/2(\alpha')} + E_m^{S-1/2(\alpha)} - \Xi_{mj}^{S(\alpha\alpha')<}(k\omega)\right)} \left\langle \left|\boldsymbol{\Psi}_q^{S(\beta)}\right\rangle \left\langle \boldsymbol{\Psi}_j^{S-1/2(\alpha')}\right| c_{k\sigma} \right\rangle \\
&\quad + \sum_{k,i,\beta',\sigma} \frac{T_{k,i,m}^{\sigma,S(\beta'\alpha)*}}{\left(\omega - \varepsilon_k - E_q^{S(\beta)} + E_i^{S(\beta')} - \Xi_{iq}^{S(\beta'\beta)>}(k\omega)\right)} \left\langle \left|\boldsymbol{\Psi}_i^{S(\beta')}\right\rangle \left\langle \boldsymbol{\Psi}_m^{S_N-1/2(\alpha)}\right| c_{k\sigma} \right\rangle
\end{aligned}
\tag{A16}
$$

The fourth order approximation (A16) differs from the second order one (A11) in the appearance of the renormalization energies $\Xi_{mj}^{S(\alpha,\alpha')<}$ and $\Xi_{iq}^{S(\beta',\beta)>}$ given by the following expressions:

$$\begin{aligned}
\Xi_{mj}^{S(\alpha\alpha\prime)<}(k\omega) &= \sum_{k',\sigma',r,\nu}\left[\frac{\left|T_{k\prime,r,j}^{\sigma\prime,S(\nu\alpha\prime)}\right|^2\langle n_{k'\sigma'}\rangle}{\left(\omega-\varepsilon_k+\varepsilon_{k'}-\mathrm{E}_r^{S(\nu)}+\mathrm{E}_m^{S-1/2(\alpha)}\right)}\right.\\
&\quad\left.+\frac{\left|T_{k\prime,r,m}^{\sigma\prime,S(\nu\alpha)}\right|^2\langle n_{k'\sigma'}\rangle}{\left(\omega-\varepsilon_{k'}-\varepsilon_k-\mathrm{E}_j^{S-1/2(\alpha\prime)}+\mathrm{E}_r^{S(\nu)}\right)}\right]\\
\Xi_{iq}^{S(\beta\prime\beta)>}(k\omega) &= \sum_{k',\sigma',u,\nu}\left[\frac{\left|T_{k\prime,i,u}^{\sigma\prime,S(\beta\prime\nu)*}\right|^2\langle c_{k'\sigma'}c_{k'\sigma'}^{+}\rangle}{\left(\omega-\varepsilon_k+\varepsilon_{k'}-\mathrm{E}_q^{S(\beta)}+\mathrm{E}_u^{S-1/2(\nu)}\right)}\right.\\
&\quad\left.+\frac{\left|T_{k\prime,q,u}^{\sigma\prime,S(\beta\nu)}\right|^2\langle c_{k'\sigma'}c_{k'\sigma'}^{+}\rangle}{\left(\omega-\varepsilon_{k'}-\varepsilon_k-\mathrm{E}_u^{S-1/2(\nu)}+\mathrm{E}_i^{S(\beta\prime)}\right)}\right]
\end{aligned}\tag{A17}$$

In the case of neglecting the spin-orbit interaction effects, $|\Psi_i^{S(\beta)}\rangle=|S,M\rangle_\beta$; $|\Psi_j^{S-1/2(\alpha)}\rangle=|S-1/2,M-\sigma\rangle_\alpha$, and $T_{k,\,i,\,j}^{\sigma,\,S(\beta,\alpha)}\equiv V_{k,\,M,\,\sigma}^{S,\,\beta,\,\alpha}$. and the indexes i and j identify with M and $M-\sigma$ respectively.

References

1. Kondo, J. *Progr. Theor. Phys.* **1964**, *32*, 37.
2. Hewson, A. C. *The Kondo Problem to Heavy Fermions*; Cambridge University Press: Cambridge, 1993.
3. Anderson, P. W. *Phys. Rev.* **1961**, *124*, 41.
4. Hirjibehedin, C. F.; Lutz, C. P.; Heinrich, A. J. *Science* **2006**, *312*, 1021.
5. Hirjibehedin, C. F.; Lin, C. Y.; Otte, A. F.; Ternes, M.; Lutz, C. P.; Jones, B. A.; Heinrich, A. J. *Science* **2007**, *317*, 1199.
6. Otte, A. F.; Ternes, M.; Loth, S.; von Bergmann, K.; Brune, H.; Lutz, C. P.; Hirjibehedin, C. F.; Heinrich, A. J. *Nat. Phys.* **2008**, *4*, 847.
7. Ternes, M.; Heinrich, A. J.; Schneider, W. D. *J. Phys. Condens. Matter* **2009**, *21*, 053001.
8. Bulla, R.; Costi, T. A.; Pruschke, T. *Rev. Mod. Phys.* **2008**, *80*, 395.
9. Otte, A. F.; Ternes, M.; Loth, S.; Lutz, C. P.; Hirjibehedin, C. F.; Heinrich, A. J. *Phys. Rev. Lett.* **2009**, *103*, 107203.
10. Spinelli, A.; Gerrits, M.; Toskovic, R.; Bryant, B.; Ternes, M.; Otte, A. F. *arXiv:1411.4415v2 [cond-mat.mes-hall] 20 Apr 2015*; 2015.
11. Bryant, B.; Spinelli, A.; Wagenaar, J. J. T.; Gerrits, M.; Otte, A. F. *Phys. Rev. Lett.* **2013**, *111*, 127203.
12. Bryant, B.; Toskovic, R.; Ferrón, A.; Lado, J. L.; Anna Spinelli, J.; Fernández-Rossier; Otte, A. F. *Nano Lett.* **2015**, *15*, 6542.

13. Spinelli, A.; Rebergen, M. P.; Otte, A. F. *J. Phys. Cond. Matt.* **2015**, *27*, 243203.
14. Cortés, R.; Tejeda, A.; Trabada, D. G.; et al. *Phys. Rev. Lett.* **2013**, *96*, 126103.
15. Lee, P. A.; Nagaosa, N.; Xiao-GangWen. *Rev. Mod. Phys.* **2006**, *78*, 17.
16. Nozières, P.; Blandin, A. *J. Physique* **1980**, *41*, 193.
17. Georges, A.; de'Medici, L.; Mravlje, J. *Annual Rev. of Condens. Matter Phys.* **2013**, *4*, 137.
18. Kotliar, G.; Savrasov, S. Y.; Haule, K.; Oudovenko, V. S.; Parcollet, O.; Marianetti, C. A. *Rev. Mod. Phys.* **2006**, *78*, 865.
19. Goldberg, E. C.; Flores, F. *Phys. Rev. B* **2008**, 77, 125121.
20. Goldberg, E. C.; Flores, F. *J. Phys. Cond. Matt.* **2013**, *25*, 225001.
21. Blesio, G. G.; Manuel, L. O.; Aligia, A. A.; Roura-Bas, P. *Phys. Rev. B* **2019**, *100*, 1–14.
22. Di Napoli, S.; Weichselbaum, A.; Roura-Bas, P.; Aligia, A. A.; Mokrousov, Y.; Blügel, S. *Phys. Rev. Lett.* **2013**, *110*, 1–5.
23. Di Napoli, S.; Roura-Bas, P.; Weichselbaum, A.; Aligia, A. A. *Phys. Rev. B* **2014**, *90*, 125149.
24. Fernández, J.; Roura-Bas, P.; Camjayi, A.; Aligia, A. A. *J. Phys. Condens. Matter* **2018**, *30*, 374003.
25. Goldberg, E. C.; Flores, F. *J. Phys.: Condens. Matter* **2017**, *29*, 055602.
26. Goldberg, E. C.; Flores, F. *Phys. Rev. B* **2017**, *96*, 115439.
27. Roura Bas, P.; Aligia, A. A. *J. Phys. Condens. Matter* **2010**, *22*, 025602.
28. Tacca, M. S.; Jacob, T.; Goldberg, E. C. *Phys. Rev. B* **2020**, *101*, 125419.
29. Tacca, M. S.; Jacob, T.; Goldberg, E. C. *Phys. Rev. B* **2021**, *103*, 245419.
30. Orgel, L. E. *An Introduction to Transition-Metal Chemistry. Ligand Field Theory*; Methuen and Co: LTD, London, 1961.
31. Zitko, R.; Pruschke, T. *New J. Phys.* **2010**, *12*, 063040.
32. Tacca, M. S. *PhD. Thesis*; Universidad Nacional del Litoral (Santa Fe Argentina) and Ulm University: Ulm, Germany, 2020.
33. Flores, F.; Chiappe, G.; Anda, E. V.; Goldberg, E. C. Chapter one. In *Rufus Ritchie, A Gentleman and A Scholar, volume 80 of Advances in Quantum Chemistry*; Sabin, J. R., Oddershede, J., Eds.; Academic Press, 2019; pp. 1–22.
34. Bardeen, J. *Phys. Rev. Lett.* **1961**, *6*, 57.
35. Haldane, F. D. M. *Phys. Rev. Lett.* **1978**, *40*, 416.

Index

Note: Page numbers followed by "*f*" indicate figures and "*t*" indicate tables.

E

Q

R

V

W

Z

FSC
www.fsc.org
MIX
Papier | Fördert
gute Waldnutzung
FSC® C083411

Kolibri 360 GmbH
Ferdinand-Jühlke-Straße 7
99095 Erfurt, Deutschland
produktsicherheit@kolibri360.de